RELIABILITY ENGINEERING

RELIABILITY ENGINEERING

Editor

Hoang Pham

Rutgers University, USA

W⊕ World Scientific

NEW JERSEY · LONDON · SINGAPORE · BEIJING · SHANGHAI · TAIPEI · CHENNAI

Published by

World Scientific Publishing Co. Pte. Ltd.

5 Toh Tuck Link, Singapore 596224

USA office: 27 Warren Street, Suite 401-402, Hackensack, NJ 07601

UK office: 57 Shelton Street, Covent Garden, London WC2H 9HE

Library of Congress Control Number: 2025006452

British Library Cataloguing-in-Publication Data
A catalogue record for this book is available from the British Library.

RELIABILITY ENGINEERING

ISBN 978-981-98-1253-0 (hardcover)
ISBN 978-981-98-1254-7 (ebook for institutions)
ISBN 978-981-98-1255-4 (ebook for individuals)

For any available supplementary material, please visit
https://www.worldscientific.com/worldscibooks/10.1142/14294#t=suppl

Desk Editors: Soundararajan Raghuraman/Steven Patt

Typeset by Stallion Press
Email: enquiries@stallionpress.com

Preface

Most products that influence our daily lives are becoming increasingly complex. Reliability and statistical analytics play a vital role in business, finance, and industry. On one hand, these tools enable organizations to achieve better results and make fact-based decisions across various domains. On the other hand, modern business is becoming more competitive and challenging due to global competition and the inherent complexity of business operations.

This book features 18 articles selected from the *International Journal of Reliability, Quality and Safety Engineering (IJRQSE)* over the past 30 years. These articles represent the major themes of the journal, focusing on reliability, quality, and safety engineering. They are among the most cited articles in the last three years, with each receiving at least 40 citations, according to Crossref data.

Article #1, authored by Lin, Zuo, and Yam, presents new sequential imperfect preventive maintenance (PM) models that consider both age reduction and hazard rate adjustment. The authors developed methods to determine the optimal PM schedule that minimizes the mean cost rate. In Article #2, Pham and Zhang developed a software reliability model based on a nonhomogeneous Poisson process (NHPP) to predict the number of remaining errors in the software. They also created a software program to facilitate the task of obtaining estimators for the model parameters. Article #3, by Coit and Liu, focuses on methods to select components and redundancy levels to maximize system reliability within k-out-of-n subsystems, subject to system-level constraints.

Wallace and Kuhn, in Article #4, analyzed software-related failures in medical devices that, while causing no deaths or injuries, led to manufacturer recalls. They provided insights into the need for formal requirements specification and improved testing of complex hardware–software systems. Article #5 by Bouti and Kadi offered a comprehensive review of failure mode and effects analysis (FMEA) for single failures in a system. They also discussed the causes and effects of each potential failure mode and proposed corrective actions to improve system performance.

Khoshgoftaar and Allen, in Article #6, presented an integrated method using logistic regression to interpret coefficients, the use of prior probabilities, and the costs of misclassifications. They also illustrated these techniques through a case study involving a major subsystem of a military real-time system. Article #7, by Bracquemond and Gaudoin, offered a comprehensive survey of discrete probability distributions used in reliability modeling to represent the discrete lifetimes of nonrepairable systems. Article #8, by Zhong, Khoshgoftaar, and Seliya, explored multiple centroid-based unsupervised clustering algorithms, including k-means, mixture-of-spherical Gaussians, and self-organizing map, for intrusion detection. They proposed a simple yet effective self-labeling heuristic for identifying attack and normal clusters within network traffic audit data.

Chen, Mizutani, and Nakagawa, in Article #9, proposed various random age replacement policies for an operating unit that functions intermittently. These policies aim to minimize the expected cost rate. In Article #10, Kim and Bai introduced a method for estimating the lifetime distribution under use conditions for constant stress accelerated life tests. This method accounts for both intrinsic and extrinsic failure modes using a mixture of two distributions to model these failure modes.

Castagliola, Celano, and Chen, in Article #11, examined the effects of estimating the unknown in-control process variance and the exact run length distribution of a statistical control chart where the in-control process variance is estimated. In Article #12, Ding and Tian presented three opportunistic maintenance optimization models, considering various preventive maintenance approaches: perfect maintenance, imperfect maintenance, and a two-level action method. Through simulation, they demonstrated that the two-level action method was the most cost-effective across different cost scenarios.

Ahmadi, Gupta, Karim, and Kumar, in Article #13, proposed a multi-criteria decision-making methodology using an AHP-enhanced TOPSIS framework to select maintenance strategies. This approach ensures the consistency and effectiveness of maintenance decisions. Xie, Gaudoin, and Bracquemond, in Article #14, discussed refining the failure rate function for discrete distributions and introduced several useful discrete reliability functions. Article #15 by Xie, Goh, and Karalmani revealed that the average time to alarm may initially increase when a process begins to deteriorate. They proposed a new procedure for setting control limits to maximize the average run length when the process is operating at a normal level.

Dhillon and Anude, in article #16, provided a comprehensive literature review on common-cause failures related to reliability and safety engineering. Li, Chan, and Yuan, in article #17, presented the failure time distribution under a δ-shock failure model and discussed approaches to optimizing system design with respect to total system cost. Article #18, authored by Djamaludin, Murthy, and Kim, offered a thorough review of warranty and preventive maintenance. They also developed a framework to study preventive maintenance actions for items sold under warranty.

Postgraduates, researchers, and engineers will undoubtedly gain valuable knowledge from this book, which provides foundational insights into the fields of reliability, quality, and safety engineering, along with their practical applications. The book is designed for graduate and advanced undergraduate-level students.

I extend my gratitude to World Scientific Publishing for their remarkable ideas and the opportunity to bring forth this special volume, which features the most cited articles published in the *International Journal of Reliability, Quality and Safety Engineering (IJRQSE)* over the past 30 years. Importantly, I would also like to thank all the authors who have contributed their research to the *IJRQSE*, advancing the field significantly over the decades.

Hoang Pham
Piscataway, New Jersey
December 2024

About the Editor

Dr. Hoang Pham is a Distinguished Professor and former Chairman (2007–2013) of the Department of Industrial and Systems Engineering at Rutgers University, New Jersey. Prior to joining Rutgers, he was a Senior Engineering Specialist with the Boeing Company, Seattle, Washington, and the Idaho National Engineering Laboratory, Idaho Falls, Idaho.

He is the editor-in-chief of the *International Journal of Reliability, Quality and Safety Engineering* and editor of the *Springer Book Series in Reliability Engineering*. Dr. Pham is the author or coauthor of 6 books and has published over 210 journal articles and 100 conference papers and edited 20 books including *Springer Handbook of Engineering Statistics* (2nd ed., 2023) and *Handbook in Reliability Engineering*. He has delivered over 50 invited keynote and plenary speeches at various international conferences and institutions. His numerous awards include the 2009 IEEE Reliability Society *Engineer of the Year Award*. He is a Fellow of the Institute of Electrical and Electronics Engineers (IEEE), the Institute of Industrial and Systems Engineers (IISE), and Asia-Pacific Artificial Intelligence Association (AAIA).

Contents

© 2025 World Scientific Publishing Company
https://doi.org/10.1142/9789819812547_0001

Chapter 1

GENERAL SEQUENTIAL IMPERFECT PREVENTIVE MAINTENANCE MODELS*

DAMING LIN

Department of Mechanical and Industrial Engineering
University of Toronto, 5 King's College Road, Toronto
Ontario, Canada M5S 3G8
E-mail: dlin@mie.utoronto.ca

MING J. ZUO

Department of Mechanical Engineering, University of Alberta
4-9 Mechanical Engineering Building
Edmonton, Alberta, Canada T6G 2G8
E-mail: ming.zuo@ualberta.ca

RICHARD C. M. YAM

Department of Manufacturing Engineering and Engineering Management
City University of Hong Kong, 83 Tat Chee Avenue, Kowloon, Hong Kong
E-mail: mery@cityu.edu.hk

This paper presents new sequential imperfect preventive maintenance (PM) models incorporating adjustment/improvement factors in hazard rate and effective age. The models are hybrid in the sense that they are combinations of the age reduction PM model and the hazard rate adjustment PM model. It is assumed that PM is imperfect: It not only reduces the effective age but also changes the hazard rate, while the hazard rate increases with the number of PMs. PM is performed in a sequence of intervals. The objective is to determine the optimal PM schedule to minimize the mean cost rate. Numerical examples for a Weibull distribution are given.

Keywords: Preventive Maintenance (PM); Sequential Policy; Minimal Repair; Improvement Factor; Optimal Policy.

1. Introduction

A preventive maintenance (PM) policy specifies how PM activities should be scheduled. One of the commonly used PM policies is called *Periodic PM*, which specifies that a system is maintained at integer multiples of some fixed period. Another PM policy is called *Sequential PM*, in which the system is maintained at a sequence of intervals that may have unequal lengths. Periodic PM is more convenient to

---

*This chapter appeared previously on the International Journal of Reliability, Quality and Safety Engineering. To cite this chapter, please cite the original article as the following: D. Lin, M. J. Zuo and R. C. M. Yam, *Int. J. Reliab. Qual. Saf. Eng*, **7**, 253–266 (2000), doi:10.1142/S0218539300000213.

schedule, whereas sequential PM is more realistic when the system requires more frequent maintenance as it ages. A common assumption used in both these PM policies is that *minimal repair* is conducted on the system if it fails between successive PM activities. A minimal repair only restores the system to the functioning state once it fails, but does not improve the overall health condition of the system. In other words, minimal repairs do not change the *hazard rate* or the *effective age* of the system.

A PM model describes the effects of a PM activity. Many publications have reported PM models. Traditional PM models assume that the system after PM is either *as good as new* (in this case, it is called *perfect PM* or simply *replacement*), or *as bad as old* (in this case, it is the same as minimal repair). The more realistic assumption is that the system after PM lies somewhere between as good as new and as bad as old. This kind of PM is called *imperfect PM*. One common approach for modelling an imperfect PM activity is to assume that PM is equivalent to minimal repair with probability p, and equivalent to replacement with probability $1 - p$.[1-6] Another approach is to model the PM effects directly by specifying how the hazard rate or the effective age changes after PM.[7-14]

Hazard rate or instantaneous hazard rate $h(t)$ is defined to be the probability of the first and only failure of an item in the next instant of time, given that the item is presently working (p. 69, Ref. 16). Based on this definition, we have the following observations:

(i) The hazard rate is only defined when the item is working.
(ii) The hazard rate at time t reflects the health condition of the item at this point of time.
(iii) The hazard rate at time t, or the health condition of the item at time t, is a consequence of the previous operating history of the item, including previous operating conditions, previous failures and repairs, and preventive maintenances that have been performed on this item.

Two popular approaches to determining the PM intervals for a sequential PM are the reliability-based method and the optimization method. In the former method, PM is performed whenever the system reliability (or the hazard rate of the system) reaches a predetermined level.[7-11] The latter method is to search for the optimal intervals to minimize cost[12-14] or maximize mission reliability.[17]

Nguyen & Murthy[12] and Nakagawa[13] propose sequential imperfect PM models. Lie & Chun[8] and Nakagawa[15] introduce adjustment/improvement factors in hazard rate and effective age after PM, respectively. Nakagawa[14] adopts adjustment/improvement factors in hazard rate and effective age[8,15] for a sequential PM policy and proposes two PM models: (i) the hazard rate in the next PM interval becomes $ah(x)$ when it was $h(x)$ in the previous PM interval, where $a \geq 1$ is the adjustment factor and $x \geq 0$ represents the time elapsed from the previous PM time, and (ii) if the effective age of a system is t right before a PM, it reduces to bt right after the PM, where $b \leq 1$ is the improvement factor in effective age. We call the first model *hazard rate adjustment model* and the second one *age reduction model*.

The hazard rate adjustment model assumes that the hazard rate right after PM reduces to 0 and then increases more quickly than it did in the previous PM interval. The age reduction model assumes that there is an effective age reduction right after PM, the effective age right after PM may be greater than zero, and the hazard rate remains as a function of the effective age. A natural extension of these two models is to combine the two concepts of age reduction and hazard rate adjustment. A more general and realistic case is that PM not only reduces the effective age to a certain value but also changes the slope of the hazard rate function. Thus, a hybrid PM model which is a combination of the age reduction model and the hazard rate adjustment model would be more appropriate. Given a certain failure rate function $h(t)$ for $t \in (0, t_1)$, the PM activity at time t_1 produces a new failure rate function $g(t)$ for $t \in (t_1, t_2)$. The hazard rate function in the next PM interval depends only on the hazard rate function in the previous PM interval and the corresponding PM activity. In other words, $g(t)$ for $t \in (t_1, t_2)$ depends on both $h(t)$ for $t \in (0, t_1)$ and the extent of the PM activity at time t_1. In this paper we propose a form of $g(t)$ using both the hazard rate concept and the effective age concept. The proposed model is

$$g(t_1 + x) = ah(bt_1 + x),$$

where $a \geq 1$, $0 \leq b \leq 1$, and $x \in (0, t_2 - t_1)$. When $a = 1$, the hybrid model reduces to the age reduction model. When $b = 0$, the hybrid model reduces to the hazard

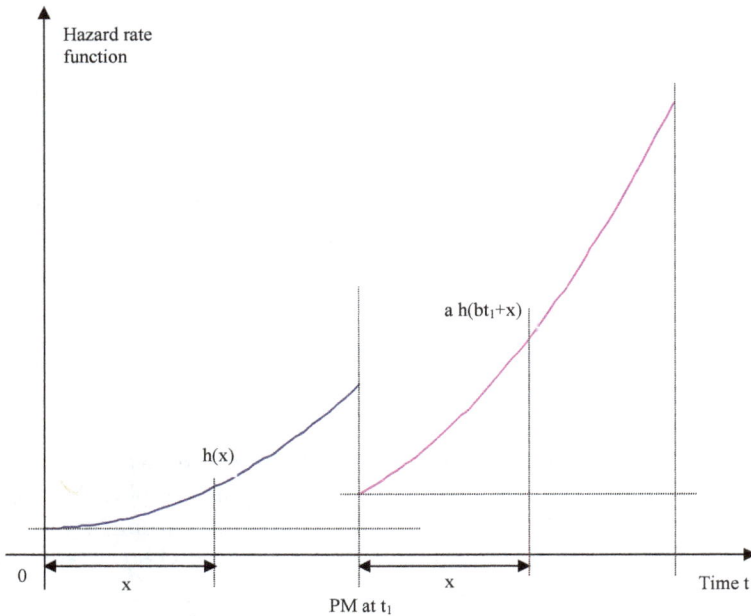

Fig. 1. Hazard rate function for the proposed hybrid model.

rate adjustment model. The hazard rate function before and after PM for a typical hybrid model is shown in Fig. 1.

Using the proposed hybrid PM model, we develop optimal PM policies through two alternatives: (i) leaving the PM intervals as decision variables, and (ii) allowing the PM intervals to be determined by the predetermined hazard rate level or reliability limit. The objective functions in both models are to minimize the mean cost rate. Numerical examples for the Weibull distribution case will be given for illustration purposes.

Notation

$h(t)$	hazard rate of the system
$H(t)$	cumulative hazard rate of the system
λ	predetermined acceptable level of hazard rate
x_k	scheduled PM intervals, $k = 1, 2, \ldots, N$
t_k	$= x_1 + x_2 + \cdots + x_k$, $k = 1, 2, \ldots, N$
y_k	effective age of the system just before the kth PM, $k = 1, 2, \ldots, N$
N	scheduled number of the interval where the system is replaced
a_k	adjustment factor in hazard rate after the kth PM, $1 = a_0 \leq a_1 \leq a_2 \leq \cdots \leq a_{N-1}$
A_k	$= \prod_{i=0}^{k-1} a_i$, $k = 1, 2, \ldots, N$
b_k	improvement factor in effective age right after the kth PM, $0 = b_0 \leq b_1 \leq b_2 \leq \cdots \leq b_{N-1} < 1$
c_m	cost of minimal repair
c_p	cost of PM
c_r	cost of replacement
C	mean cost rate of the system
d_k	$= \left[\frac{(1-b_k)^\alpha}{A_k - A_{k+1} b_k^\alpha} \right]^{\frac{1}{\alpha-1}}$

Assumptions

(1) The planning horizon is infinite.

(2) The hazard rate function of the system when there is no repair or PM, $h(t)$, is continuous and strictly increasing. We also assume that $h(t)$ is a convex function. As a result, $h'(t)$ is a monotonic increasing function. This is a reasonable assumption as the rate of deterioration increases as the system ages if there is no maintenance intervention.

(3) The times for PM, minimal repair and replacement are negligible.

(4) PM is performed at $t_1, t_2, \ldots, t_{N-1}$, and the system is replaced at t_N.

(5) The system is restored to as good as new state at replacement.

(6) The hazard rate function becomes $a_k h(bt_k)$ right after the kth PM when it was $h(t_k)$ just before the PM. After the PM, the hazard rate function is expressed as $a_k h(bt_k + x)$ for $x > 0$. Here we have $1 = a_0 \leq a_1 \leq a_2 \leq \cdots \leq a_{N-1}$ and $0 = b_0 \leq b_1 \leq b_2 \leq \cdots \leq b_{N-1} < 1$.

2. Model Formulations and Optimal Solutions

We consider the situation where a system is preventively maintained at $t_1, t_2, \ldots,$ t_{N-1} and replaced at t_N. Minimal repair is performed at failures between PMs. Replacement of the system restores the system to as good as new. The system has the hazard rate $A_k h(t)$ between the $(k-1)$th and the kth PMs, i.e., in time interval (t_{k-1}, t_k). The effective age of the system becomes $b_{k-1} y_{k-1}$ right after the $(k-1)$th PM and then becomes $y_k = x_k + b_{k-1} x_{k-1} + \cdots + b_{k-1} b_{k-2} \cdots b_2 b_1 x_1$ immediately before the kth PM; i.e., the effective age of the system changes from $b_{k-1} y_{k-1}$ to y_k in (t_{k-1}, t_k). Obviously, we have $y_k = x_k + b_{k-1} y_{k-1}$, or $x_k = y_k - b_{k-1} y_{k-1}$.

Following Ref. 14, the mean cost rate is

$$C = \frac{c_r + c_p(N-1) + c_m \sum_{k=1}^{N} A_k [H(y_k) - H(b_{k-1} y_{k-1})]}{\left[\sum_{k=1}^{N-1} (1 - b_k) y_k + y_N \right]}. \tag{1}$$

There are two main alternatives to determine the PM intervals in the literature. One is to select optimal PM intervals to minimize the mean cost rate, i.e., to leave PM intervals as decision variables in the optimization problem.[7-11] The other is to determine PM intervals by the hazard rate or reliability limit.[12-14] Both will be used in the following model formulations.

2.1. *Model 1*

In this model, the PM times $t_1, t_2, \ldots, t_{N-1}$ and the replacement time t_N are all treated as independent decision variables. Alternatively, the decision variables are N and y_k $(k = 1, 2, \ldots, N)$; i.e., the objective in this model is to determine the optimal values of N and y_k $(k = 1, 2, \ldots, N)$ to minimize the mean cost rate expressed in Eq. (1).

Following Ref. 14 and based on Eq. (1), setting the partial derivatives of the mean cost rate C with respect to y_k to zero leads to

$$A_k h(y_k) - A_{k+1} b_k h(b_k y_k) = A_N (1 - b_k) h(y_N), \qquad (k = 1, 2, \ldots, N-1) \tag{2}$$

and

$$c_m A_N h(y_N) = C. \tag{3}$$

Theorem 1. *For a fixed y_N $(0 < y_N < \infty)$, the solution to Eq. (2) with respect to y_k $(y_k > 0)$ exists if $1 - a_k b_k > 0$, $k = 1, 2, \ldots, N-1$. Further, the solution is unique if $h(t)$ is differentiable and $h'(t)$ is strictly increasing.*

Proof. If $1 - a_k b_k > 0$, we can show that the left hand side (lhs) of Eq. (2) is less than the right hand side (rhs) when $y_k = 0$:

$$A_k h(0) - A_{k+1} b_k h(0) = A_k (1 - a_k b_k) h(0) < A_N (1 - b_k) h(y_N).$$

With $1 - a_k b_k > 0$, the lhs of Eq. (2) is greater than the rhs when $y_k \to +\infty$:

$$A_k h(y_k) - A_{k+1} b_k h(b_k y_k) \geq A_k(1 - a_k b_k) h(y_k) \to +\infty \text{ as } y_k \to +\infty.$$

It should be noted that the rhs is a constant for a fixed y_N. Therefore, there exists some y_k $(y_k > 0)$ which satisfies Eq. (2). Moreover, when $h'(t)$ is strictly increasing, the derivative of the lhs of Eq. (2) with respect to y_k is

$$A_k h'(y_k) - A_{k+1} b_k^2 h'(b_k y_k) > A_k(1 - a_k b_k) h'(y_k) > 0 ;$$

i.e., the lhs of Eq. (2) is a strictly increasing function of y_k. Therefore, the solution to Eq. (2) with respect to y_k is unique.   □

Note that the condition $1 - a_k b_k > 0$ means that the hazard rate adjustment factor a_k (>1) should be smaller than the inverse of age improvement factor b_k. The extent of system improvement by PM is determined by the values of a_k and b_k: the smaller the values of a_k and b_k, the greater the improvement. It is a reasonable condition since PM tends to improve the system's health condition. In other words, the hazard rate right after a PM, $a_k h(b_k y_k)$, should be smaller than the hazard rate right before the PM, $h(y_k)$, i.e.,

$$a_k h(b_k y_k) < h(y_k) . \tag{4}$$

Because $h(t)$ is increasing and convex, we have

$$a_k h(b_k y_k) \leq a_k b_k h(y_k) .$$

Based on this equation, if we have the condition $1 - a_k b_k > 0$ satisfied, we can certainly guarantee that condition (4) is satisfied. If the condition $1 - a_k b_k > 0$ does not hold, the solution to Eq. (2) may not exist. Then the optimal y_k may be attained at boundary points or may not exist. In the following, we assume that the condition $1 - a_k b_k > 0$ holds.

Substituting each solution to Eq. (2), y_k $(k = 1, 2, \ldots, N - 1)$, into Eq. (3), we have

$$A_N h(y_N) \left[\sum_{k=1}^{N-1} (1 - b_k) y_k + y_N \right] - \sum_{k=1}^{N} A_k [H(y_k) - H(b_{k-1} y_{k-1})]$$

$$= \frac{c_r + c_p(N - 1)}{c_m} , \tag{5}$$

where each y_k $(k = 1, 2, \ldots, N - 1)$ is some function of y_N. Then, the lhs of Eq. (5) is a function only of y_N.

Theorem 2. *If $1 - a_k b_k > 0$, $k = 1, 2, \ldots, N - 1$, $h(0) = 0$, $h(t)$ is differentiable and $h'(t)$ is strictly increasing, the solution to Eq. (5) with respect to y_N $(y_N > 0)$ exists and is unique.*

Proof. Since

$$A_k h(y_k) - A_{k+1} b_k h(b_k y_k) \geq A_k(1 - a_k b_k) h(y_k) \geq 0, \qquad (k = 1, 2, \ldots, N - 1),$$

the solution to Eq. (2) with respect to y_k is zero when $y_N = 0$. Thus, if $y_N = 0$, the lhs of Eq. (5) equals zero and is smaller than the rhs of Eq. (5). On the other hand, the derivative of the lhs of Eq. (5) with respect to y_N is found to be

$$A_N h'(y_N) \left[\sum_{k=1}^{N-1} (1 - b_k) y_k + y_N \right] > 0.$$

i.e., the lhs of Eq. (5) is a strictly increasing function of y_N ($y_N \geq 0$) and equals zero at $y_N = 0$. Therefore, the solution to Eq. (5) with respect to y_N ($y_N > 0$) exists and is unique. □

Based on the above results, we obtain the following algorithm for finding the optimal PM schedule:

Step 1. Solve Eq. (2) with respect to y_k ($k = 1, 2, \ldots, N - 1$). The solutions are functions of y_N.

Step 2. Substitute the solutions in Step 1 into Eq. (5) and solve it with respect to y_N.

Step 3. Choose N to minimize $A_N h(y_N)$, where y_N is the solution obtained in Step 2.

Step 4. Compute y_k, $k = 1, 2, \ldots, N$ from expressions obtained in Steps 1 and 2 for the value of N obtained in Step 3.

Step 5. Compute x_k from $x_k = y_k - b_{k-1} y_{k-1}$, $k = 1, 2, \ldots, N$.

When the condition $h(0) = 0$ in Theorem 2 does not hold, the solution to Eq. (5) with respect to y_N ($y_N > 0$) may or may not exist. The above algorithm still works if the solution to Eq. (5) with respect to y_N ($y_N > 0$) exists. If the solution to Eq. (5) with respect to y_N ($y_N > 0$) does not exist, there are no points that make the first-order derivative with respect to y_N equate to zero. In this case, the optimal point may be achieved at a boundary point.

It should be noted that Model 1 reduces to the hazard rate adjustment model when $b_k = 0$, $k = 1, 2, \ldots, N$, and to the age reduction model when $a_k = 1$, $k = 1, 2, \ldots, N$ [see Ref. 14].

2.2. *Model 2*

In this model, PM is performed whenever the hazard rate of the system reaches the predetermined level λ and the decision variable is N. This implies that the hazard rate at time t_i ($i = 1, 2, \ldots, N$) must equal to λ; that is,

$$A_k h(y_k) = \lambda, \qquad k = 1, 2, \ldots, N. \tag{6}$$

Solving Eq. (6) with respect to y_k $(k = 1, 2, \ldots, N)$, we get the solutions as expressions of λ. Substitute these expressions into Eq. (1). Then C becomes a function of λ. Differentiating C with respect to λ and setting it to zero leads to

$$\frac{\displaystyle\sum_{k=1}^{N-1} \frac{h(y_k) - a_k b_k h(b_k y_k)}{h'(y_k)} + \frac{h(y_N)}{h'(y_N)}}{\displaystyle\sum_{k=1}^{N-1} \frac{1 - b_k}{A_k h'(y_k)} + \frac{1}{A_N h'(y_N)}} = \frac{C}{c_m}, \tag{7}$$

where y_k $(k = 1, 2, \ldots, N)$ is a function of λ. Solving Eq. (7) with respect to λ, we obtain λ as a function of N. Then we can find N which minimizes

$$\frac{\displaystyle\sum_{k=1}^{N-1} \frac{h(y_k) - a_k b_k h(b_k y_k)}{h'(y_k)} + \frac{h(y_N)}{h'(y_N)}}{\displaystyle\sum_{k=1}^{N-1} \frac{1 - b_k}{A_k h'(y_k)} + \frac{1}{A_N h'(y_N)}}, \tag{8}$$

where y_k $(k = 1, 2, \ldots, N)$ is a function of λ and thus a function of N.

Based on the above discussions, we derive the algorithm for obtaining the optimal PM schedule as follows.

Step 1. Solve Eq. (6) with respect to y_k $(k = 1, 2, \ldots, N - 1)$. The solutions are functions of λ.

Step 2. Substitute the solutions in Step 1 into Eq. (7) and solve it with respect to λ.

Step 3. Choose N to minimize the function in expression (8) where y_k $(k = 1, 2, \ldots, N)$ is the expression obtained in Steps 1 and 2.

Step 4. Compute y_k, $k = 1, 2, \ldots, N$ from expressions obtained in Steps 1 and 2 for the value of N obtained in Step 3.

Step 5. Compute x_k from $x_k = y_k - b_{k-1} y_{k-1}$, $k = 1, 2, \ldots, N$.

It can be shown that Eqs. (6) and (7) are equivalent to Eqs. (2) and (3), respectively, when $b_k = 0$, $k = 1, 2, \ldots, N$. This means that the optimal PM schedules in Model 1 and Model 2 are the same for the pure hazard rate adjustment model.

3. Numerical Examples

We consider the case where the hazard rate function of the system is Weibull, i.e.,

$$h(t) = \beta t^{\alpha - 1}, \ \alpha > 1, \ \beta > 0.$$

For Model 1, Eq. (2) becomes

$$A_k \beta y_k^{\alpha - 1} - A_{k+1} b_k \beta (b_k y_k)^{\alpha - 1} = A_N (1 - b_k) \beta y_N^{\alpha - 1}, \qquad (k = 1, 2, \ldots, N - 1).$$

Solving this equation we get

$$y_k = \left[\frac{A_N(1 - b_k)}{A_k - A_{k+1}b_k^\alpha} \right]^{\frac{1}{\alpha-1}} y_N, \qquad (k = 1, 2, \ldots, N - 1). \tag{9}$$

Substituting Eq. (9) into Eq. (3) leads to

$$y_N = \frac{[c_r + c_p(N - 1)]^{\frac{1}{\alpha}}}{\left\{ c_m\beta(1 - 1/\alpha) \left[A_N + A_N^{\frac{\alpha}{\alpha-1}} \sum_{k=1}^{N-1} d_k \right] \right\}^{\frac{1}{\alpha}}}. \tag{10}$$

Then $A_N h(y_N)$ becomes

$$\beta \frac{A_N[c_r + c_p(N - 1)]^{\frac{\alpha-1}{\alpha}}}{\left\{ c_m\beta(1 - 1/\alpha) \left[A_N + A_N^{\frac{\alpha}{\alpha-1}} \sum_{k=1}^{N-1} d_k \right] \right\}^{\frac{\alpha-1}{\alpha}}}.$$

Minimization of $A_N h(y_N)$ is equivalent to minimization of function

$$B(N) = \frac{c_r + c_p(N - 1)}{A_N^{-\frac{1}{\alpha-1}} + \sum_{k=1}^{N-1} d_k}, \qquad N = 1, 2, \ldots.$$

The optimal value of N can be found by solving inequalities $B(N+1) \geq B(N)$ and $B(N) < B(N - 1)$. These two inequalities imply that

$$D(N) \geq \frac{c_r}{c_p} \quad \text{and} \quad D(N - 1) < \frac{c_r}{c_p}, \tag{11}$$

where

$$D(N) = \frac{A_N^{-\frac{1}{\alpha-1}} + \sum_{k=1}^{N-1} d_k}{A_{N+1}^{-\frac{1}{\alpha-1}} - A_N^{-\frac{1}{\alpha-1}} + d_N} - (N - 1).$$

Thus, the optimal value N^* must satisfy inequalities (11). The optimal PM intervals are computed from $x_k = y_k - b_{k-1}y_{k-1}$ $(k = 1, 2, \ldots, N)$, where y_k $(k = 1, 2, \ldots, N)$ are given in Eqs. (9) and (10).

To compute the optimal PM schedule, we need to know the cost parameters c_m, c_p and c_r, the Weibull parameters α and β, and the adjustment factors a_k and b_k. Usually, maintenance engineers should be consulted for the determination of these parameters. For the cost parameters, we actually only need to know the cost ratios c_r/c_p and c_m/c_p. For the selection of adjustment factors a_k and b_k, please refer to Refs. 8 and 15, where the adjustment factors a_k and b_k are introduced. In this numerical example, however, we will assume the following parameter values for illustration purposes.

Table 1. Optimal PM schedules for Weibull case in Model 1.

c_r/c_p	2	5	10	20	50
N^*	1	3	5	7	11
x_1	0.5774	0.6507	0.8371	1.0822	1.5584
x_2		0.3219	0.4141	0.5353	0.7709
x_3		0.4006	0.3020	0.3905	0.5623
x_4			0.2366	0.3058	0.4404
x_5			0.3452	0.2460	0.3543
x_6				0.2005	0.2887
x_7				0.3068	0.2370
x_8					0.1954
x_9					0.1615
x_{10}					0.1338
x_{11}					0.2122

Table 1 shows the optimal PM schedules, N^*, $x_1, x_2, \ldots, x_{N^*}$, for $c_r/c_p = 2, 5,$ 10, 20, 50, where $c_m/c_p = 4$, $\alpha = 2$, $\beta = 3$ and

$$a_k = \frac{6k+1}{5k+1}, \qquad b_k = \frac{k}{2k+1}, \qquad k = 0, 1, 2, \ldots .$$

For Model 2, solving Eq. (6) with respect to y_k ($k = 1, 2, \ldots, N$), we have

$$y_k = \left(\frac{\lambda}{A_k\beta}\right)^{\frac{1}{\alpha-1}}, \qquad k = 1, 2, \ldots, N. \tag{12}$$

Substituting Eq. (12) into Eq. (7) and solving it with respect to λ, we get

$$\lambda = \beta^{\frac{1}{\alpha}}\left[\frac{c_r + c_p(N-1)}{(1-1/\alpha)c_m E(N)}\right]^{\frac{\alpha-1}{\alpha}}, \tag{13}$$

where

$$E(N) = \sum_{k=1}^{N-1}(1 - a_k b_k^\alpha)A_k^{-\frac{1}{\alpha-1}} + A_N^{-\frac{1}{\alpha-1}}.$$

Then expression (8) becomes

$$\beta^{\frac{1}{\alpha}}\left[\frac{c_r + c_p(N-1)}{(1-1/\alpha)c_m}\right]^{\frac{\alpha-1}{\alpha}}\frac{[E(N)]^{1/\alpha}}{F(N)}, \tag{14}$$

where

$$F(N) = \sum_{k=1}^{N-1}(1 - b_k)A_k^{-\frac{1}{\alpha-1}} + A_N^{-\frac{1}{\alpha-1}}.$$

Minimization of the function in expression (14) is equivalent to minimisation of function

$$Q(N) = \frac{[c_r + c_p(N-1)]^{\frac{\alpha-1}{\alpha}}[E(N)]^{1/\alpha}}{F(N)}.$$

Inequalities $Q(N+1) \geq Q(N)$ and $Q(N) < Q(N-1)$ imply that

$$W(N) \geq \frac{c_r}{c_p} \quad \text{and} \quad W(N-1) < \frac{c_r}{c_p}, \tag{15}$$

where

$$W(N) = \frac{[E(N+1)]^{\frac{1}{\alpha-1}}[F(N)]^{\frac{\alpha}{\alpha-1}}}{[E(N)]^{\frac{1}{\alpha-1}}[F(N+1)]^{\frac{\alpha}{\alpha-1}} - [E(N+1)]^{\frac{1}{\alpha-1}}[F(N)]^{\frac{\alpha}{\alpha-1}}} - (N-1).$$

Thus, the optimal value N^* must satisfy inequalities (15). The optimal PM intervals are computed from $x_k = y_k - b_{k-1}y_{k-1}$ ($k = 1, 2, \ldots, N$), where y_k ($k = 1, 2, \ldots, N$) and λ are given in Eqs. (12) and (13).

Table 2 gives the optimal PM schedules in Model 2 for the same parameters adopted in the example for Model 1.

Table 2. Optimal PM schedules for Weibull case in Model 2.

c_r/c_p	2	5	10	20	50
N^*	1	3	5	8	11
x_1	0.5774	0.7137	0.8658	1.0775	1.5230
x_2		0.3738	0.4535	0.5644	0.7978
x_3		0.2729	0.3311	0.4121	0.5824
x_4			0.2597	0.3232	0.4568
x_5			0.2092	0.2603	0.3679
x_6				0.2124	0.3002
x_7				0.1745	0.2466
x_8				0.1439	0.2035
x_9					0.1683
x_{10}					0.1395
x_{11}					0.1158

4. Concluding Remarks

In this paper we have presented a general PM model which assumes that PM not only reduces the effective age of the system but also changes the hazard rate function. Two models for obtaining the optimal PM schedules have been studied. Generally, the optimal solutions in the two models are not the same. When the general policy reduces to the hazard rate adjustment policy ($b_k = 0$), however, it can be shown that the optimal PM schedules obtained from Model 1 and Model 2 are exactly the same. From the examples, we see that the optimal solution to y_k is analytically found in the Weibull case. For a general case, however, the optimal y_k may not be analytically solved. The algorithms in Section 2 may become numerical procedures, and standard numerical approaches for solving an equation could be used in Step 1.

From Table 1, we see that x_k decreases for all k values except the last one — a result similar to that given by Nakagawa.[14] This proves again Nakagawa's comment[14]: "It would be reasonable to do frequent PM with age, but it would be better to do the last PM as late as possible because the system should be replaced at next PM." From Table 2, we observe that x_k decreases for all k values. Therefore, we may conclude that Model 1 would be more practical than Model 2.

In Model 2, the hazard rate at every PM remains the same (equals λ). The predetermined level λ is selected to minimize the mean cost rate. An alternative and reasonable way is to determine λ to meet the practical requirement of for reliability the system. That is, λ is set to be a specific level λ_0 (λ_0 is known). Then y_k ($k = 1, 2, \ldots, N$) can be computed from Eq. (6). The mean cost rate C is a function only of N. Thus, we can find N which minimizes C and finally obtain the optimal PM schedule.

Comparing our results with the results in Nakagawa,[14] we find that the optimal PM policy using our model is to do more PMs and more frequent PM than using Nakagawa's models. The reason for this is that in our model we assume that PM not only reduces the effective age to a certain value but also increases the slope of the hazard rate function. This assumption is more realistic and more general than that in Nakagawa's models. Thus, our models are more general and more flexible than Nakagawa's models. Actually, Nakagawa's models can be treated as special cases of our models.

Acknowledgment

This research was partially supported by the City University of Hong Kong and the University of Alberta. Critical comments and constructive suggestions from the editor and the referees are very much appreciated.

References

1. P. K. W. Chan and T. Downs, "Two criteria for preventive maintenance," *IEEE Transactions on Reliability* **R-27** (1978), pp. 272–273.
2. D. N. P. Murthy and D. G. Nguyen, "Optimal age-policy with imperfect preventive maintenance," *IEEE Transactions on Reliability*, **R-30** (1981), pp. 80–81.
3. T. Nakagawa, "Optimum policies when preventive maintenance is imperfect," *IEEE Transactions on Reliability*, **R-28** (1979), pp. 331–332.
4. M. Brown and F. Proschan, "Imperfect repair," *Journal of Applied Probability* **20** (1983), pp. 851–859.
5. R. A. Fontenot and F. Proschan, "Some imperfect maintenance models," in *Reliability Theory and Models* (Academic Press, 1984), pp. 83–101.

6. S.-H. Sheu and C.-T. Liou, "A generalized sequential preventive maintenance policy for repairable systems with general random minimal repair costs," *International Journal of Systems Science* **26** (1995), pp. 681–690.

7. M. A. K. Malik, "Reliable preventive maintenance scheduling," *AIIE Transactions* **11** (1979), pp. 221–228.

8. C. H. Lie and Y. H. Chun, "An algorithm for preventive maintenance policy," *IEEE Transactions on Reliability* **R-35** (1986), pp. 71–75.

9. V. Jayabalan and D. Chaudhuri, "Optimal maintenance and replacement policy for a deteriorating system with increased mean downtime," *Naval Research Logistics* **39** (1992), pp. 67–78.

10. V. Jayabalan and D. Chaudhuri, "Cost optimization of maintenance scheduling for a system with assured reliability," *IEEE Transactions on Reliability* **41** (1992), pp. 21–25.

11. A. Monga, M. J. Zuo, and R. W. Toogood, "Reliability-based design of systems considering preventive maintenance and minimal repair," *International Journal of Reliability, Quality and Safety Engineering* **4** (1997), pp. 55–71.

12. D. G. Nguyen and D. N. P. Murthy, "Optimal preventive maintenance policies for repairable systems," *Operational Research* **29** (1981), pp. 1181–1194.

13. T. Nakagawa, "Periodic and sequential preventive maintenance policies," *Journal of Applied Probability* **23** (1986), pp. 536–542.

14. T. Nakagawa, "Sequential imperfect preventive maintenance policies," *IEEE Transactions on Reliability* **R-37** (1988), pp. 295–298.

15. T. Nakagawa, "A summary of imperfect preventive maintenance policies with minimal repair," *R. A. I. R. O. Oper. Res.* **14** (1980), pp. 249–255.

16. M. Modarres, *What Every Engineer Should Know About Reliability And Risk Analysis* (Marcel Dekker, New York, 1993).

17. C. Valdez-Flores and R. Feldman, "A survey of preventive maintenance models for stochastically deteriorating single-unit systems," *Naval Research Logistics* **36** (1989), pp. 419–446.

About the Authors

Daming Lin received his B.Sc. degree in Applied Mathematics from Peking University in 1985, his M.Sc. degree in Operational Research and Control Theory from the Institute of Systems Science, Academic Sinica, China, in 1988, and his Ph.D. in Statistics from the University of Hong Kong in 1995. He worked in the Department of Mathematics at Shantou University, China from 1988 to 1999. Now he works as a postdoctoral fellow in the Department of Mechanical and Industrial Engineering at the University of Toronto. His research interests include reliability analysis, preventive maintenance, quality control, optimization techniques and Bayesian statistics.

Ming J. Zuo received his B.Sc. degree in Agricultural Engineering from Shandong Institute of Technology in 1982, and his M.Sc. degree in 1986 and his Ph.D. in 1989, both in Industrial Engineering from Iowa State University, Ames, Iowa, USA. He worked at the University of Windsor in Canada for a year in 1989. Since 1990, he has been a faculty member in the Department of Mechanical Engineering at the University of Alberta in Canada. From 1996 to 1998, he was on leave at the City University of Hong Kong. His research interests include system reliability, manufacturing systems and applied operations research. He is on the editorial board of the journal IIE Transactions on Quality and Reliability Engineering. He is also a member of IEEE and IIE.

Richard C. M. Yam received his M.Sc. degree with distinction in Fibre Science and Technology from the University of Leeds, his M.Sc. degree in Management Science from the London University Imperial College of Science & Technology and his Ph.D. in Engineering Management from the University of Warwick, UK. He is now Associate Professor and Program Leader of the M.Sc. Engineering Management Programme at the City University of Hong Kong. He is the Society Editor of the Engineering Management Journal, American Society for Engineering Management. His research interests include maintenance management system, quality management, engineering and technology management.

© 2025 World Scientific Publishing Company
https://doi.org/10.1142/9789819812547_0002

Chapter 2

AN NHPP SOFTWARE RELIABILITY MODEL AND ITS COMPARISON*

HOANG PHAM and XUEMEI ZHANG
Rutgers University

In this paper, software reliability models based on a nonhomogeneous Poisson process (NHPP) are summarized. A new model based on NHPP is presented. All models are applied to two widely used data sets. It can be shown that for the failure data used here, the new model fits and predicts much better than the existing models. A software program is written, using Excel & Visual Basic, which can be used to facilitate the task of obtaining the estimators of model parameters.

Keywords: Software Reliability; NHPP; Maximum Likelihood Estimation; Probability Distribution; Mean Value Function; Sum of Squared Errors; Reliability Prediction.

1. Introduction

Research activities in software reliability engineering have been conducted over the past two decades and more than 60 software reliability models have been proposed. Software reliability growth models (SRGMs)[1,12,15,17,24,26,28] have been proven to be successful for the estimation of software reliability and the number of errors remaining in the software. Different SRGMs make different assumptions and therefore can be applied to different situations. For example, some models assume that each time a failure occurs, the error which caused it is immediately removed and no new errors are introduced. Other models have proposed a relaxation of the above assumptions.[7,10,11,17,25,27] Presently, software applications are playing increasingly important roles in almost all fields. At the same time, software reliability research are also receiving more and more attention. New models come out continuously, which relax more assumptions of existing models and therefore are more close to the real world. Since the failure data of software products are diverse, there is no single software reliability model which can be the best for any type of data and failure phenomenon.

Among all SRGMs, a big family of stochastic reliability models based on a nonhomogeneous Poisson process, which is known as NHPP reliability models, have been widely used (like Goel & Okumoto (G–O model)[1]). Many modifications of

---

*This chapter appeared previously on the International Journal of Reliability, Quality and Safety Engineering. To cite this chapter, please cite the original article as the following: H. Pham and X. Zhang, *Int. J. Reliab. Qual. Saf. Eng*, **4**, 269–282 (1997), doi:10.1142/S0218539397000199.

these models have been investigated from which new models have been obtained. Research has been conducted to estimate the parameters of these models. The most efficient ones are maximum likelihood estimators (MLEs) and least square estimators (LSEs).

In this paper, eleven existing NHPP reliability models are summarized. A newly developed reliability model is presented and compared with other NHPP models. All models are applied to two sets of real software failure data. Results are provided to show that the new model presented in this paper fits better than the other models.

2. Software Reliability Modeling

2.1. *Software Reliability Growth Models*

Notation

$a(t)$ total errors content at time t

$b(t)$ error detection rate at time t

$\lambda(t)$ intensity function or fault detection rate per unit time

$m(t)$ the mean value function or the expected number of errors detected by time t

$R(x/t)$ reliability function of software by time t for a mission time x

$N(t)$ counting process representing the cumulative number of failures at time t

$\sum_k$ sum over k from 1 to n

SSE   sum of squared errors of a model fitting the actual data

During the testing phase of software development, each error is fixed when it is discovered. This decreases the number of errors in the software and so the error-discovery rate should decrease. In other words, the length of time between error discoveries should increase. When the error-discovery rate reaches an acceptably low level, the software is deemed suitable to be released to the customers. However, it is difficult to extrapolate from error-discovery rate in a test environment to failure rate during system operation, primarily because it is hard to extrapolate from the testing time to system operation time. Instead, the expected number of remaining errors (i.e. residual errors) provide an upper limit on the number of unique failures customers could encounter in field use.

Knowing the number of the residual errors helps determine whether or not the code is suitable for customer use and how much more testing is required if it is not. It also provides an estimate of the number of failures that customers will encounter when operating the software. This estimate helps to define the appropriate levels of support that will be required for error correction after the software is released.

Software reliability growth models are mathematical functions that describe error-detection rate. There are two major classes: concave and S-shaped models, as illustrated in Fig. 1. Concave models always bend downward while S-shaped models are first convex and then concave. This reflects their underlying assumptions that early testing is not as efficient as later testing; there is a "ramp-up" period during which the error-detection rate increase.

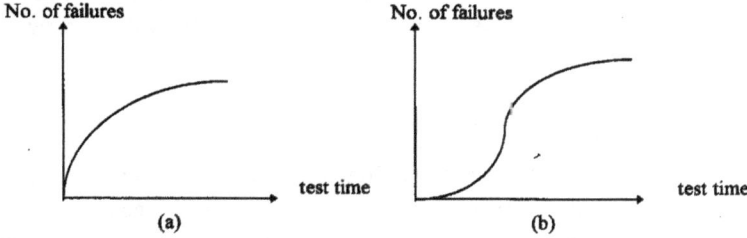

Fig. 1. Two classes of most software reliability growth models (a) Concave (b) S-shaped.

Of course, error repair can introduce new errors. Some models explicitly account for new errors during test by modifying the mathematical form of the model, while others assume new errors are accounted for by the statistical fit of the model to the data. In practice, either method works as long as the model does a good job of fitting the data.

2.2. NHPP models

The counting process $\{N(t), t \geq 0\}$ is said to be a nonhomogeneous Poisson process (NHPP) with intensity function $\lambda(t), t \geq 0$ [27], if $N(t)$ has a Poisson distribution with mean value function $m(t)$, i.e.,

$$\Pr\{N(t) = k\} = \frac{[m(t)]^k}{k!} e^{-m(t)}, \qquad k = 0, 1, 2, \ldots, \tag{1}$$

By definition, the mean value function m(t), which is the cumulative number of failures, can be expressed in terms of failure rate of the program, i.e.,

$$m(t) = \int_0^t \lambda(s) ds \tag{2}$$

where $\lambda(s)$ is also called the intensity function, or failure rate function.

The main issue in the NHPP model is to determine an appropriate mean value function (MVF) to denote the expected number of failures experienced up to a certain time point.

Software reliability $R(x/t)$ is defined as the probability that a software failure does not occur in $(t, t + x)$, given that the last failure occurred at testing time $t(t \geq 0, x > 0)$. That is,

$$R(x/t) = e^{-[m(t+x)-m(t)]} \tag{3}$$

For special cases, when $t = 0$ then $R(x/0) = e^{-m(x)}$, and $t = \infty$ then $R(x/\infty) = 1$.

2.3. A Generalized NHPP Software Reliability Model

Pham and Nordmann[14] recently present a generalized NHPP software reliability model and an analytical expression for the mean value function. They propose a generalized form for the mean value function, which can be obtained by solving the following differential equation:

$$\frac{dm(t)}{dt} = b(t)[a(t) - m(t)] \quad \text{with } m(t_0) = m_0 \tag{4}$$

This equation relates $m(t)$ to two other functions that, by their definition, possess actual physical meanings. The function $a(t)$ represents the total error content at time t and $b(t)$ represents the error detection rate. In this way, by introducing functional assumptions about $a(t)$ and $b(t)$, which are more tangible, an analytical expression for $m(t)$ can be derived. Most of the reliability models summarized here employ their own forms of the function $a(t)$ and $b(t)$, which can be interpreted as different assumptions made.

In this paper, we propose the following two functions $a(t)$ and $b(t)$

$$a(t) = c + a(1 - e^{-\alpha t}) \tag{5}$$

and

$$b(t) = \frac{b}{1 + \beta e^{-bt}} \tag{6}$$

Here the function $a(t)$ is not linearly increasing in time t, but in a exponentially way; which means the number of error increases more quickly at the beginning of the test process than at the end of it. This reflects that at the beginning more errors are introduced into the software, while at the end developers possess more knowledge and therefore introduce less errors into it. Also, the error detection rate function $b(t)$ is not a constant anymore but a non-decreasing function with a inflection S-shaped model and therefore can be used to describe a broad range of practical situation. (see Fig. 2)

From the analytical expression for the mean value function in Eq. (4),[14] we can obtain the mean value function for the new model as follows:

$$m(t) = \frac{1}{(1 + \beta e^{-bt})}[(c + a)(1 - e^{-bt}) - \frac{ab}{b - \alpha}(e^{-\alpha t} - e^{-bt})] \tag{7}$$

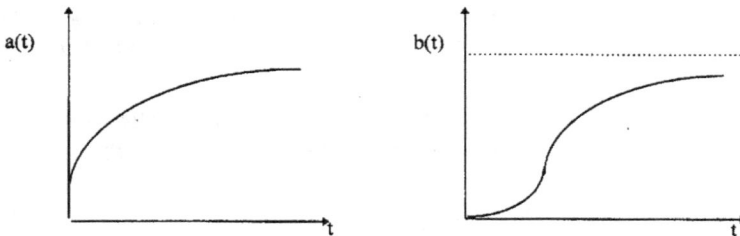

Fig. 2. The shapes of functions $a(t)$ and $b(t)$

2.4. *Summary of software reliability models*

NHPP software reliability models predict the expected number of errors. Different models use different assumptions and therefore provide different mathematical forms for the mean value function (MVF). Table 1 shows a summary of the models for the mean value function (MVF) appeared in the literature during nearly two decades.

3. Data Analysis and Models Comparisons

3.1. *Data analysis: Data set no. 1*

In this section a set of testing data is given in Table 2. The data are reported by Musa,[23] which are summarized as numbers of failures per one-hour interval of execution time and also cumulative failures. This data set belongs to the concave class.

3.2. *Data analysis: data set no. 2*

We also use another widely used data set which is from the Naval Fleet Computer Programming Center, Naval Tactical Data Systems (NTDS). It was extracted from information about failures in the development of software for the real-time, multi-computer complex which forms the core of the NTDS. The software consisted of 38 different project schedules. Each module was supposed to have 3 phases: production, test, and user. The times (in days) between failures are shown in Table 3; 26 software failures were found during the production phase, and 5 during the test phase; the last failure was found on January 4, 1971. One failure was observed during the user phase in September 1971. Again, 2 failures were observed during the test phase in 1971. In this paper, we use the first 26 data for model comparison. The reason lies in the fact that the last three of the first 26 errors in NTDS data occur almost in a cluster and there is a relatively long interval between errors before and after the cluster. Therefore the error number 26 is an unfortunate cut-off point. This data set belongs to the S-shaped class.

3.3. *Models comparisons*

Here, we apply all the models mentioned in Sec. 2.3 to the two data sets. The estimators and the curves of mean value functions and the actual failure data are obtained for each data set. First, parameters of all models are estimated and the mean value functions are obtained. Second, all the models are compared with each other and the best fitting model is found to be the new model presented in this paper.

The criterion we used here, Sum of Squared Error (SSE), is to judge the performance of the models which sum up the square of the residuals of the actual data and the mean value function ($m(t)$) of each model in terms of the number of actual failure errors at any time points.

Table 1. Summary of the software reliability models and their mean value functions.

Model Name	Model Type	$MVF(m(t))$	Comments
Goel–Okumoto (G–O)[1]	Concave	$m(t) = a(1 - e^{-bt})$ $a(t) = a$ $b(t) = b$	Also called exponential model.
Delayed S-shaped SRGM[2]	S-shaped	$m(t) = a(1 - (1 + bt)e^{-bt})$	Modification of G–O model to make it S-shaped
Inflection S-shaped SRGM[8]	Concave	$m(t) = \dfrac{a(1 - e^{-bt})}{1 + \beta e^{-bt}}$ $a(t) = a$ $b(t) = \dfrac{b}{1 + \beta e^{-bt}}$	Solves a technical condition with the G–O model. Becomes the same as G–O if $\beta = 0$
Gompertz[4]	S-shaped	$m(t) = a(b^{c^t})$	Used by Fujitsu, Numazu Works.
Pareto[5]	Concave	$m(t) = a(1 - (1 + t/\beta)^{1-\alpha})$	Assumes failures have different failure rates and failure with highest rate removed first.
Weibull[6]	Concave	$m(t) = a(1 - e^{-bt^c})$	Same as G–O when $c = 1$
Yamada Exponential[7]	Concave	$m(t) = a(1 - e^{-r\alpha(1 - e^{(-\beta t)})})$ $a(t) = a$ $b(t) = r\alpha\beta t e^{-\beta t}$	Attempt to account for testing-effort
Yamada Rayleigh[7]	S-shaped	$m(t) = a(1 - e^{-r\alpha(1 - e^{(-\beta t^2/2)})})$ $a(t) = a$ $b(t)r\alpha\beta t e^{-\beta t^2/2}$	Attempt to account for testing-effort
Yamada Imperfect Debugging model (1)[21]	S-shaped	$m(t) = \dfrac{ab}{\alpha + b}(e^{\alpha t} - e^{-bt})$ $a(t) = ae^{\alpha t}$ $b(t) = b$	Assume exponential fault content function and constant error detection rate
Yamada Imperfect Debugging model (2)[21]	S-shaped	$m(t) = a[1 - e^{-bt}]\left[1 - \dfrac{\alpha}{b}\right] + \alpha at$ $a(t) = a(1 + \alpha t)$ $b(t) = b$	Assume constant introduction rate α and the error detection rate
Pham-Nordmann[14]	S-shaped and concave	$m(t) = \dfrac{a[1 - e^{-bt}]\left[1 - \dfrac{\alpha}{b}\right] + \alpha at}{1 + \beta e^{-bt}}$ $a(t) = a(1 + \alpha t)$ $b(t) = \dfrac{b}{1 + \beta e^{-bt}}$	Assume introduction rate is a linear function of testing time, and the error detection rate function is non-decreasing with an inflexion S-shaped model.
New model	S-shaped and concave	$m(t) = \dfrac{1}{(1 + \beta e^{-bt})}[(c+a)(1 - e^{-bt})$ $\qquad - \dfrac{ab}{b-\alpha}(e^{-\alpha t} - e^{-bt})]$ $a(t) = c + a(1 - e^{-1\alpha t})$ $b(t) = \dfrac{b}{1 + \beta e^{-bt}}$	Assume introduction rate is exponential function of the testing time, and the error detection rate is non-decreasing with an inflexion S-shaped model.

Table 2. Failure in 1 hour (execution time) intervals and cumulative failures (data set no. 1).

Hour	Number of Failures	Cumulative Failures
1	27	27
2	16	43
3	11	54
4	10	64
5	11	75
6	7	82
7	2	84
8	5	89
9	3	92
10	1	93
11	4	97
12	7	104
13	2	106
14	5	111
15	5	116
16	6	122
17	0	122
18	5	127
19	1	128
20	1	129
21	2	131
22	1	132
23	2	134
24	1	135
25	1	136

The SSE of a model can be expressed as follows:

$$SSE = \sum_{k=1}^{n} [y_k - \tilde{m}(t_k)]^2 \tag{8}$$

where

y_k: total number of failures observed at time t_k according to the failure data
$\tilde{m}(t_k)$: estimated cumulative number of failures at time t_k obtained from the fitted mean value function, $k = 1, 2, \ldots, n$.

Therefore, the less the SSE value, the better the model performs.

Theoretically, we can obtain the MLEs of the parameters by solving a group of differential equations. However, in practice, the estimations may not be obtained, especially under some circumstances in which the mean value function is complex.

Table 3. Software Failure Data From NTDS (Data set no. 2).

Error No. k	Inter Failure Times x_k (days)	Cumulative Time $s_k = \sum_{i=1}^{k} x_i$
Production Phase		
1	9	9
2	12	21
3	11	32
4	4	36
5	7	43
6	2	45
7	5	50
8	8	58
9	5	63
10	7	70
11	1	71
12	6	77
13	1	78
14	9	87
15	4	91
16	1	92
17	3	95
18	3	98
19	6	104
20	1	105
21	11	116
22	33	149
23	7	156
24	91	247
25	2	249
26	1	250
Test Phase		
27	87	337
28	47	384
29	12	396
30	9	405
31	135	540
User Phase		
32	258	798
Test Phase		
33	16	814
34	35	849

This is because the differential equations are complex and the estimators cannot be expressed explicitly. We have developed a software program using Excel and Visual Basic tools to facilitate the task of finding the MLEs. Users can input their mean value function and with a single mouse click, the results of the MLEs parameters as well as the curve of mean value function fitting the actual failure data will be provided.

3.3.1. *Models comparison when appied to concaved data (Data set no. 1)*

First of all, we fit all models to data set no. 1 which is a set of concave data. Table 4 illustrates the estimators and the SSEs for all models.

From Table 4, we can see that the new model performs significantly better than the others when applied to this concave data. The SSE of the new model, which is 4071, is much lower than those of other models. It is worthwhile to notice that Yamada imperfect debugging model (1) works also significantly better with SSE value 4812.

Figure 3 gives the actual failure data as well as the curves of the mean value functions of the four best fitting models which are Yamada imperfect debugging model (1) and (2), Pham-Nordmann model and the new model. We can see that the new model is the best among all.

Table 4. Model comparison: SSEs for data set no. 1 (concave data).

Model Name	$MVF(m(t))$	SSE
Goel–Okumoto(G-O)[1]	$m(t) = a(1 - e^{-bt})$	6816
Delayed S-shaped SRGM[2]	$m(t) = a(1 - (1 + bt)e^{-bt})$	27827
Inflection S-shaped SRGM[8]	$m(t) = \dfrac{a(1 - e^{-bt}}{1 + \beta e^{-bt})}$	6428
Gompertz[4]	$m(t) = a(b^{c^t})$	200000
Pareto[5]	$m(t) = a(1 - (1 + t/\beta)^{1-\alpha})$	8609
Weibull[6]	$m(t) = a(1 - e^{-bt^c})$	5361
Yamada Exponential[7]	$m(t) = a(1 - e^{-r\alpha(1-e^{-\beta t})})$	6789
Yamada Rayleigh[7]	$m(t) = a(1 - e^{-r\alpha(1-e^{(-\beta t^2/2)})})$	69557
Yamada Imperfect Debugging model (1)[21]	$m(t) = \dfrac{ab}{\alpha + b}(e^{\alpha t} - e^{-bt})$	4812
Yamada Imperfect Debugging model (2)[21]	$m(t) = a[1 - e^{-bt}]\left[1\dfrac{\alpha}{b}\right] + \alpha at$	5069
Pham-Nordmann	$m(t) = \dfrac{a[1 - e^{-bt}]\left[1 - \dfrac{\alpha}{b}\right] + \alpha at}{1 + \beta e^{-bt}}$	5425
New model	$m(t) = \dfrac{1}{(1 + \beta e^{-bt})}[(c + a)(1 - e^{-bt}) - \dfrac{ab}{b - \alpha}(e^{-\alpha t} - e^{-bt})]$	4071*

m(t)

| the Mean Value Functions vs. the Actual Failure Data |

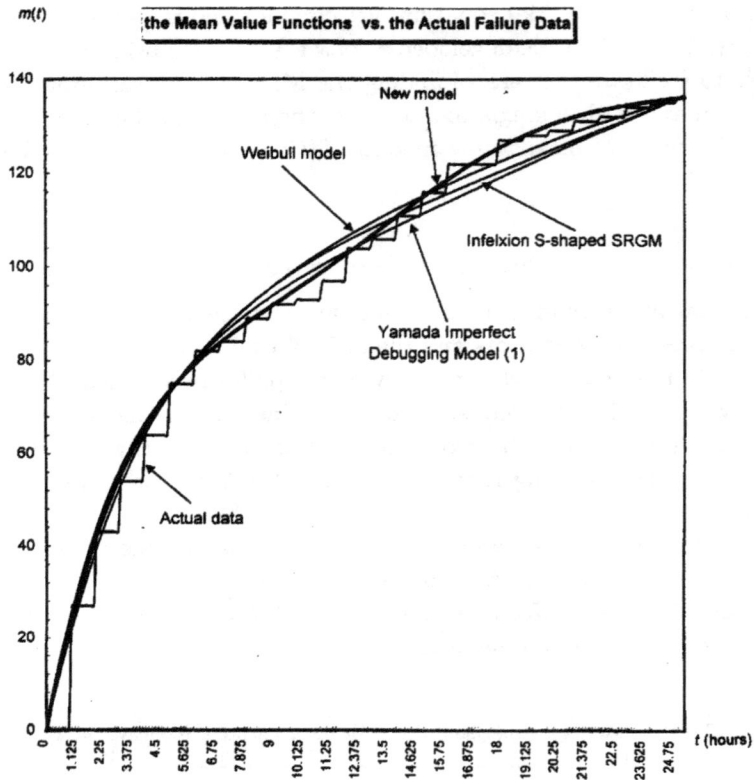

Fig. 3. The mean value functions ($m(t)$) versus the actual failure data no. 1.

3.3.2. Models comparison when appied to S-shaped data (data set no. 2)

Secondly, we fit the six S-shaped models to the data set no. 2 which is a set of S-shaped data. Table 5 gives the estimators and the SSE value for each model. From the results, we can see that the new model works better than Yamada's imperfect debugging model (1) and all the other four models. This is because the new model carries more information and fits perfectly to the S-shaped data. Therefore, we draw a general conclusion that the new model fits and predicts very well for both concave and S-shaped data sets.

Figure 4 gives the curves of the mean value functions of all the six S-shaped models and the actual failure data. We can see that the new model is the best fitting model.

Table 5. Model comparison: SSEs for data set no. 2 (S-shaped data).

Model Name	$MVF(m(t))$	SSE
G–O (Delayed) S-shaped SRGM[2]	$m(t) = a(1 - bt)e^{-bt})$	361.9
Yamada Rayleigh[7]	$m(t) = a(1 - e^{-r\alpha(1-e^{(-\beta t^2/2)})})$	1141
Yamada Imperfect Debugging model (1)[21]	$m(t) = \dfrac{ab}{\alpha + b}(e^{\alpha t} - e^{-bt})$	565.9
Yamada Imperfect Debugging model (2)[21]	$m(t) = a[1 - e^{-bt}]\left[-\dfrac{\alpha}{b}\right] + \alpha at$	629.3
Pham–Nordmann[14]	$m(t) = \dfrac{a[1 - e^{-bt}]\left[-\dfrac{\alpha}{b}\right] + \alpha at}{1 + \beta e^{-bt}}$	394.5
New model	$m(t) = \dfrac{1}{(1 + \beta e^{-bt})}\left[(c + a)(1 - e^{-bt})\right.$ $\left. - \dfrac{ab}{b - \alpha}(e^{-\alpha t} - e^{-bt})\right]$	251.5*

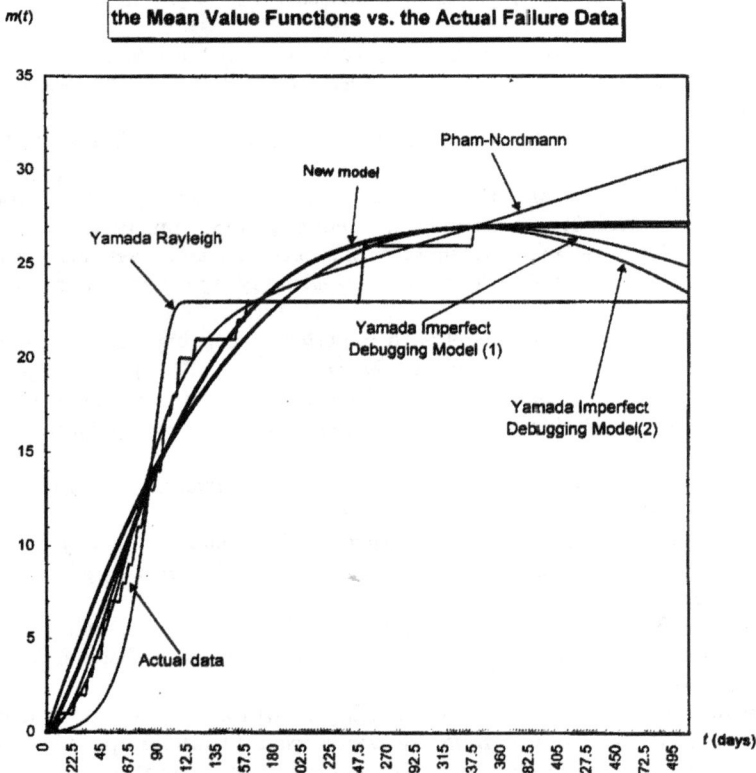

Fig. 4. The mean value functions (m(t)) versus the actual failure data no. 2.

4. Conclusions

In this paper, a new reliability model is developed and applied to two widely used data sets. It can be proven that this model fits the failure data very well. It is also compared to other existing NHPP software reliability models. The results show that the new model performs better than the others when applied to both kinds of failure data presented in this paper. A software program is developed to facilitate the process of finding the MLEs of model parameters and the mean value functions.

References

1. A. L. Goel and K. Okumoto, "Time-dependent error-detection rate model for software and other performance measures," *IEEE Transactions on Reliability*, **28** (1979) pp. 206–211.
2. S. Yamada, M. Onha, and S. Osaki, "S-shaped reliability growth modeling for software error detection," *IEEE Transactions on Reliability* **12** (1983), pp. 475–484.
3. S. A. Hossain and Ram C. Dahiya, "Estimating the parameters of a non-homogeneous poisson process model for software reliability," *IEEE Transactions on Reliability*, **42** (4) (1993) pp. 604–612.
4. D. Kececioglu, *Reliability Engineering Handbook, Vol. 2* (Prentice-Hall, Englewood Cliffs, N.J., 1991).
5. B. Littlewood, "Stochastic reliability growth: A model for fault removal in computer programs and hardware design," *IEEE Transactions on Reliability*, **12** (1981) pp. 313–320.
6. J. Musa, A. Iannino, and K. Okumoto, *Software Reliability* (McGraw-Hill, New York, 1987).
7. S. Yamada, H. Ohtera, and H. Narihisa, "Software reliability growth models with testing effort," *IEEE Transactions on Reliability*, **4** (1986) pp. 19–23.
8. S. Yamada, "Software Quality/Reliability measurement and assessment: software reliability growth models and data analysis," *Journal of Information Processing*, **14** (3) (1991) pp. 254–266.
9. H. Pham, "A software cost model with imperfect debugging, random life cycle and penalty cost", *International Journal of Systems Science*, **27** (5) (1996) pp. 455–463.
10. P. K. Kapur and V. K. Bhalla, "Optimal release policies for a flexible software reliability growth model," *Reliability Engineering and System Safety Journal*, **35** (1992) pp. 45–54.
11. Y. W. Leung, "Optimal software release time with a given cost budget," *Journal of Systems and Software*, **17** (1992) pp. 233–242.
12. H. Pham, "Software Reliability and Testing," *IEEE Computer Society Press*, 1995.
13. H. Pham, "Fault-Tolerant Software Systems: Techniques and Applications," *IEEE Computer Society Press*, 1992,
14. H. Pham and L. Nordmann, "A generalized NHPP software reliability model," *Proceedings of the Third ISSAT International Conference on Reliability & Quality in Design* **3** (1997) pp. 116–120.
15. S. Yamada and S. Osaki, "Software reliability growth modeling: models and applications", *IEEE Transactions on Software Engineering*, **11** (1985) pp. 1431–1437.
16. M. Ohba, "Software reliability analysis models," *IBM Journal of Research Development*, **28** (1984) pp. 428–443.
17. H. Pham, *Software Reliability Assessment: Imperfect Debugging and Multiple Failure Types in Software Development* (EG&G-RAAM-10737; Idaho National Laboratory, 1993).

18. M. Ohba, "Inflection S-shaped software reliability growth models," *Stochastic Models in Reliability Theory*, ed. S. Osaki and Y. Hatoyama, (Springer-Verlag Merlin, 1984) pp. 144–162.

19. M. Ohba and S. Yamada, "S-shaped software reliability growth models," *Proc. 4th Int. Conf. Reliability and Maintainability*, (1984) pp. 430–436.

20. S. Yamada and S. Osaki, "Optimum software release policies for a nonhomogeneous software error detection rate model," *Microelectronics and Reliability — An International Journal*, **26** (1986) pp. 691–702.

21. S. Yamada, K. Tokuno, and S. Osaki, "Imperfect debugging models with fault introduction rate for software reliability assessment," *International Journal of Systems Science*, **23** (12) 1992.

22. Wood, Alan, "Predicting Software Reliability," *IEEE Computer*, (1996) pp. 69–77.

23. J. D. Musa, *Software Reliability Data* (DACS, RADC, New York, 1980).

24. H. Pham and M. Pham, *Software Reliability Models for Critical Applications* (Idaho National Engineering Laboratory, EG&G-2663, 1991).

25. H. Ohtera and S. Yamada, "Optimal allocation and control problems for software-testing resources," *IEEE Transactions on Reliability*, **39** (1990) pp. 171–176.

26. P. N. Misra, "Software reliability analysis", *IBM Systems Journal*, **22** (1983), pp. 262–270.

27. N. Kareer, P. K. Kapur and P. S. Grover, "An S-shaped software reliability growth model with two types of errors," *Microelectronics and Reliability — an International Journal*, **30** (1990), pp. 1085–1090.

28. M. Ohba, X. M. Chou, "Does imperfect debugging affect software reliability growth?," *Proceedings of the 11th IEEE International Conference on Software Engineering*, (1989) pp. 237–244.

About the Authors

Hoang Pham received his Ph.D. in industrial engineering from the State University of New York at Buffalo in 1989. He is on the faculty of the Department of Industrial Engineering at Rutgers University. Prior to joining Rutgers in August 1993, Dr. Pham worked with the Idaho National Engineering Laboratory (INEL) in Idaho Falls as a Senior Engineering Specialist from July 1990 to July 1993 and the Boeing Company in Seattle from March 1989 to July 1990 as a Senior Specialist Engineer. He is the editor of 8 books and the author of over 40 refereed journal articles.

He is Editor-in-Chief of the *International Journal of Reliability, Quality, and Safety Engineering*. He is also an associate technical editor of the *IEEE Communications Magazine* and an editorial board member of the *IIE Transactions on Quality and Reliability Engineering, Journal of Computer and Software Engineering*, and *International Journal of Plant Engineering and Management*. He was the guest editor of the *IEEE Transactions on Reliability and the Journal of Systems and Software*.

He is the general chair of the Fourth International Conference on Reliability and Quality in Design in Seattle, 1998. He was conference chair of the Third International Conference on Reliability and Quality in Design in Anaheim in 1997, and general chair of the IASTED Pacific-Rim International Conference on Reliability Engineering and Its Applications. He was conference program chair of the International Conference on Reliability and Quality in Design in 1994 and 1995, and was conference program vice-chairman of the IEEE 1994 Annual Reliability and Maintainability Symposium. He was also conference program chair of the International Conference on Reliability, Quality Control, and Risk Assessment in 1992, 1993, and 1994. He has served on technical committees for many major national and international conferences.

Xuemei Zhang received her B.S. and M.S. degrees from Beijing Institute of Technology, Beijing, China. Currently she is a Ph.D. candidate in the Department of Industrial Engineering at Rutgers, the State University of New Jersey. Her research interests include software reliability, software failure data analysis, software cost modeling and tools for software reliability and cost studies.

© 2025 World Scientific Publishing Company
https://doi.org/10.1142/9789819812547_0003

Chapter 3

SYSTEM RELIABILITY OPTIMIZATION WITH
k-out-of-n SUBSYSTEMS[#]

DAVID W. COIT* and JIACHEN LIU

Department of Industrial Engineering
Rutgers University, Piscataway, NJ 08854, USA
*E-mail: coit@rci.rutgers.edu

Optimal solutions to the redundancy allocation problem are determined for systems
designed with multiple k-out-of-n subsystems in series. The objective is to select the
components and redundancy levels to maximize system reliability given system-level
constraints. The individual subsystems may use either active or cold-standby redundancy, or they may require no redundancy. Previously, optimization methods for this
problem either pertained to k-out-of-n systems consisting of a single subsystem or to
series–parallel systems ($k = 1$). Additionally, it had generally been assumed that only
active redundancy was to be used. In practice design problems can vary appreciably
from these restrictions and the design process may consider more complex system configurations. Unfortunately, available optimization algorithms are inadequate for many
of these design problems. The methodology presented here is specifically developed to
accommodate the case with k-out-of-n subsystems. Optimal solutions to the problem are
found by an equivalent problem formulation and integer programming. The methodology
is demonstrated on a well-known test problem with interesting results. The availability
of this tool fills a void and should result in more reliable and cost-effective engineering
designs.

Keywords: Redundancy Allocation Problem; Reliability Optimization; k-out-of-n
Reliability.

1. Introduction

A new problem formulation and solution method is presented to determine optimal system design configurations when a system design includes multiple k-out-of-n
subsystems that are designed with either active or cold-standby redundancy. The
redundancy allocation problem has previously been analyzed for many different system structures, objective functions and time-to-failure distributions. Generally, the
problem domain has been limited to series–parallel systems with active redundancy
or k-out-of-n systems consisting of a single subsystem. The redundancy allocation
problem involves the selection of components from among discrete choices and the
determination of a system-level configuration to maximize system reliability given
constraints on the system. Formulation of the problem to accommodate k-out-of-n

---

[#] This chapter appeared previously on the International Journal of Reliability, Quality and
Safety Engineering. To cite this chapter, please cite the original article as the following:
D. W. Coit and J. C. Liu, *Int. J. Reliab. Qual. Saf. Eng*, **7**, 129–142 (2000), doi:10.1142/
S0218539300000110.

subsystems and different redundancy options offers enhanced capabilities. More engineering design problems can be analyzed, thereby providing a better tool for designers and reliability analysts.

When formulated to consider only active redundancy and restricted to series–parallel systems, efficient optimization algorithms have been developed to determine optimal designs using dynamic programming or integer programming. These existing solution methodologies are not applicable when the system design may involve active and cold-standby redundancy in different parts of the design and when there may be subsystems where more than one component is required for the system to operate ($k_i > 1$).

A methodology is presented and demonstrated here to determine optimal solutions to this more general form of the redundancy allocation problem. The problem is solved by developing an equivalent problem formulation and the application of zero–one integer programming methods. For this problem formulation, component time-to-failure is distributed according to the exponential distribution.

1.1. *Notation*

$R(t)$ system reliability at time t depending on design vectors $\mathbf{z}$ and $\mathbf{n}$
$\mathbf{n}$ $= (n_1, n_2, \ldots, n_s)$
n_i number of components used in subsystem i
$n_{\max,i}$ upper bound for $n_i (n_i \leq n_{\max,i} \forall i)$
$\mathbf{z}$ $= (z_1, z_2, \ldots, z_s)$
z_i index of component choice used for subsystem i, $z_i \in \{1, 2, \ldots, m_i\}$
m_i number of available component choices for subsystem i
$\mathbf{k}$ $= (k_1, k_2, \ldots, k_s)$
k_i minimum number of operating components for subsystem i
s number of subsystems
T_{ij} time-to-failure of the jth available component for subsystem i
λ_{ij} component failure rate (exponential distribution parameter) for the jth available component for subsystem i, $f_{ij}(t) = \lambda_{ij} \exp(-\lambda_{ij}t)$
C, W system-level constraint limits for cost and weight
c_{ij}, w_{ij} cost and weight for the jth available component for subsystem i
t mission time (fixed)

2. Redundancy Allocation Problem

The redundancy allocation problem has been studied in great detail as an efficient means to select sound design configurations. There are two general problem classifications. For the first, there are discrete component choices with known characteristics (cost, weight, etc.). The objective for this combinatorial problem is to select which components to use and the corresponding redundancy levels. For the second general problem class, component reliability (or an exponential distribution parameter) is treated as a design variable and component cost is a predefined in-creasing function of component reliability. This paper pertains to the first type of problem. It is a difficult combinatorial problem which has been shown to be NP-hard.[1]

The design of new products involves the specification of performance require-ments, the selection of components to perform clearly defined functions and the determination of a system-level architecture. Detailed engineering specifications prescribe minimum levels of reliability, maximum weight, maximum volume, etc. If the design is to be produced within some specified budget, numerous design al-ternatives must be considered, resulting in a complex combinatorial optimization problem.

The redundancy allocation problem studied here pertains to a system of s in-dependent k-out-of-n subsystems in series. Each subsystem may be designed with one component ($k_i = n_i = 1$), parallel redundant components ($k_i = 1 < n_i$), or components in a k-out-of-n configuration. Figure 1 presents a typical example of the system configuration being considered. Within a particular subsystem, the redundant components are either active or in a standby mode. For a k-out-of-n subsystem with cold-standby redundancy, k of the original n components are fully active and susceptible to failure. As a component fails, it is detected and one of the redundant components is activated. It is assumed that the system has perfect failure detection and switching.

For each subsystem, there are m_i equivalent components that may be selected. There is an unlimited supply of each of the m_i choices. Each available component has different levels of cost, weight and other characteristics. Component time-to-failure is independent and is distributed according to the exponential distribution, $T_{ij} \sim \text{EXP}(\lambda_{ij})$. There are system-level constraints on cost and weight (and others) and the objective is to select the component choices and levels of redundancy to maximize the system reliability.

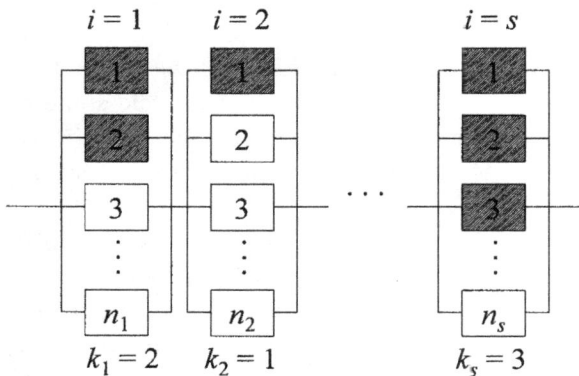

Fig. 1. System with k-out-of-n subsystems.

A general problem formulation to maximize system reliability is defined as Problem P1 and presented as follows:

Problem P1:

$$\max \quad R(t),$$

$$\text{s.t.} \quad \sum_{i=1}^{s} c_{iz_i} n_i \leq C,$$

$$\sum_{i=1}^{s} w_{iz_i} n_i \leq W,$$

$$n_i \in \{k_i, k_i + 1, k_i + 2, \ldots, n_{\max,i}\},$$

$$z_i \in \{1, 2, \ldots, m_i\}.$$

The problem has most often been formulated considering only series–parallel systems ($k_i = 1$) or k-out-of-n systems consisting of a single subsystem ($s = 1$). There has only been very limited research directed towards optimization of systems with cold-standby redundancy and k-out-of-n subsystems. Previous optimization methods also require the assumption that only one type of redundancy will be used throughout the system. That is, the system design will use only active redundancy or only cold-standby redundancy throughout the subsystem.

In practice, system design may involve some subsystems that use cold-standby and other subsystems that use active redundancy. The choice of redundancy type is often dictated by the particular failure mechanisms and technology of the part types being used. If it is not efficient or possible to detect a failed component and activate a redundant component (without significantly interrupting service), then active redundancy will be used. Furthermore, there may be a time lag associated with the switching that is unacceptable for some safety critical systems. Alternatively, if power, cost or space restrictions preclude the simultaneous operation of redundant components, then cold-standby will be used. If either redundancy strategy is possible for a particular subsystem, then cold-standby redundancy will generally be preferable when the failure detection and switching capability is highly reliable. In practice, it is unrealistic to limit a system design to exclusively one redundancy strategy.

There has been significant research activity devoted to different forms of the redundancy optimization problem. Kuo *et al.*[2] provides a comprehensive overview of system reliability optimization problem formulations and solution methodologies. For a single subsystem, there have been research findings describing optimal configurations for the k-out-of-n subsystem. For multiple subsystems in series, reliability optimization research has largely been limited to series–parallel systems. This is equivalent to k-out-of-n subsystems with k_i always set equal to one.

Pham[3] describes an optimal design method for k-out-of-n redundant systems. The optimization problems are formulated and solved to minimize expected total cost of k-out-of-n systems. The optimal number of units can be obtained for a system consisting of a single k-out-of-n subsystem. Similarly, Pham[4] also demonstrates methods to optimally determine the most economical number of components in k-out-of-n subsystems. In this case, methods are presented to determine the optimal values of k (for fixed n) and n (for fixed k) to minimize the mean total cost of k-out-of-n subsystems. Suich and Patterson[5] have also derived several cost models for a single k-out-of-n subsystem and presented a numerical solution for k and n that minimizes the total cost of a k-out-of-n system.

Pham and Malon[6] consider the problem of achieving optimal system size n, and a threshold k for k-out-of-n subsystem with competing failure modes. Methods to determine optimal k (given n); optimal n (given k); and optimal k, n are described. In each case, only one subsystem is considered.

Bai, Yun and Chung[7] describe the problem of determining the optimal number of redundant units in k-out-of-n subsystems with common-cause failures (CCFs). The mean cost rate is obtained, and the number of redundant units minimizing the mean cost rate is shown to be finite and unique. Again, only one subsystem is considered.

Chiang and Chiang[8] considered a relayed mobile communication system. Such a system can be considered as a consecutive k-out-of-n subsystem. They present equations for computing the mean number of stations needed for a successful relay and studied the optimal choice of k to minimize the mean number. Hwang and Shi[9] demonstrate that it is always better to replace a consecutive k-out-of-n subsystem by a consecutive l-out-of-n line but with k redundancy. The problem of choosing an optimal k still has no closed-form solution but is more tractable than the original problem studied by Chiang and Chiang. There has been other research pertaining to optimization problems for k-out-of-n subsystems.[10–14] For each of these, only one subsystem is considered.

Kuo *et al.*[2] provide the most comprehensive description of system reliability optimization. There are generally two classes of problems. The first is where component reliability is a continuous design variable and cost is expressed as an explicit function of component reliability. The second is where components are selected from discrete choices. Both problems are difficult and many solution strategies have been proposed for both. Generally, the first problem classification has been analyzed using problem-specific variants of nonlinear programming, often relying on the problems special structure. The second class has been addressed with integer programming, dynamic programming and heuristic approaches.

For multiple subsystems in series, the problem has been formulated to maximize system reliability and solved using dynamic programming[15,16] integer programming[17,18] and genetic algorithms.[19–21] Systems with k-out-of-n subsystems have not received significant attention, although Misra and Sharma[22] and Coit and Smith[21] do consider these systems.

Misra and Sharma[22] consider the most general case of system designs with mixed redundancies. They use a combination of direct search methods and random search.

Coit and Smith[21] develop a problem-specific genetic algorithm (GA) to analyze series–parallel systems and to determine the design configuration to maximize reliability when there are multiple component choices available for each k-out-of-n subsystem. Only active redundancy is considered. While this GA has repeatedly been demonstrated to yield good solutions, it cannot guarantee optimality, and furthermore, the final solution may be dependent on the initial population that is randomly selected.

A similar problem has been addressed by Nakashima and Yamato.[23] They consider improving the reliability of multivalue output systems by the use of n redundant subsystems. The mean loss of multivalue output systems with multichannels can be minimized by adopting k-out-of-n redundancy for each channel. The optimal k depends on the probability and loss matrices, but it can be specified in some special cases. Only active redundancy is considered.

3. System Reliability with k-out-of-n Subsystems

A system design may employ active redundancy for some subsystems and cold-standby redundancy for other subsystems. The choice of redundancy strategy may be based either on reliability benefits or practical design constraints. The choice of redundancy strategy may be dictated by the practicality of sensing a failure and activating a component in cold-standby. For other subsystems, there may already been a decision to use a single component in series. System reliability is a function of the design variables represented by $\mathbf{z}$ and $\mathbf{n}$, given a mission time t, and minimum component numbers $\mathbf{k}$. For notational brevity, system reliability will simply be expressed as $R(t)$, i.e., $R(t) = R(\mathbf{z}, \mathbf{n}|t, \mathbf{k})$.

The reliability of a system, with both active and cold-standby redundancy is given in Eq. (1). $R_i(t, z_i, n_i, k_i)$ is the reliability of the ith subsystem.

$$R(t) = \prod_{i=1}^{s} R_i(t, z_i, n_i, k_i). \tag{1}$$

A k-out-of-n subsystem with cold-standby redundancy requires k_i components to be operating, and thus, they are prone to failure. The remaining $n_i - k_i$ redundant components are not exposed to operating stress and not prone to failure. As components do fail, the failures are sensed and a redundant component is activated. Prior to subsystem failure, there are always a collective k_i operating components. The subsystem failure process can be considered as a Poisson process with rate $\lambda_{ij} k_i$. The expected time to failure of the ith subsystem can be determined by summing the expected interarrival times of the first $n_i - k_i + 1$ failure events of the Poisson process. If X is defined as the Poisson process interarrival time, then

$$MTTF_i = (n_i - k_i + 1)E[X] = (n_i - k_i + 1)\frac{1}{\lambda_{ij} k_i} = \frac{n_i - k_i + 1}{k_i}\, E[T_{ij}], \tag{2}$$

where $MTTF_i$ = mean time to failure of the ith subsystem with cold standby redundancy.

Subsystem reliability for cold-standby subsystems is the probability that there are less than or equal to $n_i - k_i$ failures observed until time t. This probability can now be computed from the Poisson distribution. Subsystem reliability for active redundancy is computed using standard techniques. System reliability is now given by the following equation:

$$R(t) = \prod_{i \in A} \sum_{l=k_i}^{n_i} \binom{n_i}{l} (\exp(-\lambda_{i,z_i} t))^l (1 - \exp(-\lambda_{i,z_i} t))^{n_i - l}$$

$$\times \prod_{i \in S} \exp(-\lambda_{i,z_i} k_i t) \sum_{l=0}^{n_i - k_i} \frac{(\lambda_{i,z_i} k_i t)^l}{l!} \tag{3}$$

A = set of all subsystems using active redundancy
S = set of all subsystems using cold-standby redundancy

4. Solution Methodology

A methodology for maximization of system reliability was developed by transforming the problem and defining new decision variables to yield an equivalent zero–one integer programming problem. This approach is based on the research and algorithms from Misra and Sharma.[24] The problem was transformed by taking the logarithm of Eq. (3) and by defining new 0–1 decision variables y_{ijp}. This linearizes the problem and allows for the use of integer programming algorithms. y_{ijp} is defined as follows:

$$y_{ijp} = \begin{cases} 1, & p \text{ of the } j\text{th component is used for subsystem } i, \\ 0, & \text{otherwise}. \end{cases}$$

The following methodology is used to determine an optimal solution to Problem P1. The resulting solution maximizes system reliability.

Step 1. Partition each subsystem i into sets A, S, N.
A = set of all subsystems limited to active redundancy
S = set of all subsystems limited to cold-standby redundancy
N = set of all subsystems which will not use additional redundancy $(n_i = k_i)$

Step 2. Compute,

$$\alpha_{ijp} = c_{ij} p \quad \text{for} \quad 1 \le i \le s, 1 \le j \le m_i, k_i \le p \le n_{\max,i}$$

$$\beta_{ijp} = w_{ij} p \quad \text{for} \quad 1 \le i \le s, 1 \le j \le m_i, k_i \le p \le n_{\max,i}$$

for $i \in A$,

$$\gamma_{ijp} = \begin{cases} -\lambda_{ij}k_i t, & p = k_i \\ \ln\left(\sum_{l=k_i}^{n_i} \binom{n_i}{l}(\exp(-\lambda_{ij}t))^l(1-\exp(\lambda_{ij}t))^{n_i-l}\right), \\ & k_i < p \leq n_{\max,i} \end{cases}$$

for $i \in S$,

$$\gamma_{ijp} = \begin{cases} -\lambda_{ij}k_i t, & p = k_i \\ -\lambda_{ij}k_i t + \ln\left(\sum_{l=0}^{n_i-k_i} \dfrac{(\lambda_{ij}k_i t)^l}{l!}\right), & k_i < p \leq n_{\max,i} \end{cases}$$

for $i \in N$,

$$\gamma_{ijp} = -\lambda_{ij}k_i t, p = k_i$$

$$n_{\max,i} = k_i$$

Step 3. Solve the following 0–1 integer program (Problem P2) using any convenient branch-and-bound or cutting plane algorithm.

Problem P2:

$$\max \sum_{i=1}^{s}\sum_{j=1}^{m_i}\sum_{p=k}^{n_{\max,i}} \gamma_{ijp}y_{ijp}$$

$$\text{s.t} \sum_{i=1}^{s}\sum_{j=1}^{m_i}\sum_{p=k}^{n_{\max,i}} \alpha_{ijp}y_{ijp} \leq C$$

$$\sum_{i=1}^{s}\sum_{j=1}^{m_i}\sum_{p=k}^{n_{\max,i}} \beta_{ijp}y_{ijp} \leq W$$

$$\sum_{j=1}^{m_i}\sum_{p=k}^{n_{\max,i}} y_{ijp} = 1 \ \forall \ i$$

$$y_{ijp} \in \{0,1\}$$

Step 4. Interpret the results. There will be exactly s y_{ijp} values equal to one in the optimal solution and the remainder will be equal to zero.

(1) For $i \in A$ and $y_{ijp} = 1$, the optimal system design uses p of the jth available component choice in active redundancy for subsystem i.

(2) For $i \in S$ and $y_{ijp} = 1$, the optimal system design uses p of the jth available component choice in cold-standby redundancy for subsystem i (i.e., k_i active component and $p - k_i$ in cold-standby).

(3) For $i \in N$ and $y_{ijp} = 1$, the optimal system design uses one of the jth available component choice.

To solve the problem, the user is required to pre-specify t, $\mathbf{k}$ and $n_{\max,i}$ and determine sets A, S and N. These decisions should be based on mission requirements and preliminary design decisions. If it is technologically possible to assign a particular subsystem to set A, S, or N but the decision is unclear, then the subsystem should be assigned to set S unless the detection and switching capabilities are suspect. In that case, the subsystem should be assigned to A. $n_{\max,i}$ is an upper-bound for n_i. Usually, $n_{\max,i}$ can be selected based on pragmatic restrictions. Otherwise, $n_{\max,i}$ can be computed as

$$n_{\max,i} = \max_j \min \left[\frac{C}{c_{ij}}, \frac{W}{w_{ij}} \right].$$

α_{ijp}, β_{ijp} and γ_{ijp} are expressed entirely as a function of specified component and problem parameters. Any problem with exponential component times-to-failure and known cost and weight measures will yield a unique set of α_{ijp}, β_{ijp} and γ_{ijp} values.

Problem P2 is linear and in the form of a standard 0–1 integer program with $\sum_{i=1}^{s} (n_{\max,i} - k_i + 1) m_i$ decision variables. Optimal solutions can be found using standard algorithms developed specifically for 0–1 integer programs.[25–27] There are many readily available automated packages (e.g., CPLEX, LINDO) to solve integer programs. The optimal solution is found using branch-and-bound or cutting plane methods that are based on a linear programming relaxation and successive application of the simplex algorithm.

For very large problems, the number of variables and/or constraints may become prohibitive and the capacities of available automated algorithms may be exceeded. In this case, it may be necessary to use other approaches for very large problems. Linear program complexity is largely dictated by the number of constraints. In the problem formulations presented here, there were only two constraints, but there may be many more in practice. The use of surrogate constraints methods[16,17] is one viable option. This involves the combination of all constraints into a single surrogate constraint creating a surrogate problem that can be readily solved using special algorithms for the knapsack problem. Then, a series of surrogate problems is solved with different constraint combinations to determine the optimal solution to the original problem.

If there are many constraints but only one component choice per subsystem ($m_i = 1$), then the specialized integer programming algorithm proposed by Misra and Sharma[28] can be used to solve Problem P2. This algorithm is designed specifically for reliability optimization problems and exploits the special structure of the problem for coherent systems.

If the problem is sufficiently large that none of these approaches are viable, then the genetic algorithm proposed by Coit and Smith[21] can be extended for this problem formulation. This approach acts directly on the Problem P1 formulation. It does not guarantee optimal solutions, although it has been demonstrated repeatedly to yield high-quality solutions that are optimal or near-optimal.

This problem has been formulated with two constraints but it can be readily expanded to accommodate additional linear constraints. The redundancy allocation problem can also be formulated to minimize system cost given a constraint for system reliability R (a reliability requirement or minimal acceptable reliability). System cost becomes the objective function and the following constraint is added:

$$\sum_{i=1}^{s} \sum_{j=1}^{m_i} \sum_{p=k_i}^{n_{\max,i}} \gamma_{ijp} \, y_{ijp} \geq \ln[R] \,.$$

5. Illustrative Example

A system reliability optimization example is provided to demonstrate the methodology. The example has been adapted from the example provided by Fyffe, Hines

Table 1. Component data for example.

Subsystem			Component Choice 1			Component Choice 2			Component Choice 3			Component Choice 4		
i	k_i	type	λ_{ij}	c_{ij}	w_{ij}	λ_{ij}	c_{ij}	w_{ij}	λ_{ij}	c_{ij}	w_{ij}	λ_{ij}	c_{ij}	w_{ij}
1	1	A	.001054	1	3	.000726	1	4	.000943	2	2	.000513	2	5
2	2	A	.000513	2	8	.000619	1	10	.000726	1	9	—		
3	1	A	.001625	2	7	.001054	3	5	.001393	1	6	.000834	4	4
4	2	A	.001863	3	5	.001393	4	6	.001625	5	4	—		
5	1	A	.000619	2	4	.000726	2	3	.000513	3	5	—		
6	2	A	.000101	3	5	.000202	3	4	.000305	2	5	.000408	2	4
7	1	A	.000943	4	7	.000834	4	8	.000619	5	9	—		
8	2	S	.002107	3	4	.001054	5	7	.000943	6	6	—		
9	3	S	.000305	2	8	.000101	3	9	.000408	4	7	.000943	3	8
10	3	S	.001863	4	6	.001625	4	5	.001054	5	6	—		
11	3	S	.000619	3	5	.000513	4	6	.000408	5	6	—		
12	1	S	.002357	2	4	.001985	3	5	.001625	4	6	.001054	5	7
13	2	S	.000202	2	5	.000101	3	5	.000305	2	6	—		
14	3	S	.001054	4	6	.000834	4	7	.000513	5	6	.000101	6	9

NOTES: A = active redundancy, S = cold-standby redundancy, units for λ_{ij} are failures/hour and Lee.[15] The system is designed with 14 subsystems. For each subsystem, there are three or four component choices. Component cost, weight and exponential distribution parameter (λ_{ij}) are provided in Table 1. The objective is to maximize system reliability at a time of 100 h given constraints for system cost ($C = 130$) and system weight ($W = 170$).

The problem was revised by randomly selecting $k_i \in \{1, 2, 3\}$ for each subsystem and by assigning each subsystem to set A or S. The original problem was for series–parallel systems ($k_i = 1$) with exclusively active redundancy. The component exponential distribution parameters in Table 1 yield the same component reliability data (for $t = 100$) as originally presented by Fyffe, Hines and Lee.[15] For each subsystem, the selection of k_i and redundancy type has been indicated in Table 1. The maximum number of components within a subsystem has been defined to be six ($n_{\text{max},i} = 6$).

For this problem, there are 244 0–1 decision variables in the reformulated problem. The number of prospective unique solutions to the problem is larger than 1.6×10^{17}. The problem was then solved on a personal computer using readily available linear programming software (Hyper-LINDO).

The optimal solution is given in Table 2. It corresponds to a system with system reliability of .4466, a system cost of 118 and a system weight of 170.

Table 2 also presents the optimal solution to the original problem with active redundancy and $k_i = 1$. The comparison in Table 2 is interesting, but it must be emphasized that by changing k_i and redundancy type, the problem is fundamentally different. For 13 of the 14 subsystems, the optimal solutions to the two problems

Table 2. Example results.

i	Optimal Solution		Solution- Fyffe[15]	
	z_i	n_i	z_i	n_i
1	3	2	3	3
2	1	2	1	2
3	4	1	4	3
4	3	3	3	3
5	2	1	2	3
6	2	2	2	2
7	2	1	1	2
8	1	3	1	4
9	3	3	3	2
10	2	4	2	3
11	1	4	1	2
12	1	2	1	4
13	2	2	2	2
14	3	4	3	2

involved the same component selection (z_i). Alternatively, the number of components used (n_i) is the same for only four of the 14 subsystems. The differences for n_i in the two optimal solutions is not unexpected. For systems with k-out-of-n subsystems, n_i must be at least as large as k_i.

6. Conclusions

A solution methodology is presented to determine optimal solutions for the redundancy allocation problem for systems consisting of multiple k-out-of-n subsystems in series. This problem had not previously been satisfactorily solved and the proposed method offers the capability to solve a greater range of engineering design problems.

Acknowledgment

David W. Coit's work was supported by NSF CAREER grant DMII-9874716.

References

1. M. S. Chern, "On the computational complexity of reliability redundancy allocation in a series system," *Operations Research Letters* **11** (1992), pp. 309–315.
2. W. Kuo, V. R. Prasad, F. A. Tillman, and C.-L. Hwang, *Fundamental and Applications of Reliability Optimization* (Cambridge University Press, Cambridge, 2000).
3. Hoang Pham, "Optimal Design of k-out-of-n Redundant Systems," *Microelectronics and Reliability* **32**(1/2) (1992), pp. 119–126.
4. Hoang Pham, "On the Optimal Design of k-out-of-n: G Subsystems," *IEEE Trans. Reliab.* **45**(2) (1996), pp. 254–260.
5. R. C. Suich and R. L. Patterson, "k-out-of-n: G systems; Some Cost Considerations," *IEEE Trans. Reliab.* **40** (1991), pp. 259–264.
6. Hoang Pham and D. M. Malon, "Optimal design of systems with competing failure modes," *IEEE Trans. Reliab.* **43**(2) (1994), pp. 251–254.
7. D. S. Bai, W. Y. Yun, and S. W. Chung, "Redundancy optimization of k-out-of-n systems with common-cause failures," *IEEE Trans. Reliab.* **40**(1) (1991), pp. 56–59.
8. D. T. Chiang and R. F. Chiang, "Relayed communication via consecutive-k-out-of-n," *IEEE Trans. Reliab.* **R-30** (1986), pp. 65–67.
9. F. K. Hwang and D. Shi, "Optimal relayed mobile communication systems," *IEEE Trans. Reliab.* **38**(4) (1989), pp. 457–459.
10. A. Lesanovsky, "Minimal size of the system with two dual modes of failure having prescribed reliability," *Microelectronics and Reliability* **33**(4) (1993), pp. 543–563.
11. A. Rushdi and K. A. Al-Hindi, "Table of the Lower Boundary of the Region of Useful Redundancy for k-out-of-n System," *Microelectronics and Reliability* **33**(7) (1993), pp. 979–992.
12. Gurov, Sergey, Utkin, Lev, Shubinsky, and B. Igor, "Optimal reliability allocation of redundant units and repair facilities by arbitrary failure and repair distributions," *Microelectronics and Reliability* **35**(12) (1995), pp. 1451–1460.
13. A. A. Charl, "Optimal redundancy of k-out-of-n: G system with two kinds of CCFs," *Microelectronics and Reliability* **34**(6) (1994), pp. 1137–1139.
14. Kalyan, Radha, Kumar, and Santosh, "Study of protean systems-redundancy optimization in consecutive k-out-of-n: F systems," *Microelectronics and Reliability* **30**(4) (1990), pp. 635–638.

15. D. E. Fyffe, W. W. Hines, and N. K. Lee, "System reliability allocation and a computational algorithm," *IEEE Trans. Reliab.* **R-17** (1968), pp. 64–69.
16. Y. Nakagawa and S. Miyazaki, "Surrogate constraints algorithm for reliability optimization problems with two constraints," *IEEE Trans. Reliab.* **R-30** (1981), pp. 175–180.
17. R. L. Bulfin and C. Y. Liu, "Optimal allocation of redundant components for large systems," *IEEE Trans. Reliab.* **R-34** (1985), pp. 241–247.
18. M. Gen, K. Ida, and J. U. Lee, "A computational algorithm for solving 0–1 goal programming with GUB structures and its application for optimization problems in system reliability," *Electronics and Communications in Japan* Part 3, **73** (1990), pp. 88–96.
19. K. Ida, M. Gen, and T. Yokota, "System reliability optimization with several failure modes by genetic algorithm," *Proceedings of the 16th International Conference on Computers and Industrial Engineering* (1994), pp. 349–352.
20. L. Painton and J. Campbell, "Genetic algorithms in optimization of system reliability," *IEEE Trans. Reliab.* **44**(2) (1995), pp. 172–178.
21. D. W. Coit and A. E. Smith, "Reliability optimization of series–parallel systems using a genetic algorithm," *IEEE Trans. Reliab.* **45**(2) (1996), pp. 254–260.
22. K. B. Misra and U. Sharma, "Reliability optimization of a system by zero-one programming," *Microelectronics and Reliability* **31**(2/3) (1991), pp. 323–335.
23. K. Nakashima and K. Yamato, "On optimal redundancy of multivalue-output systems," *IEEE Trans. Reliab.* **R-36**(2) (1987).
24. K. B. Misra and U. Sharma, "Reliability optimization of a system by zero-one programming," *Microelectronics and Reliability* **12**(3) (1973), pp. 229–233.
25. G. Nemhauser and L. Wolsey, *Integer and Combinatorial Optimization* (John Wiley & Sons, 1988).
26. G. Parker and R. Rardin, *Discrete Optimization* (Academic Press, 1988).
27. H. Taha, *Integer Programming: Theory, Applications and Computations* (Academic Press, 1975).
28. K. B. Misra and U. Sharma, "An efficient algorithm to solve integer programming problems arising in system-reliability design," *IEEE Trans. Reliab.* **40**(1) (1991), pp. 81–91.

About the Authors

David W. Coit is an assistant professor of Industrial Engineering at Rutgers University. He has a BSME from Cornell, an MBA from Rensselaer Polytechnic Institute, and M.S. and Ph.D. degrees in Industrial Engineering from the University of Pittsburgh. He previously worked for 12 years at the IIT Research Institute (IITRI), Rome, New York as a reliability engineer and engineering manager. He teaches classes in reliability, applied probability, and optimization. He is currently working on system reliability research projects as part of the Quality and Reliability Engineering (QRE) Center at Rutgers. His interests involve the prediction and optimization of system reliability when there is uncertainty about the reliability of critical components within the system. He is a member of IEEE, IIE and INFORMS.

Jiachen Liu received his B.S. and M.S. degrees in Engineering from Tsinghua University, P. R. China, in 1995 and 1998, respectively. He is currently a Ph.D. student and a research assistant of Industrial Engineering at Rutgers University. He has also worked for Honeywell Inc. His research interests include system optimization, supervisory control, and precision tracking.

© 2025 World Scientific Publishing Company
https://doi.org/10.1142/9789819812547_0004

Chapter 4

FAILURE MODES IN MEDICAL DEVICE SOFTWARE: AN ANALYSIS OF 15 YEARS OF RECALL DATA[#]

DOLORES R. WALLACE[*,†] and D. RICHARD KUHN[‡]

Information Technology Laboratory, National Institute of Standards and Technology
Gaithersburg, MD 20899, USA
[†]*drwallace@erols.com*
[‡]*kuhn1@aol.com*

Most complex systems today contain software, and systems failures activated by software faults can provide lessons for software development practices and software quality assurance. This paper presents an analysis of software-related failures of medical devices that caused no death or injury but led to recalls by the manufacturers. The analysis categorizes the failures by their symptoms and faults, and discusses methods of preventing and detecting faults in each category. The nature of the faults provides lessons about the value of generally accepted quality practices for prevention and detection methods applied prior to system release. It also provides some insight into the need for formal requirements specification and for improved testing of complex hardware-software systems.

Keywords: Assurance; Detection; Failures; Fault; Fault Model; Medical Device; Prevention; Software Quality.

1. Introduction

Henry Petroski devotes an entire book to failures in engineering and lessons to be learned.[1] In his preface, he states "the concept of failure — mechanical and structural failure in the context of this discussion — is central to understanding engineering, for engineering design has as its first and foremost objective the obviation of failure." He further states "the lessons learned from ... disasters can do more to advance engineering knowledge than all the successful machines and structures in the world."

We take license in extending Petroski's views from mechanical and structural engineering into the domain of software system failures. Many software assurance techniques, including inspections, failure modes and effects analysis, flaw

---

*This author performed the work while an employee of NIST and is now employed as a contractor at the NASA Goddard Space Flight Center.

[#] This chapter appeared previously on the International Journal of Reliability, Quality and Safety Engineering. To cite this chapter, please cite the original article as the following: D. R. Wallace and D. R. Kuhn, *Int. J. Reliab. Qual. Saf. Eng*, **8**, 351–371 (2001), doi:10.1142/10.1142/S021853930100058X.

hypothesis penetration testing, and some specification-based test methods, benefit from knowledge of the types of faults that typically occur in a given class of software. Lessons learned from failure analysis can either affirm proposed software engineering principles or help define new ones.

Several industries, including telecommunications, space, finance, and defense, were early drivers of computer technology. Within these industries, more and more systems are controlled by, or dependent on, software today than in the early years. We find a great need to examine software-based failures from many domains to gain insight about possible common causes of failures and the means to prevent them in the next system or, at the very least, to detect them before the system is released. The purpose is to reduce costs by finding and detecting problems before systems are recalled from multiple users. Loss of revenue from the customer and additional costs for fixing a faulty system after release can become exorbitant.

Systems in all industries can fail for many reasons, including acts of nature, hardware failures, human error, vandalism and software. While the distribution of failures due to specific causes may differ by industry, most experience failures attributable to these causes at some time or another. Our long-term objective is to study failure data from several industries individually and then in the aggregate, to identify the relationships to software problems. We have previously examined failures of the public switched telephone network.[2]

We focus our current study on medical devices that have been voluntarily recalled by the manufacturers due to computer software problems. Any findings may well apply to other application domains. Like most industries, the health care industry depends on computer technology to perform many of its functions, ranging from financial management and patient information to patient treatment. The use of software in some kinds of medical devices has become widespread only in the last two decades or so. Their developers had limited software experience and had to develop the expertise for avoiding preventable problems.[a] The Federal Food Drug & Cosmetic Act defines a medical device as:

> "an instrument, apparatus, implement, machine, contrivance, implant, in vitro, reagent, or other similar or related article, including a component part, or accessory which is:
>
> - recognized in the official Formulary, or the United States Pharmacopoeia, or any supplement to them,
> - intended for use in the diagnosis of disease or other conditions, or in the cure, mitigation, treatment, or prevention of disease, in man, or other animals, or
> - intended to affect the structure or any function of the body of man or other animals, and which does not achieve any of its primary intended

---

[a]From the lecture by Lynn Elliott, "When Safe Patients Means Dependable Software," in the Lecture Series on High Integrity Systems, U.S. National Institute Standards and Technology, October 1995.

purposes through chemical action within or on the body of man or other animals and which is not dependent upon being metabolized for the achievement of any of its primary intended purposes."

The problems cited in this study were found in medical devices recalled by their manufacturers either in final testing, installation, or actual use from 1983 to 1997. It is important to note that there were no deaths or serious injuries caused by these failures, nor was there sufficient information to guess at potential consequences had the systems remained in service.

Using the Food and Drug Administration (FDA) database of medical device failures, we have examined the symptoms that indicated there were problems, identified the software faults that may have caused the problems, provided some generic guidance, and assessed what could have been done to prevent or detect the classes of faults. Section 2 contains a characterization of the system failure data, while Sec. 3 provides an analysis of the software faults. Section 4 contains a synopsis of the lessons learned with Sec. 5 providing conclusions about this study and recommendations for additional work.

2. Characterization of the Data

A medical device may be as simple as a tongue depressor, but this paper is concerned only with those containing software. The study includes only those devices in the categories of anesthesiology, cardiology, diagnostics, radiology, general hospital use, and surgery. Examples of these devices are insulin pumps, cardiac monitors, ultrasound imaging systems, chemistry analyzers, pacemakers, electrosurgical devices, and anesthesia gas machines. The following highly simplified description is provided only to enable understanding of the classes selected for observed symptoms of malfunctions. A device is a system providing a service, involving one or more components. Some components may contain computer software, executing functions that produce an output either to the next function within a component or to another component of the system (e.g., a display device). The system behaves according to the values or messages it receives from the functions' output. An alarm may sound and/or the device may cease operation. A dosage rate or volume may change. Equipment may move. Measurements of various specimens or human reactions may be taken, and data may be recorded and associated with a patient's name. The failures have been observed as a response of the physical system and usually not as an obvious software fault.

2.1. *General features of the recall data*

The FDA recall data consists of the recall number, the product name, a problem description, and a cause description. The code for the recall number yields the year of the recall and the general type of device. To protect the privacy of the manufacturers, we do not publish either the recall number or the product name. Our purpose is to understand the types of software problems and to abstract generic guidance about preventing and detecting the software faults before systems are released. Over time, manufacturers may have improved their software development processes and eliminated many factors contributing to these failures. This study reinforces the need for software quality practices and provides specific guidance on how to prevent and detect faults.

For the Fiscal Years 1983–1991, there were 2,792 quality problems that resulted in recalls of medical devices, including devices that do not contain software. Of those, 165, or 6%, were related to computer software. While the second group of data from 1992–1997 is not quite complete, the results are within the same ranges. We base our study on only the software recalls. The total number of software recalls from 1983–1997 is 383. The years 1994, 1995, 1996 have 11%, 10%, and 9% of the software recalls. One possibility for this higher percentage in later years may be the rapid increase of software in medical devices. The amount of software in general consumer products is doubling every two to three years.[3]

The medical devices can be grouped into classification panels according to the primary function of the medical device. The medical devices fit into 7 panels: anesthesiology, cardiology, general hospital, diagnostic, radiology, general and plastic surgery, or other. Diagnostic includes chemistry, hematology, immunology, microbiology, pathology, and toxicology. The label "other" includes anything else such as obstetrics and gynecology, or ophthalmology for which there were not enough recalls to be grouped into their own panels. The distribution of recalls by classification panel is shown in Fig. 1. The pie wedges match the legend going clockwise, starting with anesthesiology, near the top, at 10%.

Some systems are more difficult to develop than earlier similar devices, such as in radiology where ultrasound and tomography are highly complex. The added complexity in algorithms and system interactions may have affected the failure rates for radiology.

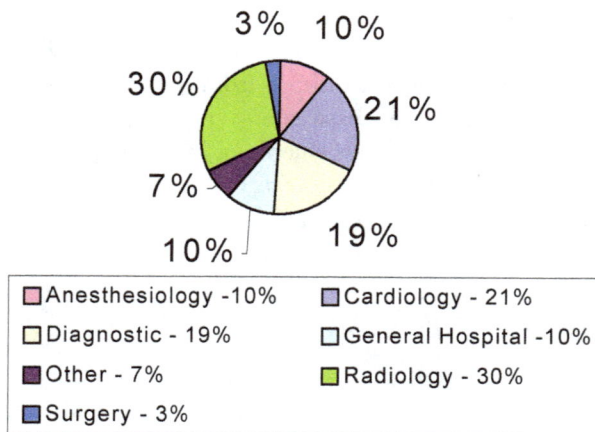

Fig. 1.   Failure distribution by device panel.

2.2. *Observed behavior signifying recall*

The problem and cause descriptions contain information on which we base our analysis. They provide observations about the system or a feature as shown in the following examples:

- An alarm failed to sound.
- Dosages were too fast, too slow or were stopped inconsistent with the data on the display unit.
- Display unit values were inconsistent with other visual outputs of the device, for example, name of patient on screen not correct.
- The system simply stopped.
- The device performed in a manner completely unplanned, when several conditions occurred simultaneously.
- Data were lost or corrupted.
- A calculation or other function was missing, or an instruction was omitted from the user manual.

For each recall, we reduced the problem description to a symptom of the failure (e.g., behavior–alarm did not sound; output — incorrect relationship with display). We next reduced the list to only the key attribute and one description, such as behavior alarm and ended with thirteen primary symptoms shown in Fig. 2. The pie wedges match the legend going clockwise, starting with behavior, near the top, at 22%.

Definitions for the thirteen primary symptoms are the following:

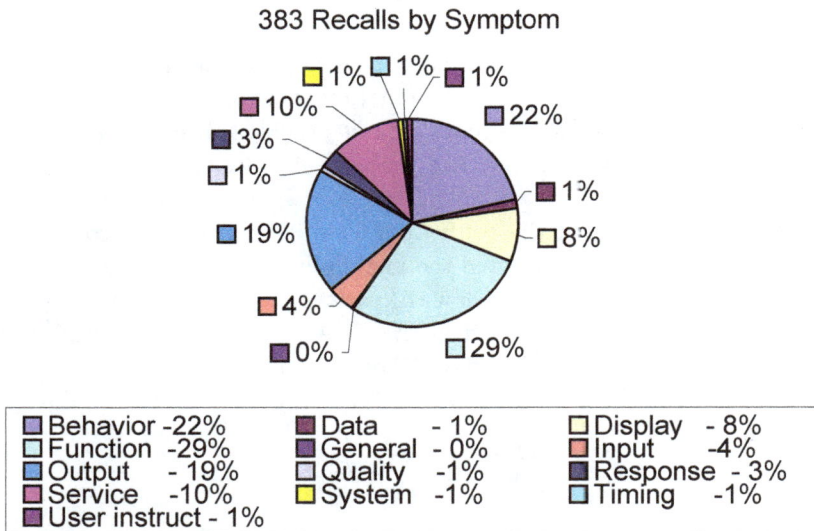

Fig. 2.   Distribution of 383 failures by symptom.

- **Behavior:** the system performs an action due to some output of some function. The action is a physical action, e.g., movement of the gantry.
- **Data:** a consequence to the data, usually corruption or loss of input data.
- **Display:** the visual display on a screen — numbers, text, or images in various formats.
- **Function:** usually a single calculation or activity; a software function in one module.
- **General:** not enough information to assign to a category.
- **Input:** the initial input (typed, sampled, read off equipment, database, file or tape, etc.) on which some operation is performed.
- **Output:** result of some function; generally an output to be used by the next function.
- **Quality:** user observations stated that "quality requirement was not met".
- **Response:** something has happened that should not, e.g., power emitted above allowed amount; manifested in some hardware function.
- **Service:** an identifiable system service involving multiple functions such as pumping, ventilating, giving medication; generally involves more than one component (module, subsystem).
- **System:** the total system.
- **Timing:** timing of the instrument or a service of the device
- **User instruction:** manual, or other descriptions for the operator/user.

3. Analyses of the Data

While the observed symptoms provide some insight about the nature of the failures of the medical devices, the vendors' determination of the software fault is important information. In many cases, the vendors did not provide this information. We were limited in determining the fault by the problem and description data; there was no mechanism for getting any further details. We want to understand the nature and reason for occurrence of the software fault and to develop lessons regarding software quality practice. We selected the final fault class terminology from several published taxonomies and reasoned how the various problems best fit, based on the problem and description as provided in the FDA database. We had no access to the manufacturers or to any other data. From this limited information, we could discern the fault type for only 342 failures. Only these 342 failures are discussed in the rest of this study.

3.1. *Fault distributions*

In many cases there could have been 2 or 3 fault types contributing to a failure. Often study of the symptom revealed the generic nature of the fault. For example, the observed behavior may indicate that two or more events had occurred at their

Table 1.   Partial list of detailed fault categories.

Accuracy; rounding	Logic; initialization
Algorithm; logic	Memory; dead code
Algorithm; rate	Missing code
Assignment	Missing information in user manual
Calculation; factor	Not enough information
Calculation; fault tolerance	Not validated; QA
Change impact; QA	Reinitialization
COTS; memory lost; size	Requirement — wrong formula
Data passing; QA	Scaling
Improper impact of change	Sequence of operations; QA
Incorrect change to counting	Transpoition
Initialization; data passing	Typo
Input; data passing	Units, calculation
Interface; parameter value	Volume

boundary values simultaneously, resulting in an incorrect or unexpected response. Possibly, the developers had not specified in the requirements that these events could occur, or the logic of the design failed to account for these simultaneous events, or the code logic was incorrect. If the first situation had been true, then the problem would have been classified as a requirements problem (e.g., omission, ambiguity, conditions not considered). While recognizing the value of better specification methods, specifically formal methods in some of these situations, we classified most of these as logic problems at the point of failure. Without additional information we could not classify some of these problems as requirements. In Table 1 the primary fault type is shown first, followed by one or more specific problems related to it, for example, "rate" following "algorithm" indicates a function performed at wrong rate in an algorithm.

We reduced the number of fault categories to the final list in Fig. 3, placing the detailed fault type into the class it best fit. For example, "incorrect change to counting" was placed under "calculation" because the error occurred in the counting algorithm and did not cause additional problems that would have fit under "change impact." In Fig. 3, the pie wedges match the legend going clockwise, starting with calculation, near the top, at 24%.

Among the fault types, logic appears very high at 43%; with further details, some of these faults might fit into other classes. This class includes possible errors such as incorrect logic in the requirement specification, unexpected behavior of two or more conditions occurring simultaneously, and improper limits. The group "data" includes units, assigned values, or problems with the actual input data. The group "other" includes problems in COTS, EPROM, hardware, resources (e.g., memory), configuration management, typos, mistakes in translating requirements into code, and quality assurance. For quality assurance, either the processes were not sufficient, or a new version was not validated.

For 1996–1997, calculation faults occur 9 times in radiology compared with 13 faults in all the panels. For 1996, logic has 4 faults in cardiology and 3 in radiology out of 11. For 1997, logic has 5 in diagnostics, but only 1 in radiology. The other

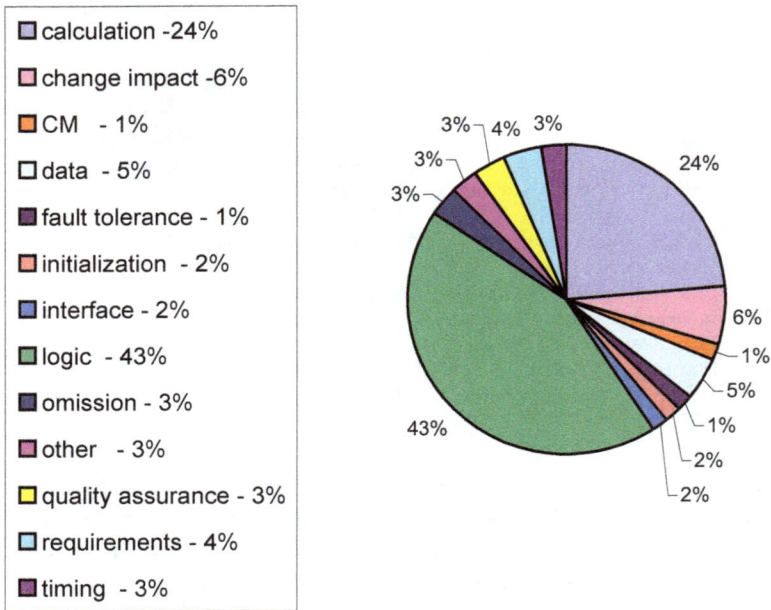

Fig. 3.   Fault class distribution.

fault classes are smaller and vary over the years. For the other years, also, the higher percentages are generally for calculation and logic. The obvious questions are "Why are logic and calculation the prevalent types?" and "What can prevent or detect them before product release?"

3.2. *Prevention and detection of faults*

These software recalls were distributed over 342 devices built by different vendors. What could have been done, individually, to prevent or detect each fault before the release of the device? We examined each fault in each of the thirteen classes and attempted to determine an answer to this question. By prevent, we mean some method applied by the development group before testing. By detect, we mean some method applied during testing or by quality assurance staff.

Obviously we cannot ascertain whether these methods were used or not. We have no evidence that more experienced companies used these more than inexperienced companies. Rather, we can indicate perhaps an affirmation that these are best practices, consistent with today's focus on process and need to be utilized.[4] Thirteen fault classes contain 342 faults. First, by each class, for each fault, we considered various techniques/methods for prevention, and then for detection. Next we reduced the results to a smaller, generic set for each fault class. While we provide descriptions of typical problems, only one problem per class is shown below with prevention or detection approaches.

Certain methods appear frequently in the complete synopses as well as in the few examples provided in this paper. We include inspection as both a prevention and detection technique, where inspection as prevention is used in a broader sense than the original Fagan inspection.[5] Glass explains this broader view which is based on practitioners' presentations in workshops and conferences.[6] In the prevention approach, then, inspection may include code reading and various static analyses. Sometimes we were specific, because the fault description warranted more specificity. When inspection appears as a detection technique, it generally means the traditional Fagan-type inspection.

While these faults occurred in medical devices, some faults occur in many other types of system. For example, truncation or rounding problems may occur in almost any software. The intent of providing details in the tables is to provide understanding of problems that may occur. Each system that may possibly have that type of error can benefit from the prevention and detection techniques.

The class *Calculation* includes many types of algorithmic problems. Attention to algorithms and computations includes such details as verifying units, operators, intervals, limits, ranges, transformations from mathematical expressions into their implementation, and others. Sometimes even verifying that the original algorithm requirement is the correct version may require significant effort. Understanding how the specific computer will handle registers and floating point values is mandatory. Verifying all the issues for a calculation may require expertise outside computer

Table 2. Some examples for calculation.

Generic Problem	Prevention	Detection
Constants or table of constants incorrectly coded.	Design, code reading to ensure correct relationship between the specified constant or table and the code.	Code reading, inspection. Unit test.
Improper handling of boundary conditions (e.g., circumstances close to limitation of the operating environment).	Assertions. Fault tolerance.	Focused inspections, code reading or walkthrough. Unit test.
Improper handling of data structure (array, bitset, list, queue, set stack, vector, etc.)	Low level design review. Code review.	Code reading or walk-through, review inspection. Unit test.
Precision problem (truncation or rounding error during I/O or calculation).	Low level design review. Code review.	Code review. Unit test.
Improper handling of abnormal conditions (e.g., wire disconnect from the device, electrical noise).	Assertion. Fault tolerance.	Unit test.
Graphical output meaningless.	Review requirements for relationship of computation output to next function.	Interface test.
Overflow.	Assertion. Fault tolerance.	Unit test.

Table 3.   Some examples for change impact.

Generic Problem	Prevention	Detection
Logic (incorrect conditions or incompatibility with sequential relationship).	Traceability analysis. Change impact analysis.	Inspection of logic relative all areas affected by change with focus on original assumptions (input values, selection criteria for a function).
No verification against original design specification.	Traceability analysis. Change impact analysis.	Inspection of proposed changes. Regression test.
Loss of correct functions over several upgrades; Reversion to defect from at least two version back.	Configuration management. Change impact analysis.	Traceability analysis. Verification against original specifications. Interface test.

science or software engineering. Often someone must verify that the algorithm is adequate for its intended use, e.g., increments used in the algorithm will be useful in the displayed output (neither too large nor too small to be meaningful). Examples for calculation appear in Table 2.

While *change impact* is not necessarily considered a fault type, these cases indicate that failure to examine the impact of changes hides other problems. In all cases, another practice, performing a traceability analysis, is a prerequisite for performing change impact analysis. The analyses identify the region the proposed change will affect. Examples for change impact appear in Table 3.

For *configuration management* (CM), that is, keeping all artifacts correctly associated with the appropriate version of the system, several problems may have been due to the incorrect exercise of CM procedures. Others may have been prevented simply by using CM. The use of tools to manage the software versions would be helpful. In some cases, the problems stem not from improper software versions, but

Table 4.   Example problems for CM.

Generic fault	Prevention	Detection
Incorrect configuration for non-domestic systems.	Use of CM tools.	Verify usage of CM tools for all changes. Verify configuration for non-domestic use.
Software incompatible with other components.	Record assumptions about all components, in the CM data.	Inspection of requirements for component interfaces. CM approval for configuring system components.
Failure to upgrade accompanying system, to match software changes.	Traceability analysis. CM tools.	Verification of changes; regression test.
Use of wrong master program when making software revision.	Use of CM tools.	Verification of appropriate master program. CM manager releases the versions.

Table 5. Some example for data.

Generic Problem	Prevention	Detection
System failed due to invalid input data.	Assertions for invalid values, checks for ranges that imply incorrect data. Design: set criteria of input data validation. Code: implementation of input data validation.	Review for completeness of data specification, and that all data specifications are included in the user instructions. Inspection: focus on data validation. Test against invalid data.
Inconsistency of data retrieved from database and that expected by the program.	Assertions on validity of data retrieved from database.	Testing focused on data retrieval.
Database corruption.	Database administration.	Error handling routine in software.

from selecting a software program that is not compatible with the hardware. This is also a problem of requirement specification; once hardware and software configurations are selected, the assumptions about each component need to be recorded as part of the CM history. Some examples for for CM appear in Table 4.

Problems in software programs can arise from input *data*. Data requirements for a program must be specified, entered in a data dictionary, and validated before the operation using the data is executed. The specification includes information such as units, acceptable range of values, the expected quantity or frequency with which values will change. The specification is published in the data dictionary of the database and in user instructions, emphasizing values that could cause program stoppage if they are out of range. Of course, the program itself may address some potential problems by containing assertions for input values or input omission, with actions to take when data are incorrect or missing. When a program is fielded, data in a database should be protected against database corruption. The software should facilitate an error-handling package to detect database corruption. Table 5 provides examples for data.

The *fault tolerance* category relates to safety-critical systems that should include facilities to handle abnormal or anomalous conditions. Fault tolerance examples appear in Table 6.

Initialization is essential for enabling programs either to begin or to perform more than one cycle of a function. Default values for variables are a necessity, and likewise, re-initialization of a variable must be established. Explicitly documenting initial conditions in requirements through the code is essential. Code reviews and code reading need to focus not on whether initialization is specified, but specified according to good programming practices. Examples for initialization appear in Table 7.

In a system, *interfaces* allow software to send and receive data (that is, interface) to physical components of the system, as well to other software modules and to users. Clearly, the requirement specification must be accurate, complete, and

Table 6.   Fault tolerance examples.

Generic Problem	Prevention	Detection
Excessive use of the program program causes failure.	Fault tolerance such as warnings to operators.	Stress/volume test. Testing against boundary and abnormal conditions.
Incorrect action due to external abnormal/unexpected condition related to power supply or other components.	Fault tolerance. Software cannot control abnormal condition external to it, but can provide a procedure in that event. Requirement needs to be written for FT.	Test against boundary, abnormal, and special conditions, Exception handling routine in software.
Incorrect action due to operator error.	Fault tolerance in design through code to protect against human error.	Inspect, review for protection against operator error. Test against unacceptable data entry.

Table 7.   Examples for initialization.

Generic Problem	Prevention	Detection
Lack of initialization of the runtime environment while the program initially or restarts.	Use assertions for initialization. For C or C++, see http://hissa.nist.gov/effProject/handbook/c++/variables.htm	Inspections; Code review. Test against initial executes conditions.
When the program executes first time, it fails to store necessary initialization values for the succeeding run.	Document initial conditions for both initial run and consecutive run. Design review.	Code review. Stress test (run the program multiple times).

Table 8.   Examples for interface.

Generic Problem	Prevention	Detection
Software does not properly interface with external device or other software component.	Trace requirements through design through code to ensure all software functions have interfaces to either another software module or to an output device or other system component or user. Examine the specification for each interface.	Inspections, reviews Integration test.

consistent. A traceability scheme provides a basis for ensuring that all interfaces are addressed and included correctly. A well-developed test plan for integration testing must be executed to verify the interfaces between devices or software components. Table 8 provides examples for interface.

Logic problems appear to be significant. While some failures of the devices did result from bad logic, the "bad" logic might have resulted from incorrect, incomplete, or inconsistent requirements or designs. Frequently, interactions among different functions might not have been considered at all or might have been neglected at

Table 9. Examples for logic.

Incomplete or incorrect control logic.	Design review. Walk through the software implementation against design.	Code Review. Inspection. Testing.
Configuration scheme for component interaction allows incorrect behavior.	Modeling. Simulation. Formal methods.	Code review. Interface analysis. Integration test. System test.
Improper handling of boundary conditions. (e.g., limits of value range).	Design review. Verify logic for all conditions, esp. at boundaries. Fault tolerance. Code review.	Code review. Inspection; Test against boundary and abnormal conditions.
Improperly handle abnormal or exceptional (e.g., power lost, multiple inoperative conditions occurred, I/O interrupt, I/O error).	Design review. Assertion. Fault tolerance. Review error recovery routines. Code review.	Code review. Inspection; Test against abnormal and exceptional conditions.
Improper data validation. (e.g., input or output data out of range).	Design review. Walk through the software implementation against design. Verify logic for data out of limits.	Code review. Inspection; Test against I/O boundary conditions.
Programming error (e.g., error in pointer, addressing, looping, indexing, subscript, memory management).	Low level design review. Code review.	Code review. Unit test.

boundary conditions of a function. Sometimes the logic might have been incorrect in the design. All of these were classified as logic problems, but it should be understood that the source of the problem could have been requirements, design, or code. Two examples include (1) "When power lost and then restored, system defaults to off status, which causes false information to operator and possible hazard to the operator" and (2) "When a second cartridge is in the other slot and detects an artifact condition, the monitor is prevented from alarming below set levels." Table 9 provides examples for logic.

The class *omission* indicates a required system function that is missing from the final implementation. Documentation is missing or not sufficient to install or operate the product. Two examples for omission are shown in Table 10.

Other faults too low in frequency to be classified separately include problems such as performance issues, I/O problems, typographical errors. Other types of faults appear in Table 11.

The role of *quality assurance* (QA) is to ensure that quality practices are defined in company standards and that they are used. Procedures are necessary for validation after modifications. The problems described in the recall data often cite that process checks were not made on the testing process and that testing was not performed after modifications. The problem descriptions do not reveal whether procedures for testing or other quality practices had been defined. Change impact analysis is a key task to ensure appropriate tests after modifications. While QA is

Table 10.   Examples of omission faults.

Vital system functions are missing.	Trace requirements through design through code, focus on all interfaces. Trace into user and test documentation. Use critical path analysis to ensure completion. Prepare system test scenarios at requirements specification and examine them for relationship to trace through code.	Inspections, reviews examining traceability of functions. System Test.
Lack of documentation, or improper documentation.	Proper release procedure. Traceability	Verify completeness by examining trace. Inspection.

Table 11.   Other types of faults.

Generic Problem	Prevention	Detection
Out of compliance with the performance standard.	Simulation. Design review. Code review.	Performance test.
Calculations associated with the "%" activity curve have been printed incorrectly. Formatting subroutine for screen display.	Code review: review special I/O routine. Understand the hardware/software requirements of the display system.	Unit testing with focus on verifying output against internal calculations.
A typographic error in software algorithm causes incompatibility between two devices.	Code reading against algorithm specifications.	Walkthrough focused on algorithms. Testing.

Table 12.   Examples for QA.

Generic Problem	Prevention	Detection
Test plan was not implemented or executed appropriately.	Software project management oversight.	Project status review. QA process checks.
Regression test was not performed on modified software.	Software project management oversight. Change impact analysis.	Project status review. QA process checks.
No validation before initial release.	Specified procedures regarding testing before product release. Software project management oversight.	Project status review. QA process checks.
No validation on software changes.	Software project management.	Project status review. QA process checks.

not a fault type, it is a process problem whose use might have prevented some of the failures. For this category, prevention techniques refer to discovering problems with QA. The responsibility for quality belongs to everyone on the project. QA examples appear in Table 12.

Some faults, such as omission, logic, and calculation, may have their genesis in the *requirements* specification. This category demonstrates the need to develop,

Table 13.   Examples of requirements faults.

Generic Problem	Prevention	Detection
Exceptional conditions were not specified in the requirement specification.	Modeling. Analysis. Traceability.	Interface analysis. Requirement review System test.
Functions missing in the requirement specification.	Modeling. Analysis. Traceability.	Interface analysis. Requirement review. System test.
Requirement specification was incorrect for its usage with other components.	Modeling. Interface Analysis. Traceability.	Requirement review. Interface analysis. Design review. System test.
Test hooks or monitors were not specified.	Requirement review. Design review.	Integration, system test.

Table 14.   Examples for timing.

Generic Problem	Prevention	Detection
Two inter-react processes are out of time synch with one another.	Simulation. Design review. Code review.	Timing analysis. Integration test.
Real time clock was not accurate.	High quality real time operating system. Fault tolerance.	Timing analysis. System test.
Scheduled event did not occur due to timer failure.	High quality real time operating system. Fault tolerance.	Timing analysis. System test.

verify and validate a requirement specification, in some cases uses formal methods. The document specifying the product requirements is critical to the completeness and correctness of the software of the final product. The review of the requirements may require experts with different types of expertise to ensure that the requirements call for the right functions, appropriate algorithms, correct interfaces, function interaction, and other aspects. Examples for requirements appear in Table 13.

Timing, or synchronization, is vital to the execution of real-time applications. Examples for timing appear in Table 14.

4. Lessons Learned

The information about the software faults that caused these system failures provides valuable lessons and affirmation of quality practices. These concern development procedures, assurance practices during development and maintenance activities, and testing or assurance strategies. Methods to prevent and detect faults should focus on logic and calculation errors. For logic, methods should address improved handling of various conditions, assumptions, and interactions among functions. Attention must be given to the details of calculations, such as verifying that the correct algorithm has been specified in the first place or that the programmed operators and increments are correct. The lessons addressed below are based on problems that were observed in this study, that is, they stood out as prevalent problems for this set of data and are related to the faults indicated in the fault tables in Sec. 3. Therefore the practices suggested in this paper will likely vary in other domains. Studies of other domains may provide a variation of the lessons learned here along with a roadmap for selecting the best quality strategy within a company or domain from more general guidance on quality practices. Other guidance discussing general good practices on software development and assurance includes the Capability Maturity Model, and NIST documents on life cycle development and assurance, and verification and validation.[4,7,8]

4.1. *Development and maintenance*

While software development processes are already well defined by such models as the CMM, this study indicates particular practices which would help prevent the faults that led to these specific failures. For example, training in the characteristics of the computer on which the device will reside might have prevented some of the computation errors concerning registers. Training in the application domain concerning how the outputs of functions interact and will be used by the operator might have prevented wrong interval size which produced unusable charts. Attention to details, that is, checking and verifying one's work as related to the specifications for that work, might have prevented several problems. A member of the software team with experience in the application domain may have caught several problems. Many logic faults stemmed from misunderstanding of how various functions interact, that is, under certain conditions, and in some cases, that they would interact at all. A traceability map, used regularly, can identify inconsistencies or incompleteness. The following list highlights some of the practices recommended for development and maintenance tasks:

- Complete specification of requirements, with emphasis on conditions and interactions of functions. Formal methods may be considered for highly complex systems.
- Traceability of the development artifacts: requirements to design (high, low levels) to code to user documentation and to all test documentation, especially location of source of faults. The analysis should be conducted forward and backward.
- Traceability and configuration management of all changes to the product as result of any assurance activities.
- Software configuration management.
- Change impact analysis.
- Expertise in the application domain by at least one person involved with quality practices such as requirements analysis, inspections, testing.
- Daily attention to details of the current process, the mapping to results of the previous process, and personal reviews of one's work.
- Training.

4.2. *Assurance practices*

The quality of software is the responsibility of everyone involved in its development. Practices listed above for development and maintenance are a few enabling factors in establishing an environment in which this responsibility is recognized. Other tasks fall into the category of quality assurance, but may be performed by the persons engaged in development of the software artifacts or by those separated organizationally under some quality assurance name. Every artifact of development processes needs to be scrutinized. The list of techniques supporting this scrutiny is long, and again, published elsewhere. Instead we focus on the few techniques whose value is indicated by the faults causing the failures of these devices. The inspection technique, as per Glass,[6] can be perceived as a variety of techniques that examine artifacts, ranging from requirements to design to code to test cases. Such techniques may include code reading, formal inspection meetings, review by programmer using various analytic techniques, and focused inspections. Porter and Votta describe scenario-based inspections in which participants looked for certain classes of errors.[9] To focus on a class of errors, the inspectors need to have some idea of the prevalent classes of errors of the product they are examining. The following list summarizes these suggestions:

- Focused review, inspection of the artifact against the types of faults characteristic of the domain, and the vendor's history.
- Traceability analysis, especially focused on completeness.
- Mental execution of potentially troublesome locations (e.g., an algorithm, a loop, an interface).
- Code reading.
- Recording of fault information from the assurance activities and better usage of this information.
- Recording, during development and quality assurance activities, of the symptoms that indicated there are faults.
- Checklists, questions, methods designed to force those symptoms to manifest themselves.
- Formal or informal proof of algorithm correctness.
- Use of simulation in complex situations where several interactions may occur, especially involving several components of the system.

4.3. *Testing*

How thorough was the testing applied to the devices that were recalled? One way to study this question is to look at what conditions are required to trigger the faults that remained after release. That is, is the fault manifested in a single condition, or two or more conditions? Some of the failures (109 out of the complete set of 342) contained sufficient detail to determine what level of testing would be required to detect the fault. For example, one problem report said that "if device is used with old electrodes, an error message will display, instead of an equipment alert." In this case, testing the device with old electrodes would have detected the problem. Another indicated that "upper limit CO2 alarm can be manually set above upper limit without alarm sounding." Again, a single test input that exceeded the upper limit would have detected the fault.

Other problems were not so easily manifested. One noted that "if a bolus delivery is made while pumps are operating in the body weight mode, the middle LCD fails to display a continual update." In this case, detection would have required a test with the particular pair of conditions that caused the failure: bolus delivery while in body weight mode. One vendor's description of a failure manifested on a particular pair of conditions was "the ventilator could fail when the altitude adjustment feature was set on 0 meters and the total flow volume was set at a delivery rate of less than 2.2 liters per minute."[b] Only three of 109 failures indicated that more than two conditions were required to cause the failure. The most complex of these involved four conditions and was presented as "the error can occur when demand dose has been given, 31 days have elapsed, pump time hasn't been changed, and battery is charged." The remaining 233 failures did not contain sufficient detail to make a judgment on the number of test conditions required to demonstrate a fault; many described the cause as simply "software error." It is significant however, that of the 109 reports that are detailed, 98% showed that the problem could have been detected by testing the device with all pairs of parameter settings.

Medical devices generally have a relatively small number of input variables, each with either a small discrete set of possible settings, or a finite range of values. Nevertheless, testing all possible combinations of settings may not be practical. For example, consider a device that has 20 inputs, each with 10 settings, for a total of 10^{20} combinations of settings. The few hundred test cases that can be built under most development budgets will of course cover less than a tiny fraction of a percent of the possible combinations. But the number of *pairs* of settings is in fact very small, and since each test case must have a value for each of the ten variables, more than one pair can be included in a single test case. Algorithms based on orthogonal latin squares are available that can generate test data for all pairs (or higher order combinations) at a reasonable cost. One such method makes it possible to cover all pairs of values for this example using only 180 test cases.[9] This level of test effort should be practical for most devices in the categories reviewed in this report.

Testing is part of the general quality practices, with unit, integration, and system testing all conducted. The failures in this study indicated specific test strategies might have been useful in detecting problems before the systems were delivered.

---

[b]The policy of the National Institute of Standards and Technology is to use metric units of measurement in all its technical papers. In this document however, works of authors outside NIST are cited which describe measurement values in certain non-metric units, and it is not appropriate to provide converted values.

Many failures were recognized by behavior of the system, for example, a part moved unexpectedly, or medication was provided at an incorrect rate. Most of these resulted from logic faults, so test cases in complex systems should attempt to drive these symptoms to appear. In some cases, the systems were updated versions, so previous test histories may also have been helpful. The list summarizes these points:

- Test cases aimed at manifesting prevalent symptoms observed by device operators.
- Stress testing.
- Change impact analysis and regression testing.
- SCM release of versions only with evidence of change impact analysis, regression testing; validation of changes.
- Integration testing focused on interface values under varying conditions.
- System testing under various environmental circumstances, with some conditions, input data incorrect or different from expected environmental conditions.
- Recording of test results, with special recording of all failures and their resolution, by failure and symptom of the system, and by fault type of the software.

5. Conclusions

This study yielded information affirming use of quality practices and identifying approaches for using fault and failure information to improve development and assurance practices. The nature of several faults indicates that known practices may not be used at all or may be misused. An important conclusion is that the use of many generally accepted quality practices, rather than use of a "silver bullet" is significant toward reduction of system failures. Questions remain for further research:

- If the practices were not used, what can be done to make them more readily usable?
- If the practices were used, why did they fail to prevent or detect the fault?
- What methods not yet generally accepted may help to prevent some faults and subsequent failures?

The analysis in this study demonstrates that different application domains may have different prevalent fault classes and different characteristic failure symptoms. Suggestions for improvement of assurance practices include:

- gathering failure and fault data,
- understanding the types of faults that are prevalent for a specific domain, and,
- developing prevention and detection approaches specific to these.

The subject of this study, failures of medical devices, is dealing with a relatively young industry, often new to adding microprocessors to devices.[c] As experience

---

[c]A medical device manufacturer adding software to a device for the first time called one author during preparation of this paper.

with software development and complexity of the software grow, the prevalent fault classes may change. In domains with a long history of software, the classes may also differ. In newer applications such as Electronic Commerce, which rely on newer technologies, operating systems, and languages, we would anticipate perhaps new fault classes for the domains as well as for the underlying software technologies. Data collection and analysis can help to identify the most prevalent faults and the areas where better methods are needed to prevent and detect them before system delivery.

This paper has shown that valuable lessons can be learned from system failures involving software. Some lessons may apply specifically to the application domain of study while some apply universally. It is important to continue this research on failures using modern technologies in various domains. The authors may be contacted by anyone willing to supply data.

Acknowledgments

The authors are grateful to the Food and Drug Administration (FDA) for making this data available to us. Our analyses and conclusions do not reflect any analyses or conclusions by the FDA. We appreciate the reviews and suggestions of Dr. Larry Reeker and the efforts of Mark Zimmerman and Michael Koo for their technical support.

References

1. H. Petroski, *To Engineer Is Human* (Vintage Books of Random House, Inc., New York, 1992).
2. D. R. Kuhn, "Sources of Failure in the Public Switched Telephone Network," *Computer* **30**(4) (April, 1997).
3. W. Gibbs, "Software's Chronic Crisis," *Sci. Am. (Int.Ed.)* **271**(3) (September, 1994), pp. 72–81.
4. Paulk *et al.*, "Capability Maturity Model, Version 1.1," *IEEE Software* (July, 1993), pp. 18–27.
5. M. E. Fagan, "Design and Code Inspections to Reduce Errors in Program Development," *IBM Systems Journal* **15**(3) (1976), pp. 219–248.
6. R .L. Glass, "Inspections — Some Surprising Findings," *Communications of the ACM* **42**(4) (April, 1999), pp. 17–19.
7. D. R. Wallace and L. M. Ippolito, "A Framework for the Development and Assurance of High Integrity Software," NIST SP 500–223, (National Institute of Standards and Technology, Gaithersburg, MD 20899) (December, 1994). http://hissa.nist.gov/publications/sp223/
8. D. R. Wallace, L. Ippolito and B. Cuthill, "Reference Information for the Software Verification and Validation Process," NIST SP 500–234 (National Institute of Standards and Technology, Gaithersburg, MD 20899, April, 1996). http://hissa.nist.gov/VV234/
9. A. Porter *et al.*, "An Experiment to Assess the Cost-Benefit of Code Inspections in Large Scale Software Developments," *Proceedings of the Ninth Annual Software Engineering Workshop* (National Aeronautics and Space Administration Goddard Space Flight Center, Greenbelt, MD 20771, December 1994).

About the Authors

Dolores R. Wallace is a system analyst for SRS Information services at the Software Assurance Technology Center at GSFC, NASA where she conducts research and writes standards and guidance in software assurance. Prior to joining SATC, Ms. Wallace spent almost 18 years at the U.S. National Institute of Standards and Technology (NIST) where she lead the Software Reference Fault and Failure Data Project, served as the project leader for the High Integrity Software System Assurance project, developed guidance on software verification and validation, and conducted research in related software quality topics.

Her publications on software verification and validation include NIST SP 500–234, Reference Information for the Software Verification and Validation Process, V&V articles in IEEE Software, the Encyclopedia of Software Engineering (Wiley) and the IEEE Tutorials on Software Requirements Engineering and Software Engineering. She is co-author, Software Quality Control, Error Analysis, and Testing, Noyes Data Corporation, 1995 and co-chair of the IEEE STD 1012–1998, Software Verification and Validation. She has published papers on software experimentation and other software engineering topics. She received the 1994 U.S. Department of Commerce Bronze Medal Award. While at NIST she served on the Industrial Advisory Board for the IEEE Computer Society's Software Engineering Body of Knowledge Project. Currently she serves on the editorial board of the American Society for Quality's Software Quality Professional, the Wiley Software Engineering Encyclopedia, and the Board of Directors for the Center for National Software Studies. She is a member of the ACM and the IEEE Computer Society. She has an MA in mathematics from Case Western Reserve University. She may be contacted at drwallace@erols.com.

D. Richard Kuhn is a computer scientist in the Computer Security Division at the National Institute of Standards and Technology. His primary technical interests are in information security, software assurance, and open systems. He has published more than 30 papers in these areas and holds a patent on implementation of role based access control in multi-level secure systems. He received the Excellence in Technology Transfer award from the Federal Laboratory Consortium in 1998 for work in role based access control, and a Bronze Medal from the Department of Commerce in 1990 for contributions to open systems standards. He is a senior member of the Institute of Electrical and Electronics Engineers and has served on a variety of government and industry working groups.

From 1996 to 1999 he was manager of the Software Quality Group at NIST, and from 1994 to 1995 served as Program Manager for the Committee on Applications and Technology of the President's Information Infrastructure Task Force. Before joining NIST in 1984, he worked as a systems analyst with NCR Corporation and the Johns Hopkins University Applied Physics Laboratory. He received an MS in computer science from the University of Maryland at College Park, and an MBA from the College of William and Mary. He may be contacted at Kuhn1@aol.com.

© 2025 World Scientific Publishing Company
https://doi.org/10.1142/9789819812547_0005

Chapter 5

A STATE-OF-THE-ART REVIEW OF FMEA/FMECA*

ABDELKADER BOUTI and DAOUD AIT KADI

*Department of Mechanical Engineering, Laval University
Ste Foy, Québec, Canada G1K 7P4*

The Failure Mode and Effects Analysis (FMEA) documents single failures of a system, by identifying the failure modes, and the causes and effects of each potential failure mode on system service and defining appropriate detection procedures and corrective actions. When extended by Criticality Analysis procedure (CA) for failure modes classification, it is known as Failure Mode Effects and Criticality Analysis (FMECA). The present paper presents a literature review of FME(C)A, covering the following aspects: description and review of the basic principles of FME(C)A, types, enhancement of the method, automation and available computer codes, combination with other techniques and specific applications. We conclude with a discussion of various issues raised as a result of the review.

Keywords: FMEA; FMECA; Dependability; Quality Assurance; Teamwork; Integration; Automation.

1. Introduction

The Failure Mode and Effects Analysis (FMEA)[34] is a "bottom-up" technique, based on inductive logic to analyze the system's dependability.[46] It starts at the lowest system indenture level specified and iterates upwards (i.e., it looks at all the failure modes of a component and analyzes their effects on overall system and on various subsystem operations). Its objectives are as follows: to provide a systematic examination of possible system single failures; analyze the effects of each failure mode on system operation; record the causes; and specify appropriate detection procedures and corrective actions.[200] FMEA is generally extended by a Criticality Analysis procedure (CA) which ranks the failure modes by evaluating their criticalities. The combination of FMEA and CA procedures corresponds to the Failure Mode, Effects and Criticality Analysis (FMECA).[233]

The FMEA was developed in the early 1950s and used in the aeronautics field to analyze aircraft safety.[58] Since then, it has been the focus of considerable research as reviewed by Dhillon.[58] It was used as a powerful dependability[a] analysis tool in

---

[a]Dependability is a discipline composed of reliability, availability, safety, and maintainability.

---

*This chapter appeared previously on the International Journal of Reliability, Quality and Safety Engineering. To cite this chapter, please cite the original article as the following: A. Bouti and D. Ait Kadi, *Int. J. Reliab. Qual. Saf. Eng*, **1**, 515–543 (2001), doi:10.1142/10.1142/S0218539394000362.

a wide range of industries, such as the aerospace, nuclear and electronic industries in the 1960s and the rail and automotive industries in the 1970s.[50,114] In the 1980s, the use of FMEC(A)[b] was extended to other areas such as bio-systems,[5–190] software[182] and robotics.[33]

Its importance was recognized by both governmental and international organizations resulting in standards such as the MIL-STD-1629A published by the US Department of Defense in 1980 and IEC-812[34] prepared by the International Electrotechnical Commission in 1985. In addition, several industrial companies, have developed their own procedures and guidelines for performing FME(C)A.[50,58,114]

The present paper presents a review of research and development work carried out on FME(C)A over the last two decades. The review covers the following aspects: description and review of the basic principles of FME(C)A, types, method enhancement, automation and available computer codes, combination with other techniques and specific applications. The paper is organized as follows: Sec. 2 proposes a procedure for elaborating an FME(C)A and the method types are outlined in Sec. 3. Section 4 reviews a number of papers related to the method and the final section concludes with a discussion of the various issues raised.

2. The FME(C)A Procedure

The MIL-STD 1629A and IEC-812[34] standards provide a comprehensive account of the FME(C)A method, describing its basic principles and an operational procedure consisting of the following steps (see Fig. 1):

Step 1: Build-up an interactive FME(C)A team.
Step 2: Definition and modeling of system under consideration.
Step 3: Identification of failure modes and associated causes and effects.
Step 4: Evaluation of criticality factor. (i.e., only in the case of FMECA procedure.)
Step 5: Definition and planning of corrective actions.
Step 6: Documentation of analysis results.

2.1. FME(C)A: An activity oriented teamwork

The first step in conducting an FME(C)A is to build up multi-disciplinary and multi-functional teamwork.[50,57,146,147,189] The team's members are generally selected from different departments involved in the various life cycle phases of the system under consideration. Furthermore, when necessary, suppliers and customers can also be integrated into the group, animated by a leader selected by participants to ensure the group coordination and supervision, based on criteria such as human rapport, thoughtfulness, skills, competence, and availability. Two problems that often arise during FME(C)A sessions, are the disparity in analytical capability between team members and the difficulty in maintaining continuity of teamwork over the life cycle

[b]FME(C)A acronym specifies FMEA and FMECA methods.

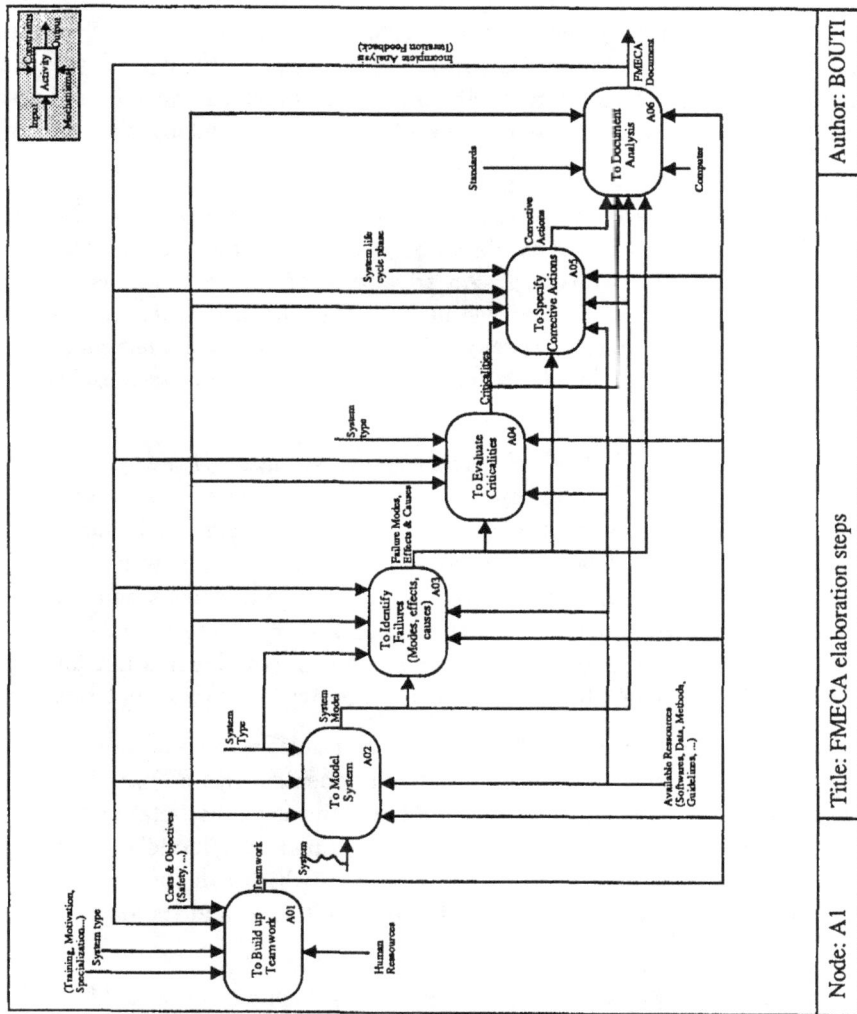

Fig. 1. The main activities of FME(C)A procedure.

| Node: A1 | Title: FMECA elaboration steps | Author: BOUTI |

phases of system. Very few authors have paid much attention to the importance of the teamwork factor in FME(C)A elaboration (see Table 1).

2.2. *System definition and modeling*

Prior to performing a failure analysis, the teamwork must clearly understand the normal operation of the system through its definition and modeling, being both fundamental and crucial.[16] System definition involves determining all basic system knowledge such as the type, mission, operation phases, environment, life cycle, ..., etc. System modeling is based upon specific techniques which can vary from a team to another, or even from one system to another, as illustrated by the following models:

(i) *Causal model*[9]: The behavior of the whole system is described in terms of the cause-and-effect behavior which is exhibited between inputs, outputs and the internal state of the components involved. An advantage of such a model is that the cause-and-effect behavior can be modeled numerically. However, this requires that the model builder must correctly specify all possible causes and their effects. In addition, having built such a model it may be difficult to insure its reusability or maintainability.

(ii) *Structural models*[8,100,101,114,131,137]: The physical structure of the system is represented by the organization of its components and their connections. The behavior of individual components is described independently of any particular system. The complete system operation can be generated by analyzing interactions between components within the specified structure. An advantage of such a model is that components can often be reused in different systems within the same field. However, it is neither suitable for systems with a large number of components nor for those whose behavior cannot easily be synthesized from individual component behavior.

(iii) *Functional models*[16,20,156,188]: The team understands system operation by analyzing the presence and function of the various components in system. So, functional models represent system structure in terms of functions and functional interconnections. One of the major advantages of exploiting functional models comes from their ability to focus on a particular part of system, which can then be analyzed in detail. Thus, the complexity of the model to be analyzed and its ambiguity are reduced.

It should be noted that a combination of the previous approaches (Multi-View modeling) is possible and recommended in the case of FME(C)A of particular systems.[79,156,177]

2.3. *Failure modes, effects and causes identification*

Based on knowledge provided by system modeling, the team performs the following tasks[20]:

(i) *Failure Mode Analysis (FMA)*: lists the failure modes of each component or function within the system. This is done by using a set of ground rules on system operation, operation phases and interactions with environment.

(ii) *Failure Cause Analysis (FCA)*: identifies the causes of each identified failure mode. This is based on the five generic causes[128] (Methods, Means, Materials, Manpower and Milieu) and other tools such as the cause-effect (Ishikawa) diagram.

(iii) *Failure Effects Analysis (FEA)*: determination of the effects of each failure mode on components within the same indenture level (local effects) or within higher ones (global effects).[233]

2.4. *Criticality factor evaluation*

To quantify the effect of each failure mode on system operation, several evaluation methods of criticality factor are applied. In Refs. 112–114, authors have used the Risk Priority Number (RPN) method to calculate the criticality factor or RPN, for each cause recorded, either adding or multiplying Severity, Occurrence and Detection ratings (S, O, D). These ratings are generally estimated with an integer interval ranging from 1 to 10. For each failure mode the Risk Priority Sum can be calculated by adding all RPNs of failure causes of a given failure mode.

Wei[233] calculated the criticality factor (C_r) for one failure mode and effect, for a single component, based upon the combination of the following values: (1) the component's failure rate λ; (2) the probability β that a specified failure effect will occur, provided the failure mode occurs; and (3) the failure mode ratio α that the failure will involve a specified failure mode, given that the component fails. C_r is then calculated from the following formula:

$$C_r = K_A . K_E . \alpha . \beta . \lambda . t . 10^6$$

where

K_A is an operational factor adjusting λ for the difference between operating stresses from component analysis vs those from measurement;
K_E an environmental factor adjusting λ for the difference between environmental stresses from component analysis vs those from measurement, and;
t, the required operating time period.
C_r is evaluated under a given severity classification defined by the MIL-STD 1629A as: catastrophic loss, critical loss, marginal loss and minor loss.

Other expressions for C_r proposed in Refs. 1, 17, 217 depend mainly on the context of analysis. For example in the context of quality assurance,[17] C_r was modified by taking into account the screening factor γ in C_r evaluation by redefining λ, α, and β.

2.5. *Corrective actions*

Based upon failures classification, the teamwork specifies the necessary corrective actions to eliminate failures or to reduce their effects.[15,23,233] The corrective actions may affect the system design, manufacturing process, control plan, ..., etc; depending on human, technological and economic constraints. Once the corrective actions are established, the teamwork iterates the procedure by taking into account the system modifications.

2.6. Documentation of analysis results

The FME(C)A results can be summarized using:

(1) Standard tabular form[34,233] (see Fig. 2): this is composed of a set of columns whose number vary according to the application. Despite the advantages of providing detailed, descriptive information on each failure mode, the tabular form has analytical limitations in that its format does not lend itself to efficient accountability and traceability of information.[6]

(2) Matrix form[101,131] (see Fig. 3): this was introduced for the first time by Barbour[6] in 1977 in the aerospace field for analyzing electrical and electronic systems. It summarizes FMEA results in a matrix form whose vertical lines represent failure effects and horizontal lines correspond to the system inputs, outputs and components. Symbols for characteristic failure modes are used to indicate the relationship between component failure mode(s) and the generated failure effect. Its major objectives[86] are as follows: to accelerate the FMEA procedure, integrate it in the design phase, reduce its costs and make FMEA easy to computerize with easier access to information. However, it becomes rather complicated when the number of inputs, outputs, parts, failure modes and effects increase. Moreover, other desired information (e.g., causes, corrective actions) are not highlighted.

3. FME(C)A Types

Depending on system life cycle, the main types of FME(C)A are:

(i) *FME(C)A-Product*[21,50]: performed during the product design phase, it allows one to confirm the production means and technology choices, to refine the schedule of conditions, to analyze the options of subcontracting or integration, to compile and to classify the critical characteristics of the product, and to insure that all product specifications exist and everyone has an operational definition.

(ii) *FME(C)A-Process*[21,50,175]: focuses on the means of production in order to: (i) assist the technicians to identify potential failures related to process activities, (ii) establish actions necessary for failure prevention, (iii) evaluate the process activities, (iv) increase the product quality, (v) improve the process productivity, and (vi) organize process environment for safety by identifying critical parameters in the process. The conditions pushing the manufacturer to perform FME(C)A-Process are generally identified from critical experiences on similar processes such as: highly warranty costs, inconvenient reliability, complexity of assembly and manufacturing procedures, unsatisfied customer and high cost production.

Project:_____ Version:_____ Date:_____

System:_____ Subsystem:_____ Teamwork leader:_____

Comp.	Funct.	Fail. Mode	fail. Cause	Local Effect	Global Effect	S	O	D	RPN	Corr. Actions

Fig. 2. Example of FMECA Tabular worksheet.

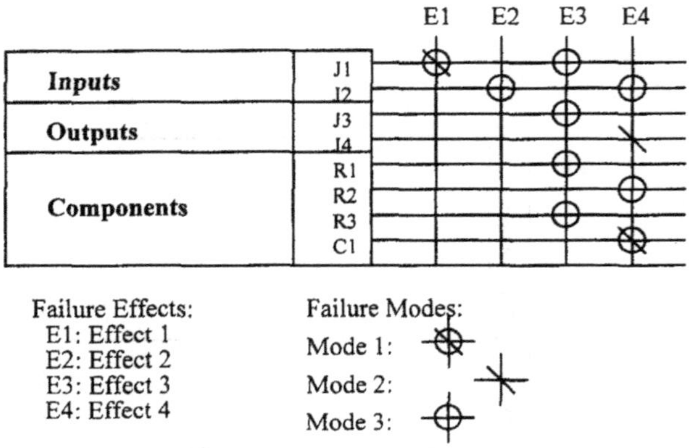

Fig. 3. Example of FMEA Matrix worksheet.

(iii) *FME(C)A-Quality Assurance*[17,21,128]: this is the procedure by which each potential failure is analyzed for early detection and prevention. Its purpose is therefore to provide early criteria for quality inspection and test planning.[18]

4. A Review of Relevant Literature

The works on FME(C)A can be conveniently classified into the following categories: (1) Description and review of the basic principles; (2) Performance evaluation and comparison with other techniques; (3) Technique enhancement; (4) Automation and computer codes; (5) Combination with other techniques; and (6) Specific applications. This classification was followed to build-up Table 1.

4.1. *Descriptions and reviews of the basic principles*

Several authors have provided a detailed description of the FME(C)A procedure and its basic principles (see Table 1). Dhillon[58] has introduced briefly the FMEA; presented an extensive list of references; enumerated documents developed by the US DoD, NASA and IEEE and outlined some ways by which FMEA helps the user. Bouti et al.[21] have discussed the basic principles of FMECA and its types and presented a framework showing the relationships between the different FMECAs. Wei[233] has argued that the FMECA is an essential task in design, proposed a unified approach to perform it, identified the application criteria of such a technique on a given system and enumerated its advantages. The role of FMECA as a technique to support other activities such as Built In Test (BIT) analysis, safety analysis or Logistic Support Analysis has been dealt with by Luthra[139] who suggested an approach to improve FMECA. Other FME(C)A surveys can be found elsewhere.[50,61]

4.2. *Performance evaluation and comparison with other techniques*

For performance evaluation of FME(C)A, many different approaches[50,214] have been used. Evans[71] asserts that "there are many places where much of this analysis is

Table 1. Classification of FME(C)A references.

Paper Category	References
Description and reviews	1, 4, 6, 8, 9, 16-17, 19, 21, 23, 25, 34, 50, 53, 57-59, 61, 64, 65, 71, 72, 77-79, 82, 87, 97, 103, 104, 107, 108, 111-113, 115, 117, 120, 121, 128, 130, 135, 137, 139, 144, 146, 147, 151, 159, 164, 165, 182, 188, 189, 192, 197-201, 207, 218, 227, 233, 243, 250
Performance evaluation or comparisons with other techniques	1, 25, 50, 71, 96, 97, 112-114, 130, 146, 147, 167, 214
FME(C)A Enhancement	6, 20, 28, 70, 75, 79, 104, 112-114, 159, 171, 201, 247
Automation	9, 32, 37, 38, 41, 51, 55, 63, 85, 86, 90, 100, 101, 105, 110, 112, 126, 131, 132, 134, 137, 148, 150, 158, 161, 162, 176, 177, 178, 185, 186, 187, 188, 195, 196, 207, 229, 230 232, 242, 248, 249
Combination with other techniques	5, 18, 42, 44, 46, 54, 61, 79, 98, 106, 125, 139, 143, 156, 194, 220, 222, 225, 237, 249
Specific applications	
•Aerospace & Aeronautics	8, 10, 11, 13, 32, 62, 83, 118, 122, 127, 133, 134, 145, 146, 163, 166, 170, 183, 196, 224
•Electrical/Electronic/Power domains	7, 11, 13, 24, 28, 29, 36, 40, 48, 51, 52, 80, 81, 91, 92, 94, 96, 106, 116, 119, 120, 122, 132, 134, 136, 138, 139, 149, 150, 157, 158, 161, 162, 175, 181, 183, 185, 194, 204, 208, 209, 211, 213, 215, 226, 234, 238, 240, 243, 247
•Automotive/Mechanical/Hydraulic domains	2, 4, 8, 12, 45, 47, 50, 79-81, 88, 95, 102, 109, 112-114, 123, 140, 141, 146, 152, 155, 157, 161, 168, 172, 203, 212, 213, 244
•Nuclear domain	3, 26, 27, 30, 43, 60, 160, 202, 210, 242
•Chemical/Petroleum area	31, 37, 65, 68, 77, 87, 89, 99, 107, 143, 170, 184, 206, 228, 231, 240, 24
•Software domain	39, 74, 93, 135, 179, 180, 182, 217, 236, 240, 246
•Biomedical/biotechnological /Industrial human behavior areas	5, 153, 190, 191, 239, 241
•Telecommunication	133, 134, 146, 173, 211
•Robotic/Control systems	33, 76, 84, 179, 210, 215, 221, 225, 231, 236
•Submarine/Plastics/Foundry domains	31, 56, 73, 128, 215, 223, 232, 235
•Quality Assurance	17, 41, 48-50, 57, 69, 82, 95, 103, 117, 118, 120, 121, 124, 128, 144, 154, 169, 173, 192, 193, 197-199, 218, 232
•Others	14, 20, 22, 23, 35, 66, 67, 129, 142, 156, 174, 205, 216, 219

pointless, even though for some projects it is very worthwhile." Hecht[97] agrees with this fact and proposes some actions such as the revision of the MIL-STD 1629A, to revitalize FMECA. Similarly, and according to Agarwala,[1] the MIL-STD 1629A fails in criticality analysis by not including (in criticality numbers computation) the contributions from the lower severity classifications of a component failure mode. The author suggests modifying the β definition and its evaluation.

Dale & Shaw[50] have provided an evaluation of the use of FMEA in the UK motor industry concluding with the following points: (i) though a number of suppliers are now using the technique to improve the product quality, its initial use was generated by requests from the customer; (ii) FMEA is generally considered as a team activity but still considered by engineers as a hard slog; (iii) the lack of FMEA computerized tools; (iv) the existence of time constraints, poor organization, inadequate training and lack of management commitment. Other drawbacks were pointed out by McKinney,[146] including: lack of defined failure causes, reckless and improper severity classification, no data sources, failure to recognize deficiencies and failure modes of earlier systems of similar design, lack of recommendations pertaining to the operation and system support, and narrow scope of analysis.

Suokas[214] carried out a detailed evaluation of four methods used in hazard identification: Hazard and Operability Study (HAZOP), Action Error Analysis (AEA), Management Oversight and Risk Tree (MORT) and FMEA. The evaluation was based on the descriptions of accidents and disturbances as indicators of the real accident contributors. Among the author's suggestions is the use of several techniques and levels of depth in hazard identification. Similarly, Pitblado[170] provided a review on other hazard identification methods such as checklists, codes, HAZOP, and FMEA; frequency methods such as fault tree analysis (FTA) and event tree analysis; and a wide range of consequence calculation models.

Hecht[96] presented FMEA as a reliability engineering activity that supports fault-tolerant design, testability, safety, logistic support and related functions. He identified some problems with FMEA for digital avionics and proposed a promising approach to capture adequate information. Finally, Lawson[130] outlined the benefits of FMECA method and illustrated its successful application in numerous projects. Other papers with relevant evaluations and comparisons are listed in Table 1.

It should be noted that it is difficult to compare FME(C)A with other techniques because of the qualitative nature of its results and the various authors' experience with its use.

4.3. Technique enhancement

Great effort has been made to improve various aspects of the FME(C)A (see Table 1).

Barbour[6] proposed a modification to the tabular form of FME(C)A to overcome what he saw as its major drawback (i.e., the non-efficiency of accountability and traceability of information). As an alternative he proposed a matrix format (see Sec. 2).

The calculation of criticality number is considered by many other authors as another major problem that the team can encounter during FMECA procedure. In this context, Fleming et al.[75] have proposed a methodology dealing with the application of fuzzy logic to RPN calculation.[70] For each failure considered, the relationship linking severity (S), occurrence (O) and detection (D) ratings as well as the required corrective action (CoA) is modeled using a rule in the form of IF (S is s_1) & (O is o_1) & (D is d_1) THEN (CoA is a_1). S, O, D and CoA are linguistic variables and s_1, o_1, d_1 and a_1 their respective linguistic values. The main advantage of this approach is its flexibility because it allows the team to model vagueness and uncertainty in the selection of S, O, and D ratings. Similarly, Kara-Zaitri *et al.*[112–114] have proposed an "advanced FMEA methodology", which in addition to RPN calculation (Local and Global Priority Indices) offers a graphical approach to summarize the results. An advantage of such an approach is that a probabilistic method is used to permit the team to express Priority Indices in terms of a common scale and therefore to make rational comparisons. Ohlef *et al.*[159] have developed a mathematical model involving a general statistical approach to the FMEA of a system (see also Ref. 171). A benefit of the mathematical model is that the team is not required to make any intuitive assumptions or simplifications on the complexity of the system block diagram, or to decide which components should be examined for criticality.

In assessing the performance of FMEA during the analysis of mechanical and hydraulic systems, Gandhi & Agrawa[79] presented a FMEA method based on di-graph and matrix approach. The method takes into account the structural as well as functional interaction of the system. Advantages of the proposed methodology are its efficient computer processing and its application not only at the design stage but also during the operational phase.

Other practical problems with the use of MIL STD 1629A are also addressed by Sexton,[201] who proposed an alternative cost-saving methodology.

Finally, as discussed below, the use of computer to automate FME(C)A and its integration with other methods are considered by many authors as an essential enhancement of the method.

4.4. *Automation and available computer codes*

Numerous computer codes were developed during the last two decades.[61,178] (see also Table 2) The automation objectives are: quick and accurate elaboration; avoidance of repetitive, tedious and time-consuming tasks; concentrates efforts on other tasks necessary for FME(C)A success; easy memorizing, recovering and up-dating of FME(C)As; integrating FME(C)A in Computer Aided activities (such as CAD/CAE) environment.

To automate the FMEA different approaches were considered. According to Ref. 9, there are three primary approaches: numerical simulation, expert systems and causal reasoning. They depend mainly on analysis objectives, system type, and available resources.

One of the first non-clerical automated FMEA tool was developed by Legg in 1978 and named FUME (FMEA Using Matrix Effects).[131] It checks a matrix FMEA for consistency and calculates failure probabilities. Its purpose was to study maintainability and reliability of electrical and electronic systems without performing any circuit analysis. However, it has limitations as regards to the number of failure effects, required for each analysis, that could be defined. To avoid FUME limitations, it was revised and expanded in a novel version called FUMER[100] in 1981.

Table 2. Examples of FME(C)A computer codes.

ACRONYM	PURPOSE
AMDEC	Generates FMECA worksheets. Developed in France in 1988 on PC in dBase 3.
AMIDEC	Edits FMECA reports, calculates failure criticalities, and studies failure propagation. Developed in France.
APACHE	Represents thermohydraulic systems with equations. Developed in France.
IFME	Models effects of failures in an engineering system, and uses the model for: behavior simulation, FMECA, fault diagnosis, ..etc. IFME is an object oriented software developed in the UK[229]
IRAS	Evaluates the reliability of electronic circuits in various ways including FMECA[24].
KARA-ZAITRI software	Generates a matrix form of FMEA and evaluates criticalities. Developed in UK in 1992 on PC in Turbo Basic.[113]
LARA	Generates tabular worksheet of FMECA. Developed in France in 1989 on PC[63].
ORACLE	Analyzes system and generates automatically FMEA-Matrix reports. Developed in USA in 1983 on Honeywell DPS 870 in FORTRAN VI.
Q-PAK	Performs functional analysis and evaluates failure criticalities. Developed in France.
LIDA	Manages and processes FMECA for the projects TELECOM 2, SPOT 2, and HELIOS[196]. Developed by the CNES and MATRA SPACE (France).

Similarly, in 1985 Goddard et al.[85,86] developed an interactive FORTRAN program called FEADS (Failure Effects And Data Synthesis) to automate the advanced Matrix Technique. FEADS provides all information required by the MIL-STD-1629A and specialized reports in accordance with the MIL-HDBK-472, Procedure 5, and BIT assessment. It uses an analyst file defining system and can be applied during design, maintainability, testability, logistics or reliability studies in the electronic field.

Renault and Sligos[37] developed AMDAO ("Analyse des Modes de Défaillance Assistée par Ordinateur") which integrates a relational database describing the functional analysis and FMECA of a given system.

With the advent of Artificial Intelligence (AI) techniques, several approaches based on AI technology were developed. A main cause of such an infatuation is the possibility of automating the reasoning portion of FME(C)A.[105,161,162,188] To illustrate this statement several works are presented below.

FADES,[242] an Expert System (ES) for performing FTA and FMEA of complex connected systems, uses a graphical editor to describe system by drawing its components and linking them together. The knowledge base created is used to simulate qualitatively the system behavior. By inducing all possible failures in the system and determining their effects, a set of facts is built-up and used to generate fault trees and FMEA tables. FURAX,[150] another example of ES for automatic generation of reliability models including FMEA and FTA, was designed to process electrical or flow networks problems on the basis of their physical descriptions. The knowledge-based system generator used to develop FURAX is the multi-purpose tool SPIRAL. FIABEX[185] is also an ES developed by CEP Systems for the automation of reliability analysis tasks including FMECA. FMEAssist,[32] a workshop software used during a space station development, assists engineers during the FMEA elaboration. The wide variety of design information required to support FMEA can be modeled by FMEAssist in a network of different disciplines and design data views generated by a database to the knowledge-based translation tool. The system designs are displayed graphically, allowing the engineers to perform the failure analysis. Russomanno et al.[186–188] discussed the application of Blackboard (BB) architecture to automate the FMEA procedure. BB model is functionally decomposed into a set of knowledge sources (KSs), each containing knowledge associated with an activity of FMEA. During FMEA generation, KSs read from and writes into BB, developing incrementally the FMEA worksheet. Price et al.[176] described an application of the model-based technology to automate the method. Bell[9] discussed the use of a causal reasoning approach to automate FMEA. They presented the Multi-Purpose Causal tool (MPC) which is applicable to different systems (electronics, hydraulics, ..., etc.). MPC consists of three major parts: a causal model, a graphics interface, and causal reasoning modules (including FMEA). After a graphical system description and the construction of its causal model; MPC generates, in an interactive way, the FMEA. An advantage of the MPC tool is the possibility of its integration in a CAD/CAE environment.

Other works on AI techniques contributing to FMEA automation are discussed in Ref. 187 and other references on FME(C)A computing are listed in Table 1.

4.5. *Combination of FME(C)A with other techniques*

The FME(C)A is an inductive, static, and qualitative method allowing the identification of single failures only. Thus, for failure analysis completeness, it can be combined with other techniques. To do this, Courties et al.[46] proposed an approach to select and combine adequate failure analysis methods including FMECA. Dale & Shaw[50] enumerated other techniques that can be used in conjunction with FMEA, especially those applied in quality assurance.

Concerning the methodologies combining FME(C)A with other methods, in addition to those cited by Dussault,[61] others exist (see Table 1), including the following:

The Integrated Critical Paths Analysis (ICPA),[222] combines FMEA, Sneak Circuit Analysis, Components Sensitivity Analysis and Critical Paths Analysis. Its main purpose is to evaluate reliability and safety of Hardware/Software systems. ICPA evaluates software effects on a man-hardware system, improves the system design and identifies possible changes.

The Failure Combined Method (FCM),[98] was developed initially for safety studies of aeronautical and space systems particularly for Concorde and Airbus airplanes, then subsequently improved by "Électricité de France" for the study of nuclear plants. It is an inductive four-step methodology based on FMEA.

The FMECA-BIT analysis integrated technique,[42,139] combines FMECA and BIT analysis. Failure modes and their detection procedures are the basis of various tests designed for BIT analysis. Its objective is to improve maintainability by detecting faults quickly and consequently isolating and announcing them properly.

The Functional Sneak Circuit Analysis (FSCA),[106] combines FMEA and Sneak Circuit Analysis to determine system response to latency conditions appearing in the absence of hardware failures.

Bieda & Hoelscher[12] presented a four-step methodology in the automotive field that integrates various reliability methods during the design phase such as: functional and reliability block diagrams; Reliability Prediction Analysis; FTA and FMECA. Bouti et al.[22] proposed another methodology based on Context diagram and SADT tool and integrating FMEA, FTA and Preliminary Risk Analysis. Niel et al.[156] presented a functional approach to integrate the concept of operational safety during the specification phase of automated manufacturing systems. The approach is based on SADT, FMEA, simulation, and state transition machines. Uguccioni et al.[223] combined FMEA, HAZOP, FTA and Sensitivity Analysis to evaluate system availability and to identify critical components during the design phase. Similarly, De Luca & Uguccioni[56] presented a Decision Tree approach based upon FMEA, FTA and Monte-Carlo techniques. Duchemin et al.[60] developed a two-step availability analysis method called DISCO (Determination of Importance Sensitivity of COmponents). It combines FMEA, FTA and eventually Markov graphs. Zwissler[250] discussed the integration problem of logistics and engineering development planning. The types of engineering analysis covered are: system reliability and maintainability allocation; prediction and measurement; FMECA; BIT effectiveness; and availability/operational effectiveness modeling. Aspects covered from the logistic side are: repair-level analysis; maintenance task analysis; spares and support equipment quantity determination; and life cycle cost analysis of design alternatives. Krause et al.[124] introduced approaches for the integration of Quality Function Deployment (QFD) and FMEA methods as well as feedback with system components for computer-aided product development. Integration is based upon information models representing product, process and factory information.

To summarize, FME(C)A can be combined with a wide range of engineering methods.

4.6. *Specific applications*

The applications summarized below highlight the widespread use of FME(C)A in different domains (see also Table 1).

Reifer[182] discussed the possible use of FMEA as a means to produce more reliable software packages. FMEA identifies critical functions and potential hazards in requirements and design of software. Van Baal[224] highlighted the application of hardware/software FMEA to analyze airplane safety with special attention to software components analysis. He concluded that the same methodology can be applied to both software and hardware. Wetterlind & Lively[236] discussed an application of FMEA and FTA methods to software for controlling robots. They pointed out the benefit of applying FMEA and FTA to software. The latter is the encouragement of communication and cooperation between software developers and industrial personnel.

To prevent problems caused by human errors in manufacturing processes, Nakajo[153] examined the use of FMEA to identify latent human errors existing within the work system before hand. A procedure for applying the method to this identification problem was defined, based on over 1000 empirical errors. In a similar context, Whittingham & Reed[239] described an FMEA based methodology for safety assessment. The purpose was to identify and reduce human errors in common industrial operations. To refine such a methodology, the authors proposed the use of Human Reliability Analysis to provide human rate data with Cost Benefit Analysis.

One new potential application of FME(C)A is in clinical engineering. In this context, Sahni & Kim[190] presented FMECA as one of the most crucial activities required during the development of a biomedical device.

Assezat[5] has applied the method to analyze the reliability of a biological wastewater treatment plant. Plant operation is simulated by a Markov model where each state corresponds to a possible failure mode identified by FMEA and checked by means of FTA.

Other new domains where FME(C)A has been used are robotics and control systems. Carrier *et al.*[33] presented a reliability study of a complete robotics unit designed for the operation and repairs of the internal upper structures of a reactor. FMEA, FTA and state graphs have been used to perform such a task. Raheja[179] proposed an approach based on FMECA to predict missing requirements that could help in reducing the downtime of intelligent controls. Fowler & Roche[76] applied FMEA and FTA for reliability analysis of a blowout preventer and hydraulic control system. Goddard,[85-86] by applying FMEA on embedded real-time control systems, pointed out that FMEA provides a cost effective way of assessing the impact of both hardware and software failures in such a type of system.

Several applications in the electronic area are available. Davies *et al.*[51] proposed a two-step methodology based on FMEA to aid reliability improvement of CMOS VLSI integrated circuits. Strandberg & Andersson[211] used FMEA and FTA to evaluate the reliability and maintainability performance of AXE switching systems in the design phase. Lehtela[132] outlined the application of computer-aided FMEA to analyze the reliability of electronic circuits. Ludwig[138] performed the technique on a general sensor signal conditioner system. Prasad[175] showed how FMEA is useful in improving the manufacturing reliability and final quality of an integrated circuit package.

FME(C)A has also been applied in areas where safety is fundamental such as submarine, nuclear, petrochemical, and aerospace systems. Caputo et al.[31] developed a methodology, based on FMEA and FTA to evaluate the mission availability of a remotely controlled submarine system designed to repair damaged pipelines in deep waters. To identify the main requirements of maintenance systems for timely detection of defects and damages, Facciolli[73] analyzed structural components of underwater tunnels by using FMEA. Uguccioni et al.[223] presented an application of FMEA to an underwater pumping station. Takagawa et al.[215] used it to analyze design reliability of a deep submergence research vehicle "Shinkai 6500" and the electric equipment installed. In the nuclear field, Cambi et al.[30] applied FMEA for reliability and safety analysis of a magnet system of a NET (Next European Torus) type machine. Ollivier & Buende[160] outlined a reliability and availability procedure based upon FMEA, for analysis of the first wall of the NET during the design phase. To improve safety of a nuclear plant, Stone & McBride[210] conducted FMEA on its control systems. Altieri et al.[3] presented an FMECA based approach to analyze risks to the cooling circuit of a reactor. In petrochemical industry, Cirilo-Bilbao et al.[36] outlined an application of the method to analyze dangers in the petroleum field. Roberts[184] described the use of hazard and risk analysis techniques including FMEA to gas industry installations. Grassick et al.[89] demonstrated how FMEA and FTA have been used to help design North Sea Oil wells. Freeman[77] discussed the AIChE Center for Chemical Process Safety Guidelines for chemical process quantitative risk analysis. FMEA is clearly highlighted and identified as a method to analyze chemical hazards. In the aerospace domain, McCrea[145] presented an application of FMEA to a space shuttle. The purpose was to meet the shuttle fail-safe program requirements during the design phase.

Several papers have focused on the use of FME(C)A in the automotive, plastics and other industries. Dale & Shaw[50] reported a state-of-the-art study on application of FMEA in the automotive field. Suhner et al.[212,213] presented a method exploiting FMECA data to generate automatically the detailed knowledge of an expert diagnostic system for complex automotive systems (see also Refs. 115, 142, 174). In the plastics industry, Webber[232] and Welsch[235] discussed the use of computer aided FMEA in injection molding shops in order to assess quality products and to reduce wastage.

In quality assurance area, FME(C)A has been adopted by several manufacturers as a tool to improve quality of processes and products.[17,41,48,198,199]

To summarize FME(C)A is widely used both in civil and military applications, by both academic and industrial staffs.

5. Conclusions

It is obvious that an appreciable literature exists on FME(C)A. This reflects primarily its accessibility and its applicability to a wide range of domains. The basic concepts are now well recognized as illustrated by the number of standards, books, papers, technical journals and academic reports.

Most of these papers, however, tend to provide descriptions and reviews of the basic method, discussing the technique and its potential applications in either the civil or military domain, restricting descriptions to the procedure based on MIL-STD 1629A and IEC-812. Few deal with the aspects of method oriented teamwork, author/reader cycle, or the representation of systems. Some authors are critical of the basic procedure and have offered suggestions to assess it. As stated by McKiney,[146] the first and possibly the most significant deficiency discovered was the exclusion of system description. If the analyst is unable to describe the system, how can he analyze it? Nevertheless, most authors support its use by highlighting its advantages.[130] Some researchers have proposed enhancements to the technique in an attempt to make it more efficient. One way to do this is by automation as outlined by several papers. Several FME(C)A software packages have been developed for this purpose and described in the literature. Unfortunately, few of these are operational or automate the whole FME(C)A procedure. Others are in the design or prototype stage. Also various methodologies integrating FME(C)A with other techniques have been proposed.

The review illustrates the interest of both academic and industrial (civil and military) sectors in the method. This is highlighted by detailed industrial applications and academic studies.

Efficient use of FME(C)A requires the total adherence of human factor (mangers, technicians, customers, suppliers, ..., etc.), the use of such a technique at the appropriate moment, training of team members, development of multi-disciplinary and multi-functionally teamwork, and its automation.

FME(C)A has been found to be flexible enough to support other activities (Reliability, Maintainability, Availability, Safety, Quality Assurance, diagnostic, testability, ..., etc.). It can be used during different system life cycle phases. Consequently several types of FME(C)A have been identified: FME(C)A-Product, FME(C)A-Process, and FME(C)A-Quality Assurance.

Finally it should be mentioned that more researches and works are required in some aspects of the method. For example in terminology, system representation, criticality evaluation, corrective actions choice, adaptation to new engineering strategies such as concurrent engineering, and automation. AI-based techniques seem to be a promising way to automate the procedure.

6. Acknowledgments

The authors would like to thank Dr. Tandjaoui Baazi and Pr. John Dickinson for the revision of the paper write-up and to the referees for their numerous remarks and suggestions on the previous version of this paper.

References†

1. A. S. Agarwala, "Shortcomings in MIL STD 1629: A guidelines for criticality analysis", in *Proc. Ann. Rel. & Maint. Symp.* (1990), pp. 494–496.

---

†Due to the wide scope of FME(C)A the authors would like to apologize in advance for overlooking any pertinent papers and not including others.

2. B. Al-Najjar, "On the selection of condition based maintenance for mechanical systems", in *Operational Reliability and Systematic Maintenance*, Eds. K. Holmberg & A. Folkeson (Elsevier, 1991), pp. 153–173.

3. D. Altieri *et al.*, "Safety and reliability analysis of the cooling system of dry-storage wells for irradiated fuel assemblies", *PSA '87* **2**, 570–575.

4. Anon., "FMEA in action", *Automotive Engineer* **13**, 26–28 (1988).

5. C. Assezat, "Probabilistic reliability analysis for biological wastewater treatment plants', *Water Science and Technology* **21**, 1813–1816 (1989).

6. G. L. Barbour, "Failure Modes and Effects Analysis by Matrix Method", in *Proc. Ann. Rel. & Maint. Symp.* (1977), pp. 114–119.

7. P. S. Baur, "Ensuring instrument reliability without overkill: Where users draw the line", *InTech* **33**, pp. 37–40 (1986).

8. S. Bednarz and D. Marriott, "Efficient analysis for FMEA", in *Proc. Ann. Rel. & Maint. Symp.* (1988), pp. 168–171.

9. D. Bell, "Using causal raisoning for automated FMEA", in *Proc. Ann. Rel. & Maint. Symp.* (1992), pp. 343–345.

10. N. Berstein and E. Kreimer, "Failure mechanism analysis", in *Proc. 7th. Int. Conf. on Rel. and Maint.* (1990), pp 105–111.

11. W. W. Bhagat, "R & M through Avionics/electronics integrity program", in *Proc. Ann. Rel. & Maint. Symp.* (1989), pp. 216–220.

12. J. Bieda and D. A. Hoelscher, "Comprehensive design reliability process for the automotive component industry via the integration of standard reliability methods", SAE Technical Paper Series, vol. 910357 (1991).

13. W. Billerbeck, "Minimizing spacecraft power loss due to single point failure", in *Proc. Intersociety Energy Conversion Engng.* (1984), pp. 345–356.

14. E. F. Bjoro, "Up date safety analysis of an advanced energy system facility", in *Proc. Ann. Rel. & Maint. Symp.* (1982), pp. 102–106.

15. M. Blanke, "A bridge from analysis of failure modes to fault handling design", in *Proc. SAFEPROCESS'94.*

16. K. G. Blemel, "Functional analysis — a methodology", in *Proc. Reliab'85* (1985), pp. 3B/1/1–3B/1/10.

17. Z. Bluvband *et al.*, "FMECA-What about the quality assurance task?", *Proc. Ann. Rel. & Maint. Symp.* (1989), pp. 242–247.

18. Z. Bluvband and A. Barel, "Testability analysis module of FMECA processor", in combined *Proc. of the 1990 and 1991 Leesburg Workshops on Rel. and Maint. Computer-Aided Engng. in Concurrent Engng.* (1992), pp. 109–116.

19. Z. Bluvband, "Availability growth aspects of reliability growth", in *Proc. Ann. Rel. & Maint. Symp.* (1990), pp. 522–527.

20. A. Bouti, D. Ait Kadi D and K. Dhouib, "Automated manufacturing systems failure analysis based on a functional reasoning", in *Proc. 10th Int. Conf. of CAD/CAM, Robotics and Factories of the Future: CARs & FOF'94*, ed. M. Zaremba (1994), pp. 423–429.

21. A. Bouti *et al.*, "L'AMDEC: un outil pour l'amélioration de la conception et de la fabrication des produits", in *Proc. CANCAM'93*, pp. 45–46.

22. A. Bouti, D. Ait Kadi and K. Dhouib, "A qualitative modeling methodology to dependability assessment of manufacturing systems", in *Proc. ICCIM'93*, pp. 446–452.

23. A. Brall, "A model for success in implementing an R&M program by a supplier of manufacturing machinery", in *Proc. Ann. Rel. & Maint. Symp.* (1994), pp. 59–64.

24. A. D. Brombacher *et al.*, "IRAS, an interactive reliability analysis for electronic circuits", *IEEE Tran. on Rel.* **34**, 507–509 (1985).

25. R. C. Bromley, E. Bottomley, "Failure modes, effects and criticality analysis (FMECA)", *IEE Colloquium on Masterclass in Systems Engineering*, Part Two (1994), pp. 1/1–7.

26. R. Buende, "Reliability and availability assessments for the next European torus", *Fusion Tech.* **14**, 197–217 (1988).

27. R. Buende, "Reliability and availability issues in NET", *Fusion Engng. and Design* **11**, N 1–2, 139–150 (1989).

28. D. J. Burns and R. M. Pitblado, "A modified HAZOP methodology for safety critical system assessment", in *Proc. of the First Safety-Critical Systems Symposium* (1993), pp. 232–45.

29. L. Cahier and R. Digout, "Functional analysis method adapted to complex electronic systems study", in *Proc. 7th. Int. Conf. on Rel. & Maint.* (1990), pp. 625–631.

30. G. Cambi *et al.*, "Reliability and safety analysis of magnet system for a NET type machine", in *Proc. IEEE 13th Symp. Fusion Eng.*, vol 1 (1989), pp. 709–712.

31. G. Caputo *et al.*, "Submarine pipeline automatic repair system. An approach to the mission availability assessment", in *Proc. Ann. Rel. & Maint. Symp.* (1992), pp. 126–130.

32. J. R. Carnes and D. E. Cutts, "FMEASSIST: A knowledge-based approach to failure modes and effects analysis", in *Proc. 3rd. Conf. on AI for Space Appl.*, part 1 (1987), pp. 187–191.

33. R. Cartier *et al.*, "La fiabilité appliquée à la robotique: opération I.S.I.S", *EDF Bulletin de la direction des études et recherches, Série A*, vol. 1 (1987), pp. 75–81.

34. CEI, "Technique d'analyse de la fiabilité des systèmes: procédure AMDE", CEI-812 (1985).

35. Y. L. Cheng *et al.*, "On establishment of I-O tables in automation of a fault tree synthesis", *Rel. Engng. & System Safety* **40**, N 3 (1993), pp. 311–318.

36. D. Cirilo-Bilbao *et al.*, "Application of FMECA to danger studies in the petroleum field", in *Proc. 7th. Int. Conf. on Rel. & Maint.* (1990), pp. 638–641.

37. L. Claude, "Un outil pour une pratique interactive des analyses de modes de défaillance", in *Proc. 7th. Int. Conf. on Rel. & Maint.* (1990), pp. 632–637.

38. P. Clavel and D. Wetter, "The Micro-computer: An essential tool for the study quality engineer", in *Proc. 7th. Int. Conf. on Rel. & Maint.* (1990), pp. 138–143.

39. J. M. Cloarec, "FMECA application to hardware–software interfaces", in *Proc. 7th. Int. Conf. on Rel. & Maint.* (1990), pp. 648–653.

40. A. W. Cockerill and M. Lavoie, "RAM analysis helps cut turbine generator systems costs", *Power Engng.* **94**, 27–29 (1990).

41. A. O. Coker *et al.*, "Computer based FMEA for quality management", *Quality Assurance* **15**, pp. 89–94 (1989).

42. R. E. Collett, "Integration of BIT effectiveness with FMECA", in *Proc. Ann. Rel. & Maint. Symp.* (1984), pp. 300–305.

43. R. C. Colley *et al.*, "Sensing loop performance monitoring in the safety systems of nuclear power stations", *Instrumentation in the Power Ind. Proc.*, vol. 34 (1991), pp. 267–269.

44. P. Cordat and M. Giraud, "AMDE par les réseaux de Petri", in *Proc 6th. Int. Conf. on Rel. & Maint.* (1988), pp. 58–62.

45. R. Coudray, "Une aide au retour d'expérience. Méthodologie d'étude de fiabilité mécanique pour un dispositif d'exploitation satisfaisante (AMDEC)", *Revue Française de Mécanique* **N 4**, 241–249 (1991).

46. J. Courties *et al.*, "A methodology for the selection of dependability methods", in *Proc. 7th. Int. Conf. on Rel. & Maint.* (1990), pp. 138–143.

47. J. S. Coutinho, "Design and logistics requirements for combat resilience", in *Proc. Ann. Rel. & Maint. Symp.* (1986), pp. 494–498.
48. C. Craig, "Advanced Quality Planning issues", in *Proc. of the Ann. Holm Conf. on Elect. Contacts* (1989), p 101.
49. R. T. Crossfield and B. G. Dale, "Developing a skip lot zero defect sampling plan. What are the main issues?", *Int. J. of Vehicle Design* **12**, N 5–6, 489–501 (1991).
50. S. B. G. Dale and P. Shaw, "Failure mode and effects analysis in the U.K. motor industry. A state of the art study", *Quality & Rel. Engng. Int.* **6**, 179–188 (1990).
51. M. S. Davies *et al.*, "Two-step methodology for CMOS VLSI reliability improvement: Step one", *Quality & Rel. Engng. Int.* **4**, N 4, 317–329 (1988).
52. R. S. Dawod & M. S. Grover, "Using FMEA for economic power transformer operation", in *Proc. 15th INTER-RAM Conf. for the Electric Power Ind.* (1988), 8pp. *supl.*
53. D. J. Dawson, "Failure mode, effect and criticality analysis", *NATO ASI Series: Computer and Systems Sciences* N 3, 55–74 (1983).
54. J. Deckers and H. Schabe, "FMEA and fault tree analysis complete each other", *Qualitaet und Zuverlaessigkeit J.* **39**, N 1, 47–50 (1994).
55. J. Deckers and H. Schabe, "Computer-aided generation of FMECA", *Qualitaet und Zuverlaessigkeit J.* **37**, N 6, 366–369 (1992).
56. F. De Luca and G. Uguccioni, "An event tree-fault tree approach to the reliability analysis of a phased underwater mission", in *Safety & Reliability'92*, pp. 572–585.
57. R. J. Dika and R. L. Begley, "Concept development through teamwork. Working for quality, cost, weight and investment", SAE technical paper, N 910212, 1991, 12p.
58. B. S. Dhillon, "Failure modes and effects analysis — bibliography", *Microelectronics & Reliability* **32**, N 5, 719–731 (1992).
59. S. Donat and M. A. Munich, "Practical planning and implementation of FMEA", *Qualitaet und Zuverlaessigkeit J.* **36**, N 3, 150–152 (1991).
60. B. Duchemnin *et al.*, "Availability studies of some PWR systems: An application of reliability methodology based on the disco method", in *Proc. Int. Conf. Nuclear Power Plant Aging, Avail. Factor and Rel. Analysis* (1985), pp. 547–552.
61. H. B. Dussault, "Automated FMEA — status and future", in *Proc. Ann. Rel. & Maint. Symp.* (1984), pp. 1–5.
62. M. Ebrahimi and J. Perry, "FMEA–IMRL correlation expert system (FICX)", in *Proc. of the 4th. Ann. A.I. and Advanced Computer Tech. Conf.* (1988), pp. 494–498.
63. ECOPOL, "LARA: Logiciel d'assistance à la réalisation d'AMDEC", tech. paper, 1989.
64. B. Edenhofer and A. Koster, "Systems analysis: The solution to the optimum use of FMEA", *Qualitaet und Zuverlaessigkeit J.* **36**, N 12, 699–704 (1991).
65. J. N. Edmondson, "Current state of risk assessment", *Institution of Chemical Enginners Symp. Series* N 120, 509–526 (1990).
66. G. W. Edwards, "Fault tolerant analysis for STS payloads", in *Proc. Ann. Rel. & Maint. Symp.* (1982),pp. 476–478.
67. J. E. Eisman Jr., "Numbers game with the logistics support analysis record (LSAR)", in *Proc. Ann. Rel. & Maint. Symp.* (1989), pp. 328–331.
68. C. Eley, "Compliance audit checklist for hazard chemicals", *Hydrocarbon Processing* **71**, N 8 (1992), 6p.
69. D. Engelhardt and K. H. Klein, "More efficient through total quality management. II", *Elektronik* **42**, N 23, 158–161 (1993).
70. C. Enrique Pelaez and J. B. Bowles, "Using fuzzy logic for system criticality analysis", in *Proc. Ann. Rel. and Maint. Symp.* (1994), pp. 449–55.

71. P. E. Evans, "Why FMECA?", *IEEE Tran. on Rel.* **38, N 2**, 161 (1989).

72. P. E. Evans, "Failure mode, effect and criticality analysis", in *Proc. Conf. Record-MIDCON'87*, vol. 11 (1987), pp. 168–171.

73. R. Facciolli, "Reliability approach to establish IMR criteria for underwater tunnels", in *Proc. of the 12th Int. Conf. on Offshore Mechanics and Arctic Engng.*, part 2 (1993), pp. 409–416.

74. K. C. Ferrera *et al.*, "Software reliability from a system perspective", in *Proc. Ann. Rel. & Maint. Symp.* (1989), pp. 332–336.

75. P. V. Fleming *et al.*, "Application of fuzzy reasoning to failure mode and effect analysis", in *Proc. of the 12th Ann. Symp. of the Soc. of Rel. Eng.* (1991), pp. 288–299.

76. J. H. Fowler and J. R. Roche, "System safety analysis of well control equipment", in *Proc. of the 25th Ann. Offshore Tech. Conf.*, part 3 (1993), pp. 427–439.

77. R. A. Freeman, "CCPS guidelines for chemical process quantitative risk analysis", *Plant/Operations Progress* **9** 231–235 (1990).

78. M. Gadoin and M. Coindoz, "An essential tool in large system program development for reliability construction", in *Proc. 5th. Int. Conf. on Rel. & Maint.* (1986), pp. 246–251.

79. O. P. Gandhi and V. P. Agrawal, "FMEA — a digraph and matrix approach", *Rel. Engng. & Syst. Safety* **35**, 147–158 (1992).

80. G. C. Gant, "Reliability in turbine governing system", in *Proc. Diesels, Gas Turbines and their Systems — the Needs for Rel. Conf.* (1985), pp. 1–12.

81. G. C. Gant, "Reliability & maintainability in a turbine governing system", in *Proc. Inst. of Mech. Engng.*, part A, vol. 201, A1 (1987), pp. 29–37.

82. W. Geiger, "FMEA — indispensable for planning a QA system", *Qualitaet und Zuverlaessigkeit J.* **36, N 8**, 468–472 (1991).

83. W. Gericke, "AMDEC en vue de la détection, la localisation et la correction des défauts", in *Proc. 6th Int. Conf. on Rel. & Maint.* (1988), pp. 714–717.

84. P. L. Goddard, "Validating the safety of embedded real-time control systems using FMEA", in *Proc. Ann. Rel. and Maint. Symp.* (1993), pp. 227–230.

85. P. L. Goddard *et al.*, "Automated advanced matrix FMEA-A logistics engineering tool", in *Proc. 19th Ann. Int. Logistics Syst. Symp.* (1984), 9pp.

86. P. L. Goddard *et al.*, "Automated advanced matrix FMEA technique", *Proc. Ann. Rel. & Maint. Symp.* (1985), pp. 77–81.

87. R. K. Goyal, "FMEA, the alternative process hazard method", *Hydrocarbon Processing* **72, N 5**, 95–99 (1993).

88. C. Granger, "Less stressful way with piston design", *Machinery & Prod. Engng.* **148, N 3775**, 46–49 (1990).

89. D. D. Grassick *et al.*, "Risk analysis of single and dual string gas lift completions", *J. Petroleum Tech.* **42, N 11**, 5p (1990).

90. C. Guidal, "Computerized FMECA — an overview of FMEGEN program", in *Proc. 5th Int. Conf. on Rel. & Maint.* (1986), pp. 163–167.

91. R. H. Gusciora, "Applying failure mode and effects analysis to a solid state relay: A case history", in *Proc. Relay 35th Conf.* (1987), pp. 9.1–9.7.

92. D. D. Hall, "First generation electronic R & M CAE", *IEEE Tran. on Rel.* **36, N 5**, 495–498 (1987).

93. F. M. Hall *et al.*, "Hardware/Software FMECA", in *Proc. Ann. Rel. & Maint. Symp.* (1983), pp. 320–327.

94. S. Hanninen, "Analysis of failures and maintenance of large electrical motors", *Rel. and Safety of Processes and Manufacturing Syst.*, eds. Y. Malmen & V. Rouhiainien, pp. 343–354.

95. M. S. Heaphy, "CI — impact on supplier requirements", *Ann. Quality Congress Tran.* **43**, 184–188 (1989).

96. H. Hecht, "Problems with failure modes and effects analysis for digital avionics", in *Proc. IEEE-AIAA The Digital Avionics Systems Conf.* (1986), pp. 695–700.

97. H. Hecht, "Comment on why FMECA", *IEEE Tran. on Rel.* **38, N 4**, 402 (1989).

98. F. Hedin *et al.*, "The failure combination method", in *Proc. Ann. Rel. & Maint. Symp.* (1981), pp. 163–171.

99. G. J. Helmstetter, "Anaerobic liquid form in place gasketing compounds", *Automotive Sealing*, SAE specials pub., N 921 (1992), pp. 25–34.

100. S. A. Herrin, "Maintainability applications using the FMEA technique", in *Proc. Ann. Rel. & Maint. Symp.* (1981), pp. 212–217.

101. S. A. Herrin, "System interface FMEA by matrix method", in *Proc. Ann. Rel. & Maint. Symp.* (1982), pp. 111–116.

102. P. A. Hogan *et al.*, "Automated fault analysis for hydraulic systems 2. Applications", in *Proc. of the Institution of Mech. Eng.*, part I, vol. 206, N 14 (1992), pp. 215–224.

103. B. Holmes, "FMEA — the number's up", *Quality Today*, July, 22–23 (1990).

104. C. L. Human, "The graphical FMECA", in *Proc. Ann. Rel. & Maint. Symp.* (1975), pp. 298–303.

105. J. E. Hunt *et al.*, "Automating the FMEA process", *Intelligent Syst. Engng.* **2, N 2**, 119–132 (1993).

106. T. Jackson, "Integration of sneak circuit analysis with FMEA", in *Proc. Ann. Rel. & Maint. Symp.* (1986), pp. 408–414.

107. H. C. Jacobs, "Improve Process safety reviews", *Hydrocarbon Processing* **68, N 7**, 67–72 (1989).

108. S. R. Jakuba, "FMEA: A tool for reliability planning and risk evaluation", in *Proc. Spring National Design Engng. Show & Conf.* (1987), pp. 89–95.

109. P. Jeswani and J. Bieda, "Predective process for spring failure rates in automotive parts applications", SAE technical paper, N 910356 (1991), 5p.

110. M. I. Johnson, "Desk-top computer database technique for integrated R & M analysis", in *Proc. Ann. Rel. & Maint. Symp.* (1991), pp. 476–480.

111. W. E. Jordan and G. C. Marshall, "Failure mode effects and criticality analysis", in *Proc. Ann. Rel. & Maint. Symp.* (1972), pp. 30–37.

112. C. Kara-Zatri *et al.*, "A smart FMEA package", in *Proc. Ann. Rel. & Maint. Symp.* (1992), pp. 414–420.

113. C. Kara-Zatri *et al.*, "An improved FMEA methodology", in *Proc. Ann. Rel. & Maint. Symp.* (1991), pp. 248–252.

114. C. Kara Zaitri, "Advanced FMEA modeling", *GSI4* (1993), pp. 265–274.

115. M. Kempf and V. Wahl, "Failure diagnosis and failure mode and effects analysis", *Informatik Forschung und Entwicklung* **7, N 4**, 103–209 (1992).

116. R. L. Kenyon and R. J. Newell, "FMEA technique for microcomputer assemblies", in *Proc. Ann. Rel. & Maint. Symp.* (1982), pp. 117–119.

117. G. Kersten, "FMEA — an efficient method for preventive quality assurance", *VDI-Z J.* **132, N 10**, 201–202, 205–207 (1990).

118. P. P. Kirrane,"The use of FMEA technique to reduce costs", *Advances in Systems Engng. for Civil and Military Avionics Conf. Proc.* (1991), pp 11.1–11.5.

119. K. Kitagawa *et al.*, "Reliability for magnetrons for microwave ovens", *J. Microwave Power Electromagn. Energy* **21, N 3**, 149–158 (1986).

120. B. Klein, "Integrated quality assurance methods IV", *Technica* **41, N 13**, 41–48 (1992).

121. B. Klein, "Quality assurance methods in product development II", Technica **40, N 24**, 14–22 (1991).

122. G. C. Klein, "Preliminary failure modes, effects and criticality analysis of the NiMH cell for aerospace batteries", in *Proc. of the 28th Intersociety Energy Conversion Engineering Conference*, vol. 1 (1993), pp. 207–217.

123. W. E. Klein and V. R. Lali, "Model-OA wind turbine generator failure mode and effects analysis", in *Proc. Ann. Rel. & Maint. Symp.* (1990), pp. 337–340.

124. F. L. Krause *et al.*, "Methods for quality driven product development", *CIRP Annals* **42, N 1**, 151–154 (1993).

125. F. J. Kreuze, "BIT analysis and design reliability", in *Proc. Ann. Rel. & Maint. Symp.* (1983), pp. 328–332.

126. P. Kukkal *et al.*, "Database design for FMEA", in *Proc. Ann. Rel. & Maint. Symp.* (1993), pp. 231–239.

127. R. W. Kunkel Jr. and M. D. Burdeshaw, "Degraded states vulnerability analysis of aircraft", in *Proc. of the Summer Computer Simulation* (1993), pp. 409–412.

128. V. S. LaFoy, "FMEA: A step toward total quality assurance", *Modern Casting* (May, 1990), pp. 29–31.

129. J. R. Lance *et al.*, "Safety analysis of an advanced energy system facility", in *Proc. Ann. Rel. & Maint. Symp.* (1980), pp. 522–527.

130. D. J. Lawson, "Failure mode, effect and criticality analysis", *NATO ASI Series, Series F: Computer and Systems Sciences* **N 3**, 55–74 (1983).

131. J. J. M. Legg, "Computerized approach for matrix form FMEA", *IEEE Tran. on Rel.* **27**, 254–257 (1987).

132. M. Lehtela, "Computer-aided failure mode and effects analysis of electronic circuits", *Microelectronics & Reliability* **30, N 4**, 761–773 (1990).

133. C. Le Mene and S. Robichez, "A self-consistent method for the management of FMECA application to space telecommunication and observation systems", combined *Proc. of the 1990 and 1991 Leesburg Workshops on Rel. and Maint. Computer-Aided Engng. in Concurrent Engng.* (1992), pp. 63–69.

134. C. Le Mene and S. Robichez, "Development and exploitation of results of FMECA software", in *Proc. 7th. Int. Conf. on Rel. & Maint.* (1990), pp. 658–663.

135. N. G. Leveson, "Software safety: Why, what, and how?", *Computing surveys* **18, N 2**, 125–163 (1986).

136. E. I. Linares, "Predicting design availability of a 400 MW repowred unit", in *Proc. Int. Power Generation Conf., ASME* (1992), pp. 1–10.

137. J. A. Lind, "Improved methods for computerized FMEA", in *Proc. Ann. Rel. & Maint. Symp.* (1985), pp. 213–216.

138. D. L. Ludwig, "Improve sensor system reliability/performance utilizing FMECA techniques", in *Proc. ISA Aerospace Instr. Symp.*, vol. 35 (1989), pp. 599–606.

139. P. Luthra, "FMECA: An integrated approach", in *Proc. Ann. Rel. & Maint. Symp.* (1991), pp. 235–241.

140. H. Malkki *et al.*, "Life cycle costing and condition monitoring applied in a pump system", in *Rel. & Safety of Processes and Manuf. Syst.*, eds. Y. Malmen & V. Rouhiainien (1991), pp. 300–310.

141. H. Malkki *et al.*, "Experience of transferring life cycle costing to manufacturing industry", in *Proc. Int. Power Generation Conf., ASME* (1991), pp. 1–8.

142. Marteel *et al.*, "Maintenance knowledge design methodology ", *IFIP Trans. B, Appl. Tech.* **B-11**, 251–259 (1993).

143. E. Martinez *et al.*, "Knowledge elicitation and structuring for a real time expert system for monitoring a butadiene extraction system", *Computers & Chemical Engng.* **16, N Suppl** S345–S352 (1992).

144. W. Masing, "Trends in quality policy", *Qualitaet und Zuverlaessigkeit J.* **36, N 3**, 141–145 (1991).

145. T. McCrea, "The shuttle processing contractors (SPC) reliability program at the Kennedy space center; the real world", in *Proc. Ann. Rel. & Maint. Symp.* (1992), pp. 376–378.

146. B. T. McKinney, "FMECA the right way", in *Proc. Ann. Rel. & Maint. Symp.* (1991), pp. 253–259.

147. J. A. McLinn, "Improve the use and implementation of FMEA", in *Proc. IASTED Int. Conf. on Rel. Qual. & Risk Ass.* (1993), pp. 202–206.

148. M. Melis *et al.*, "Computerized safety and reliability assessment: From computer codes development to expert system techniques application", *Rel. Engng. & System Safety* **25, N 2**, 175–181 (1989).

149. A. Moretti, "Continuity and electrical safety of LV user systems", *Elettrotecnica* **77, N 12**, 1149–1152 (1990).

150. R. Moureau, "FURAX: Expert system for automatic generation of reliability models for electrical or fluid networks", in *Proc. 7th. Int. Conf. on Rel. & Maint.* (1990), pp. 375–380.

151. J. Murdoch *et al.*, "Logic modeling of dependable systems", in *Proc. SAFECOMP'92*, pp. 111–116.

152. G. J. Nagy, "Organizing the engineer's toolbox, simulation and development in automotive simultaneous engng.", SAE Special Pub., **N 973**, 930836 (1993), pp. 83–88.

153. T. Nakajo, "A method of identifying latent human errors in work systems", *Quality and Rel. Engng. Int. J.* **9, N 2**, 111–119 (1993).

154. C. Nedess and J. Nickel, "Quality costs assist FMEA risk evaluation", *Qualitaet und Zuverlaessigkeit J.* **38, N 2**, 114–118 (1993).

155. J. J. Nelson *et al.*, "Reliability models for mechanical equipment", in *Proc. Ann. Rel. & Maint. Symp.* (1989), pp. 146–153.

156. E. Niel *et al.*, "A contribution for integrating the operational safety concept in CIM", *Int. J. of Computer Integrated Manufacturing* **5, N 6**, 349–360 (1992).

157. D. R. Noak and K. Wood, "System modelling for safety and fault analysis", *IEE Colloquium on Computer Aided Engineering of Automotive Electronics* (1994), pp. 3/1–7.

158. Ogino *et al.*, "An intelligent CAD-CAT system for design and testing", *J. Japanese Society for Artificial Intelligence* **5, N 4** 492–501 (1990).

159. H. Ohlef *et al.*, "Statistical model for failure mode and effects analysis and its application to computer fault tracing", *IEEE Tran. On Rel.*, 16–22 (1978).

160. G. Ollivier and R. Buende, "Reliability and Availability analysis during the design of an innovative component of a thermonuclear plant", in *Proc. 7th. Int. Conf. on Rel. & Maint.* (1990), pp. 642–647.

161. A. R. T. Ormsby *et al.*, "Towards an automated FMEA assistant", in *Proc. 5th Int. Conf. on A.I. in Engng.* (1991), pp. 739–751.

162. A. R. T. Ormsby and M. H. Lee, "A qualitative circuit simulator", *IEE Colloquium on A.I. in Simulation* (1991), pp. 5. 1–3.

163. X. P. Ostrander, "Space shuttle orbiter: A reliability challenge and achievement", in *Proc. Ann. Rel. & Maint. Symp.* (1982), pp. 468–473.

164. R. Ostvik, "A structure and some tools for assurance of industrial availability performance", in *Operational Reliability and Systematic Maintenance*, eds. K. Holmberg & A. Folkeson (Elsevier, 1991), pp. 247–259.

165. H. Ozog *et al.*, "Hazard identification and quantification", *Chemical Enging. Progress* (April, 1987), 55–64.

166. D. L. Palumbo, "Using failure modes and effects simulation as a means of reliability analysis", in *Proc. IEEE/AIAA 11th Digital Avionics Systems Conference* (1992), pp. 102–107.

167. D. M. Perkins, "Benefits of fault tree and failure modes effects analysis — more questions to ask your control supplier", in *Proc. of Food & Pharmaceutical Industries Symp.* (1992), pp. 21–24.

168. J. C. Perrot, "Failure factors analysis of machine components", in *Proc. 7th. Int. Conf. on Rel. & Maint.* (1990), pp. 703–705.

169. T. Pfeife, J. Spiekermann and T. Zenner, "Consistent avoidance of errors by using FMEA", *Qualitaet und Zuverlaessigkeit J.* **39, N 3**, 285–6, 288, 290, 292–3 (1994).

170. R. M. Pitblado, "Hazard identification and risk control: Review and recommendations", in *Proc. 14th Australian Chem. Engng. Conf.* (1986), pp. 271–276.

171. J. T. Pizzo and R. M. Adib, "Probabilistic FMECA", in *Proc. Ann. Rel. & Maint. Symp.* (1994), pp. 390–395.

172. J. K. Plastiras, "Intersystem common cause analysis of a diesel generator failure", *Risk Analysis* **6, N 4** 463–476 (1986).

173. P. E. Plsek, "FMEA for process quality planning", *ASQC Quality Congress Tran.*, 484–489 (1989).

174. A. Poon *et al.*, "Next generation TPS architecture", *AUTOTESTCON* (1990), pp. 51–61.

175. S. Prasad, "Improving manufacturing reliability in IC package assembly using FMEA technique", in *Proc. IEEE/CHMT'90 IEMT Symp.* (1990), pp. 356–360.

176. C. J. Price *et al.*, "Model based approach to the automation of failure mode effects analysis for design", in *Proc. of the Institution of Mech. Eng.*, part D, *J. of Automobile Engng.* **206, N 4**, 285–291 (1992).

177. C. J. Price and J. E. Hunt, "Automating FMEA through multiple models", in *Research and Development in Expert Systems VIII*, vol. I, eds. M. Graham & R. W. Milne, *Proc. of Expert Systems'91*, pp 25–39.

178. RAC, *Reliability & Maintainability Software Tools* (DoD Relibility Analysis Center, 1994).

179. D. Raheja, and G. Raheja, "Maintainability analysis for intelligent controls", in *Proc. 3rd. Int. Symp. Intell. Control* (1988), pp. 385–388.

180. D. G. Raheja, "Software reliability growth process — a life cycle approach", in *Proc. Ann. Rel. & Maint. Symp.* (1989), pp. 52–55.

181. U. Rauterberg, "Modern relays: Preprogrammed reliability", *Siemens Components J.* **25, N 4**, 127–130.

182. D. J. Reifer, "Software failure mode and effects analysis", *IEEE Tran. on Rel.* **28, N 3**, 247–249 (1979).

183. G. Reiss and C. Amy, "Spacecraft electrical power systems lessons learned", *Proc. Intersociety Energy Conversion Engng.* (1988), pp. 785–788.

184. P. Roberts, "Application of hazard and risk analysis to gas making and gas storage facilities", *Institution of Chemical Enginners Symp. Series*, **N 120**, 527–539 (1990).

185. E. Robinet, J. A. Agostini, "Une application de l'outil FIABEX l'AMDEC", in *Proc. 7th Int. Conf. on Rel. & Maint.* (1990), pp. 654–657.

186. D. J. Russomanno *et al.*, "A blackboard model for an expert system for FMEA", in *Proc. Ann. Rel. & Maint. Symp.* (1992), pp. 483–490.

187. D. J. Russomanno *et al.*, "Viewing computer-aided failure modes and effects analysis from an artificial intelligence perspective", *Integrated CAE* **1, N 3**, 209–228 (1994).

188. D. J. Russomanno *et al.*, "Functional reasoning in a failure modes and effects analysis (FMEA) expert system", in *Proc. Ann. Rel. & Maint. Symp.* (1993), pp. 339–347.

189. M. J. Safoutin and D. L. Thurston, "Communications-based technique for interdisciplinary design team management", *IEEE Trans. on Engng. Manag.* **40, N 4**, 360–372 (1993).

190. A. Sahni and J. Kim, "Detailed FMECA of a pacing lead", *ASQC Quality Congress Tran.* (1990) pp. 404–407.

191. A. Sahni, "Using failure mode and effects analysis to improve manufacturing processes", *Medical Device and Diagnostic Industry* **15, N 7**, 47–51 (1993).

192. J. B. Sandberg, "Reliability for profit ... not just regulation', *Quality Progress* **20, N 8**, 51–54 (1987).

193. J. S. Sarazen, "The tools of quality, part II: Cause and effect diagrams", *Quality Progress*, July, 59–62 (1990).

194. D. S. Savakoor *et al.*, "Combining sneak circuit analysis and failure modes and effects analysis", *Proc. Ann. Rel. & Maint. Symp.* (1993), pp. 199–205.

195. R. S. Jr. Schabowsky *et al.*, "Automated reliability model construction", in *Coll. of Technical Papers — AIAA Guidance, Navigation and Control Conf. Proc.* (1986), pp. 364–374.

196. B. Schietecatte and J. F. Gajewski, "LIDA: Software for FMECA (reliability studies)", in *Proc. 7th. Int. Conf. on Rel. & Maint.* (1990), pp. 788–796.

197. J. Schmidt *et al.*, "FMEA — an opportunity for middle sized companies", *Qualitaet und Zuverlaessigkeit J.* **36, N 1**, 27–30 (1991).

198. W. Schuler, "FMEA: The secret of the first two columns, or checklists as rational aids", *Qualitaet und Zuverlaessigkeit J.* **36, N 8**, 474–479 (1991).

199. H. Scuhlz and W. Hahner, "Modified FMEA as an instrument of product development", *Qualitaet und Zuverlaessigkeit J.* **36, N 8**, 480–485 (1991).

200. F. Sevick, "Current and future concepts in FMEA", *Proc. Ann. Rel. & Maint. Symp.* (1981), pp. 414–421.

201. R. D. Sexton, "An alternative method for preparing FMECA", in *Proc. Ann. Rel. & Maint. Symp.* (1991), pp. 222–225.

202. S. C. Sharma *et al.*, "Substation feeder reliability evaluation, modeling & simulation", in *Proc. Ann. Pittsburgh Conf.*, vol. 1, part 2 (1986), pp. 439–444.

203. K. Simola *et al.*, "Systematic analysis of operating experiences — an application to motor operated valve failure and maintenance data", in *Operational Rel. and Systematic Maintenance*, eds. K. Holmberg & A. Folkeson (Elsevier, 1991), pp. 129–150.

204. R. L. Smith and J. F. Cartier, "Strategy of the reliability evaluation of long lived products", in *Rel. Key to Industrial Success Proc.* (1987), pp. 57–62.

205. A. Soren, "Testing — an integrated part of the product development process", in *Operational Rel. and Systematic Maintenance*, 11–29 (1991).

206. C. A. Stacklin, "Expert control accident mitigation", *Hydrocarbon Processing* **68, N 11**, 71–76 (1989).

207. B. Stamenkovic and S. Holovac, "Failure modes effects and criticality analysis: The basic concepts and automation", *RELECTRONIC'88*, vol. 1, pp. 355–362.

208. M. Steinke, "Implementation of the standard for safety related solid state controls for household electric ranges UL 858A", *IEEE Tran. on Ind. App.* **28**, 239–250 (1992).

209. C. N. Stoll and R. E. Schmidt, "Development of a reliable and maintainable electronic countermeasures ECM pod: A case study", in *Proc. Ann. Rel. & Maint. Symp.* (1992), pp. 253–258.

210. R. S. Stone and A. F. McBride, "Safety implications of control systems", *Nuclear Engng. & Design'89*, pp. 107–111.

211. K. Strandberg and H. Anderson, "Dependability testing of switching systems in the design phase", in *Proc. Int. Conf. on Comm.* (1988), pp. 841–845.

212. M. C. Suhner *et al.*, "Utilisation de l'AMDEC pour l'aide au diagnostic de pannes de systèmes complexes automobiles", in *Proc. Canadian Conf. on Ind. Aut.* (1992), pp. 4.21–4.24.

213. M. C. Suhner *et al.*, "Use of FMECA for fault diagnosis of complex automotive systems", *Safety & Reliability'92*, pp. 598–606.

214. J. Suokas and P. Pyy, "Evaluation of the validity of four hazard identification methods with event descriptions", *Valtion Teknillinen Tutkimuskeskus*, Tutkimuksia 518 (1988), 63p.

215. S. Takagawa *et al.*, "Electric application for deep submergence research vehicle", *Tran. of the Institute of Elect. Eng. Japan*, part D, **110-D, N** 5, 437–446 (1990).

216. Thatcher *et al.*, "An integrated predictor of system reliability", in *Proc. Ann. Rel. & Maint. Symp.* (1971), pp. 254–260.

217. Ph. Thireau, "Methodology of software effects errors analysis (SEEA) on a high safety software critical analysis", in *Proc. 5th Int. Conf. on Rel. & Maint.* (1986), pp. 111–115.

218. H. Tlach, "FMEA — a strategic element of the quality management system", *Qualitaet und Zuverlaessigkeit J.* **38, N** 5, 278–280 (1993).

219. G. Toye, "Failure modes and effects recognition in the tertiary mirror assembly design", *Proc. of SPIE*, vol. 1340 (1990), pp. 383–393.

220. Y. Tsuji *et al.*, "Approachability analysis (APA)", in *Proc. Ann. Rel. & Maint. Symp.* (1981), pp. 210–216.

221. O. Tsukamoto *et al.*, "Present technologies and technical trends for improvement of reliability of control system", in *Trans. of the Institute of Electrical Engineers of Japan*, part D, **114-D, N** 3, 228–239 (1994).

222. F. Tuma, "Software/Hardware integrated critical path analysis (ICPA)", in *Proc. Ann. Rel. & Maint. Symp.* (1980), pp. 384–387.

223. G. Uguccioni *et al.*, "Design optimization of new plants by reliability engineering methodologies: Application to a subsea pumping station", in *Proc. of the 6th EuReDatA Conf.* (1989), pp. 489–508.

224. J. B. J. Van Baal, "Hardware/Software FMEA applied to airplane safety", in *Proc. Ann. Rel. & Maint. Symp.* (1985), pp. 250–255.

225. J. Van Belle and M. Van Overmeire, "Reliability and safety study of an hydraulic robot", *Reliability'91*, pp. 405–415.

226. D. Van der Scheer, "The way of product development by Holec (switchgear)", *Elektrotechniek J.* **67, N** 5, 407–410 (1989).

227. H. J. Vliegen and H. H. Van Mal, "Rational decision making: Structuring of design meetings", *IEEE Tran. on Engng. Manag.* **37, N** 3, 185–190 (1990).

228. T. V. Vo, F. A. Simonen and H. K. Phan, "Risk-based inspection priorities for PWR high-pressure injection system components", *Tran. of Amer. Nucl. Soc.* **69**, 284–285 (1993).

229. N. Ward, "IFME, analysis of safety critical systems, based around an object oriented qualitative model", in *Proc. Object-Oriented Software Conf.* (1993), 3pp.

230. B. Warnez, "Informatisation de la rédaction et de l'utilisation des AMDEC", in *Proc. 6th. Int. Conf. on Rel. & Maint.* (1988), pp. 316–318.

231. J. A. Warren, "Process control system", SAE technical paper series, 1987, paper 871551.

232. J. Webber, "FMEA: Quality assurance methodology", *Industrial Management and Data Systems J.* **90, N** 7, 21–23 (1990).

233. B. C. Wei, "A unified approach to FMECA", in *Proc Ann. Rel. & Maint. Symp.* (1991), pp. 260–271.

234. J. M. Weiss *et al.*, "Justification of response time testing requirements for pressure and differential pressure sensors", in *Proc. Inst. in the Power Ind.* vol. 34 (1991), pp. 271–273.

235. D. Welsch, "Applied computer aided FMEA in injection moulding shops", *Kunststoffe German Plastics* **83**, N **2**, 107–111 (1993).

236. P. Wetterlind and M. W. Lively, "Ensuring software safety in robot control", in *Proc. Fall Joint Computer Conf.* (1987), pp. 34–37.

237. C. P. Whetton, "Sneak analysis of process systems", *Process Safety and Environmental Protection: Transactions of the Institution of Chemical Engineers*, part B, **71**, N **3**, 169–179 (1993).

238. M. A. White *et al.*, "Preliminary design of an advanced Stirling system for terrestrial solar energy conversion", in *Proc. 25th IECEC*, pp. 297–302 (1990).

239. R. B. Whittingham and J. Reed, "Identification and reduction of critical human error using FMEA approach", *Reliab'89*, part 2, pp. 5A/1/1–5A/1/8.

240. B. C. Wilkins *et al.*, "Relevance of central hydraulics to petroleum installations", *Petroleum Review* **42**, N **497**, 30–35 (1988).

241. G. Willis, "Failure modes and effects analysis in clinical engineering", *J. of Clinical Engng.* **17**, N **1**, 59–62 (1992).

242. C. Wood, "FADES: A tool for automated fault analysis of complex systems", *A.I. in Nuclear Power Plants*, ed. P. J. Haapanen, vol. 2 (1990), pp. 77–89.

243. C. B. Xourafas and S. G. Krishnasamy, "Prediction of distribution line service reliability by probability method", in *Proc. 1st Int. Symp. on Prob. Methods Applied to Electric Power Systems* (1986), pp. 195–202.

244. K. Yamada, "Reliability activities at Toyota motor company", *Rep. Stat. Appl. Res., JUSE*, **24**, N **3** (1977).

245. H. Yanagi *et al.*, "Dynamical absorption measurement of helium impurties on a charcoal bed at 77 K", in *Proc. 11th. Int. Conf. on Magnet Tech.*, vol. 2 (1990), pp. 1387–1390.

246. W. D. Yates and D. A. Shaller, "Reliability engineering as applied to software", in *Proc. Ann. Rel. & Maint. Symp.* (1990), pp. 425–429.

247. J. Yuan and S. Chou, "Boolean algebra method to calculate network system reliability indices in terms of a proposed FMEA", *Rel. Engng* **14**, N **3**, pp. 193–203 (1986).

248. J. Y. Zang, P. Q. Li and Y. P. Yao, "Computer aided FMECA and FTA analysis", in *Proc. ICRMS'94* (1994), pp. 556–562.

249. D. Zemva *et al.*, "PC based expert system used for MTBF and FMECA analysis of computer system", in *Proc. Rliab'87*, vol. 2, pp. 5B/5/1–5B/5/6.

250. B. J. Zwissler, "Tool P. T. integration logistics and engineering development tasks", *IEEE Proc. of the Nat. Aer. & Elect. Conf.*, vol. 3 (1990), pp. 1195–1199.

About the Authors

Daoud Ait Kadi was born in 1949. He received a degree in Mechanical Engineering from École Mohammadia d'Ingénieurs, Rabat, Morocco in 1973, an M.Sc. from École Polytechnique de Montreal in 1980 and a Ph.D from Université de Montreal in 1985.

From 1973 to 1977 he worked as a mechanical engineer in the chemical company (OCP) in Morocco. In 1985, he joined the Engineering Department of the Université de Québec à Trois Rivières as a professor in industrial engineering. In 1990, he joined the Mechanical Engineering Department of Laval University. His principal research interests are reliability, maintainability, production management, FMSs, and operations research. He is a senior member of IIE and a member of IEEE.

Abdelkader Bouti was born in 1964. He received an engineering degree in Electronics Engineering in 1988 from the Constantine University, Algeria and a DEA in Control Engineering in 1989 from Institut National Polytechnique de Grenoble, France. Presently, he is a Ph.D student at the Laval University. His research interests are the modeling of industrial systems, dependability analysis, and the application of CASE and AI techniques for the automation purposes. He is a member of the IASTED.

Chapter 6

LOGISTIC REGRESSION MODELING OF SOFTWARE QUALITY*

TAGHI M. KHOSHGOFTAAR
Department of Computer Science and Engineering
Florida Atlantic University, Boca Raton, Florida, USA
E-mail: taghi@cse.fau.edu

EDWARD B. ALLEN
Department of Computer Science and Engineering
Florida Atlantic University, Boca Raton, Florida, USA
E-mail: edward.allen@computer.org

Reliable software is mandatory for complex mission-critical systems. Classifying modules as *fault-prone*, or not, is a valuable technique for guiding development processes, so that resources can be focused on those parts of a system that are most likely to have faults.

Logistic regression offers advantages over other classification modeling techniques, such as interpretable coefficients. There are few prior applications of logistic regression to software quality models in the literature, and none that we know of account for prior probabilities and costs of misclassification. A contribution of this paper is the application of prior probabilities and costs of misclassification to a logistic regression-based classification rule for a software quality model.

This paper also contributes an integrated method for using logistic regression in software quality modeling, including examples of how to interpret coefficients, how to use prior probabilities, and how to use costs of misclassifications. A case study of a major subsystem of a military, real-time system illustrates the techniques.

Keywords: Software Process; Process Measures; Software Metrics; *Fault-prone* Modules; Software Reuse; Spiral Life Cycle; Software Quality Modeling; Logistic Regression.

1. Introduction

For mission-critical systems to be highly reliable, embedded software must also be highly reliable. Software faults are not due to wear and tear during operations, but result from mistakes in design and coding decisions during development and enhancement. Early classification of modules as *fault-prone* or not is a valuable technique for guiding development processes, so that resources can be focused to remove faults before the software becomes operational. Software quality models are a tool for doing this.

Logistic regression offers easier interpretation compared to other classification techniques.[1] There are few prior applications of logistic regression to software

---

*This chapter appeared previously on the International Journal of Reliability, Quality and Safety Engineering. To cite this chapter, please cite the original article as the following: T. M. Khoshgoftaar and E. B. Allen, *Int. J. Reliab. Qual. Saf. Eng*, **6**, 303–317 (1999), doi:10.1142/S0218539399000292.

quality models in the literature,[2,3] and none that we know of that account for prior probabilities and costs of misclassification. Previous software quality modeling classification studies by other researchers have used uniform prior probabilities and equal costs for all kinds of misclassifications. In a preliminary case study, prior probabilities and costs of misclassification improved software quality models based on nonparametric discriminant analysis.[4] A contribution of this research is confirmation that the same principles apply to logistic regression.

Another contribution of this research is presentation of an integrated method for using logistic regression in software quality modeling, including examples of how to interpret logistic regression coefficients, how to use prior probabilities, and how to use costs of misclassifications.

A case study was based on a large subsystem of a tactical military system, the Joint Surveillance Target Attack Radar System, JSTARS. It is an embedded, real time application. The objective of our software quality model was to predict at the beginning of integration whether each module will be considered *not fault-prone* or *fault-prone* at the end of the current development cycle. With such predictions, one can focus review, integration, and testing resources on high risk parts of the system. The independent variables were measures of development processes prior to integration.[5]

The remainder of this paper presents general principles, a case study, and conclusions. The case study is a step by step example of how to prepare data and how to build, interpret, and validate a model.

2. Applying Logistic Regression

Logistic regression is a statistical modeling technique where the dependent variable has only two possible values.[6] Independent variables may be categorical, discrete, or continuous. In software quality modeling, we usually consider a module as an "observation"; a module is a member of one group or the other. Our dependent variable is *Class*, which has only two possible values. In this paper, we will use the groups *not fault-prone* and *fault-prone*; other groups may be used in other circumstances.

A classification model predicts membership in one group or the other. However, since a model is not likely to be perfect, some modules will probably be misclassified by the model, compared to actual group membership. Being consistent with terminology in our previously published work,[7] Type I errors misclassify modules that are actually *not fault-prone* as *fault-prone*. Type II errors misclassify modules that are actually *fault-prone* as *not fault-prone*.

There are several possible strategies for encoding categorical independent variables. For binary categorical variables, we encode the categories as the values zero and one. Discrete and continuous variables may be used directly. Let x_j be the jth independent variable, and let x_i be the vector of the ith module's independent variable values.

2.1. *Building a model*

We designate a module being *fault-prone* as an "event." Let p be the probability of an event, and thus, $p/(1-p)$ is the odds of an event. The logistic regression model has the form

$$\log\left(\frac{p}{1-p}\right) = \beta_0 + \beta_1 x_1 + \cdots + \beta_j x_j + \cdots + \beta_m x_m , \qquad (1)$$

where log means natural logarithm and m is the number of independent variables. Let b_j be the estimated value of β_j. This model can be restated as

$$p = \frac{\exp(\beta_0 + \beta_1 x_1 + \cdots + \beta_m x_m)}{1 + \exp(\beta_0 + \beta_1 x_1 + \cdots + \beta_m x_m)} , \qquad (2)$$

which implies each x_j is assumed to be monotonically related to p. Fortunately, most software engineering measures do have a monotonic relationship with faults that is inherent in the underlying processes.

Given a list of candidate independent variables and a threshold significance level α, some of the estimated coefficients may not be significantly different from zero. Such variables should not be included in the final model. The process of choosing significant independent variables is called "model selection." Stepwise logistic regression is one method of model selection that uses the following procedure.

Initially, estimate a model with only the intercept. Evaluate the significance of each variable not in the model. Add to the model the variable with the largest chi-squared p value which is better than a given threshold significance level. Estimate parameters of the new model. Evaluate the significance of each variable in the model. Remove from the model the variable with the smallest chi-squared p value whose significance is worse than a given threshold significance level. Repeat until no variables can be added or removed from the model. Tests for adding or removing a variable are based on an adjusted residual chi-squared statistic for each variable, comparing models with and without the variable of interest.[6]

We calculate maximum likelihood estimates of the parameters of the model, b_j, using the iteratively reweighted least squares algorithm. Other algorithms are also available to calculate maximum likelihood estimates. The same algorithm is used both in the stepwise procedure and for the final model. The estimated standard deviation of a parameter can be calculated, based on the log-likelihood function.[8] These calculations are provided by commonly available statistical packages, such as SAS.[9]

The odds ratio ψ_j is a statistic that indicates the relative effect on the odds of an event by a one unit change in the jth independent variable.[6] For example, suppose x_j is a binary variable with values zero or one. Let $p(1)$ be the probability of an event when $x_j = 1$, and let $p(0)$ be the probability of an event when $x_j = 0$; other things being equal.

$$\psi_j = \frac{p(1)/(1-p(1))}{p(0)/(1-p(0))} . \qquad (3)$$

Thus, the odds of an event for an observation with $x_j = 1$ is ψ_j times the odds of an event for $x_j = 0$. The odds ratio is estimated by

$$\hat{\psi}_j = e^{b_j} . \tag{4}$$

The odds ratio illustrates how straightforward interpretation is one of logistic regression's advantages.

2.2. *Prior probabilities*

We often model software development as a process that produces modules which are random samples from a large population of modules that might have been developed. From a Baysian viewpoint, our knowledge of the population is embodied in prior probabilities of class membership, that is, "prior" to knowing the attributes of any modules in the sample. Logistic regression calculates the probability of being *fault-prone* based on module attributes, but this is not enough. A decision rule that minimizes misclassifications should also take into account the overall proportions of the underlying populations for each group as well.[10]

Let π_{fp} be the prior probability of membership in the *fault-prone* group, and let π_{nfp} be the prior probability of membership in the *not fault-prone* group. We want to choose prior probabilities that are appropriate for each set of modules that we classify. When a large fit data set is representative of the population, we choose the prior probabilities, π_{fp} and π_{nfp}, to be the proportion of fit modules in each group. Otherwise, we make adjustments according to our knowledge about the data set. If we do not have information about the population of modules, we choose the uniform prior, $\pi_{fp} = \pi_{nfp} = 0.5$. Most software engineering classification models in the literature by other researchers use the uniform prior.[2,11-14] Choosing prior probabilities for software quality models is a contribution of this research.

There are several module sets of interest: the fit data set, the validation data set, and application data sets. When judging quality of fit, we use priors based on the fit data set. When validating a model with an independent data set, we consider whether or not its proportion of *fault-prone* modules should be similar to the fit data set's. If so, we use prior probabilities based on the fit data set. When applying the model to other projects or subsequent releases (application data sets), we may adjust prior probabilities based on our knowledge of project attributes and plans.

2.3. *Costs of misclassifications*

In software engineering, the cost for acting on each type of erroneous prediction will depend on the process improvement technique that uses the prediction.[4] For example, suppose we apply reliability enhancement processes, such as additional reviews, to modules identified as *fault-prone*. The cost of a Type I misclassification is to waste time on additional reviews of a module that is actually *not fault-prone*. The cost of a Type II misclassification is the lost opportunity to review a *fault-prone* module and detect its faults earlier in development. A fault that might have been discovered during a review will end up being discovered later in the project, when the cost of fixing it is much greater.

Let C_I be the cost of a Type I misclassification, and let C_{II} be the cost of a Type II misclassification. Logistic regression calculates the probability of being *fault-prone* based on attributes of the modules. A classification rule that minimizes the number of misclassifications may include prior probabilities, but this is not enough. Not all misclassifications are equivalent. Some types cost more than others. An optimal classification rule should minimize the expected cost of misclassifications (ECM), given by

$$\text{ECM} = C_I \pi_{nfp} \Pr(fp|nfp) + C_{II} \pi_{fp} \Pr(nfp|fp), \tag{5}$$

where $\Pr(fp|nfp)$ and $\Pr(nfp|fp)$ are the Type I and Type II misclassification rates, respectively. When costs of misclassification are unknown or unimportant, we choose equal costs, $C_I/C_{II} = 1$.

2.4. *Classification*

Given a logistic regression model, a module can be classified as *fault-prone* or not, by the following procedure:

1. Calculate $\hat{p}/(1 - \hat{p})$ using

$$\log\left(\frac{\hat{p}}{1 - \hat{p}}\right) = b_0 + b_1 x_1 + \cdots + b_j x_j + \cdots + b_m x_m. \tag{6}$$

2. Assign the module by a classification rule that minimizes the expected cost of misclassification.

$$Class(x_i) = \begin{cases} \textit{fault-prone} & \text{if } \dfrac{\hat{p}}{1 - \hat{p}} \geq \left(\dfrac{C_I}{C_{II}}\right)\left(\dfrac{\pi_{nfp}}{\pi_{fp}}\right), \\[2ex] \textit{not fault-prone} & \text{otherwise}. \end{cases} \tag{7}$$

This rule minimizes the expected cost of misclassification (Eq. (5)) as shown in Johnson and Wichern,[10] and generalizes when priors or costs are unknown or unimportant. We have found that this rule is valuable in software engineering applications.

3. Case Study

The Joint Surveillance Target Attack Radar System, JSTARS, was developed by Northrop Grumman for the U.S. Air Force in support of the U.S. Army.[15,5] The system performs ground surveillance, providing real time detection, location, classification, and tracking of moving and fixed objects. The system was developed under the spiral life cycle model.[16] Enhancements are currently being implemented.

We call each prototype of a spiral life cycle a "build." Successive versions of modules are created as development progresses. The "baseline version" of a build has all planned functionality implemented but not necessarily integrated and tested. The "ending version" is the one released for operational testing or the one accepted by the customer. Development of planned enhancements is done between the ending version of the prior build and the baseline version of the current build.

The following sections illustrate our method for applying logistic regression to software quality modeling. The major steps are to prepare data sets, to build a model, to interpret the model, and to validate the model. Then the model is ready for application to similar projects or subsequent spiral life cycle iterations.

We have also successfully applied the same method to other industry projects, which used product and/or process attributes as independent variables.

3.1. *Prepare data*

This major step consists of the following detailed steps: collect configuration management data, retrieving source code if necessary; analyze source code; collect problem reporting system data; calculate variables; and prepare *fit* and *test* data sets.

Collect configuration management data. The subject of this case study was the set of FORTRAN modules from a major subsystem of the final build, totaling 1643 modules and accounting for 38% of FORTRAN modules in JSTARS. Each module was a source file with one compilation unit, such as a subroutine. The project's configuration management system identified module versions for the baseline and ending versions of the system. Only modules under study that existed in both of these versions were selected.

Analyze source code. The configuration management system also retrieved archived source code for this case study. A source code analysis tool determined whether declarations or executable statements were changed from one version to the next. Changes to comments were not considered. Future work will consider software product metrics.

Collect problem reporting system data. The project uses a problem reporting system to track and control modifications to software, documents, and other software development objects. The primary raw data for the case study were Software Trouble Reports (STR). STRs are generated as problems are discovered by reviews, integration, and testing. Whenever source code is changed, the reason for the change and the version of the affected module is recorded in the STR.

One of the STR attributes is its "activity." We categorized activities of interest into four general reasons that a module was modified.[5] FAULTS means code changed due to developer errors; REQUIREMENT means code changed due to unplanned requirements changes; PERFORMANCE means code changed due to inadequate speed or capacity; and DOCUMENTATION means code changes were mandated during documentation changes. Our study included only those STRs that resulted in modification to declarations or executable statements of the code.

Calculate variables. Let *faults* be the number of FAULTS STRs that caused updates to a module's source code between the baseline version and the ending version of the build. This is essentially the number of faults discovered during integration and testing.

The dependent variable of the model was defined as class membership:

$$Class = \begin{cases} not\ fault\text{-}prone & \text{if } faults < threshold, \\ fault\text{-}prone & \text{if } faults \geq threshold, \end{cases} \qquad (8)$$

where *threshold* was chosen according to project-specific criteria. After discussions with project engineers, a classification *threshold* of two faults was selected. In other words, the project engineers considered approximately one quarter of the modules to be *fault-prone*. Another threshold might be appropriate for another project. *Class* is known at the time of the ending version.

The cumulative numbers of STRs that affected code prior to the baseline version are attributes of a module's process history as follows. Let *BaseFlts*, *BaseReq*, *BasePerf*, and *BaseDoc* be the number of FAULTS, REQUIREMENT, PERFORMANCE, and DOCUMENTATION STRs, respectively.

Since each selected STR resulted in changes to code, and changes to code are opportunities for faults, a statistical relationship to *faults* is plausible. A fault may be caused by various kinds of human mistakes. Each STR's reason entails different kinds of mental processes to implement a change to code. Diagnosis and correction of a fault occurs in the context of familiar requirements, specifications and designs. Analysis of new or changed requirements involves creating a new design that balances minimal disruption to the existing design, conformance to requirements, and flexibility for the future. Improving performance depends on an analysis of run-time behavior, rather than functionality. Diagnosing and correcting implementation discrepancies discovered during documentation revisions entails a detailed view of the code. Thus, these baseline variables are related to various aspects of software engineering which may affect the occurrence of faults.

We define categorical variables to model reuse from the prior build. New modules were created during development of the current build.

$$IsNew = \begin{cases} 1 & \text{if module did not exist in ending version of prior build,} \\ 0 & \text{otherwise.} \end{cases} \qquad (9)$$

Pre-existing modules with some changed code were reused with modifications. If a module had no code changed between the ending version of the prior build and the baseline version of the current build, then it was reused as an object.

$$IsChg = \begin{cases} 0 & \text{if no changed code since prior build}, \\ 1 & \text{otherwise}. \end{cases} \tag{10}$$

Since modules with a long history may be more reliable, we define the age of a module in terms of the number of builds it has existed.

$$Age = \begin{cases} 0 & \text{if module is new}, \\ 1 & \text{if module was new in the prior build}, \\ 2 & \text{otherwise}. \end{cases} \tag{11}$$

Our data did not include information on whether a module's age was more than two builds.

This project had data to support calculating the above variables. Other projects might have data for other process variables.

Prepare data sets. Data splitting is a technique for evaluating model accuracy when data on a similar subsequent project is not available. We impartially divided the modules into *fit* and *test* data sets. The *fit* data set was used to build a model, and the *test* data set was used to validate it. The *fit* data set had two thirds of the modules (1096), and the *test* data set had the remaining third (547). (Other split ratios may be appropriate in other studies.)

3.2. *Build model*

This major step consists of the following detailed steps: select significant independent variables; and estimate parameters of the final logistic regression model.

Select significant independent variables. The candidate independent variables represent the history of each module, known at the time of the baseline version, namely, the cumulative number of STRs for each reason, and its reuse from previous builds (*BaseFlts, BaseReq, BasePerf, BaseDoc, IsNew, IsChg, Age*). Independent variables were selected in the order shown in Eq. (12) below, using stepwise logistic regression at the $\alpha = 0.15$ significance level. In this study, we did not consider interactions among independent variables. This is a topic for further research.

Estimate parameters of the final logistic regression model. Recall that p is the probability that a module is a member of the *fault-prone* class. Logistic regression estimated the following model based on the *fit* data set.

$$\log\left(\frac{\hat{p}}{1-\hat{p}}\right) = -1.611 + 0.247 \; BaseFlts + 0.728 \; IsNew + 0.779 \; BaseReq$$

$$-0.560 \; Age + 0.457 \; IsChg. \tag{12}$$

Table 1 shows each estimated parameter b_j, its standard deviation s_j, and its odds ratio ψ_j. This model with these coefficients is only applicable to our case

Table 1. Model.

Variable	Coeff. b_j	Std. Dev. s_j	Odds Ratio ψ_j
Intercept	−1.611	0.337	—
BaseFlts	+0.247	0.032	1.280
BaseReq	+0.779	0.206	2.180
IsNew	+0.728	0.323	2.070
IsChg	+0.457	0.242	1.580
Age	−0.560	0.187	0.571

Other variables were not significant at 15% level

study project. The modeling method is generally applicable, but each software development project must build and calibrate its own model.

3.3. *Interpret the model*

This major step interprets the following aspects of a model: the significance of selected independent variables, the signs of coefficients, and the odds ratios of the independent variables.

Interpret the significance of selected variables. *BaseFlts* was the first variable selected, implying that it was more closely correlated to *Class* than the other variables. This is also indicated by fact that it has the smallest standard deviation relative to the size of the coefficient. Its relationship to *Class* may be due to several factors. For example, code changed in fixing faults are opportunities for more mistakes. Moreover, the underlying software attributes that cause human errors are probably not removed by fixing bugs.

The significance of *BaseReq* confirms conventional wisdom and earlier research on the relationship between enhancements and faults.[17] Unplanned requirements changes often disrupt the development process, resulting in more faults to be discovered later.

IsNew, *IsChg*, and *Age* were also significant. New modules have not been tested as much as older modules, and changed modules are often changed by someone other than the original designer, risking misunderstandings.

BasePerf and *BaseDoc* were not included in this model at significance level $\alpha = 0.15$, and thus, are not strongly related to faults discovered during integration and testing.

Interpret the sign of coefficients. Since p was defined as the probability of being *fault-prone*, a negative intercept b_0, means that if we ignore the other variables, we expect a typical module to be *not fault-prone*, which is consistent with the small proportion of *fault-prone* modules in the *fit* data set. A positive coefficient

$b_1, \ldots, b_m$, means that larger values of that variable were associated with worse reliability. *Age* was the only variable with a negative coefficient. This implies that older modules were more reliable, because the probability of being *fault-prone* was less. This was expected since they have had more testing and operational use.

Interpret the odds ratio of variables. The odds ratio for *IsChg* was 1.580. This indicates that the odds of being *fault-prone* for changed modules was about 58% more than the odds for unchanged modules.

The odds ratios for *BaseFlts* and *BaseReq* indicate the influence of an additional baseline STR on the likelihood of faults during integration and testing. Suppose modules A and B had the same process history except B had one more REQUIRE-MENT STR than A. The odds ratio of *BaseReq* was 2.180. Therefore, the odds of B being *fault-prone* would be more than twice the odds of A. Since the odds ratio of *BaseReq* was larger than that of *BaseFlts*, we conclude that a requirements change prior to the baseline version had a greater expected impact on the likelihood of faults during integration and testing than fixing a bug.

3.4. *Validate the model*

This major step consists of these detailed steps: select priors ratio and cost ratio; classify each module in the *test* data set; and compare the predicted values with the actual values.

Select the ratio of prior probabilities and the ratio of costs of misclas-sification. We considered both uniform and nonuniform priors. For nonuniform priors, we assumed the probability of *fault-prone* modules in the *test* data set was the same as the proportion in the *fit* data set. We considered a wide range of mis-classification cost ratios. The *fit* data set had 809 *not fault-prone* modules and 287 *fault-prone modules.*

Classify each module in the *test* data set. Once parameters are estimated, one can input values for the independent variables into Eq. (12) and then apply the decision rule in Eq. (7) to classify a module. Validation of a model should be done with an independent *test* data set which also contains values for both the dependent variable and independent variables. Validation tells us the level of accuracy to expect when applying the model to a similar data set at the beginning of integration, when the actual class memberships are not known.

Compare the predicted values with the actual values. The model was ap-plied to the *test* data set. Table 2 and Fig. 1 compare the actual with the predicted classifications for a range of values of C_I/C_{II}.

Since only about one quarter of the modules in the fit data set were actually *fault-prone*, the case of uniform priors ($\pi_{nfp}/\pi_{fp} = 0.5/0.5$) and equal misclassifi-cation costs ($C_I/C_{II} = 1/1$) was similar to the case of ($\pi_{nfp}/\pi_{fp} = 809/287$) and

$(C_I/C_{II} = 1/3)$. For this data, uniform priors and equal costs were not a satisfactory classification rule for practical use because the Type II misclassification rate (74.13%) is so large that a reliability enhancement effort would miss most of the *fault-prone* modules.

Table 2. Validation.

			Misclassifications		
π_{nfp}/π_{fp}	C_I/C_{II}	$(\pi_{nfp}/\pi_{fp})(C_I/C_{II})$	Type I	Type II	Overall
0.5/0.5	1/1	1.000	23	106	129
			5.69%	74.13%	23.58%
809/287	1/1	2.829	5	123	128
			1.24%	86.01%	23.40%
809/287	1/2	1.409	13	116	129
			3.22%	81.12%	23.58%
809/287	1/3	0.940	24	104	128
			5.94%	72.73%	23.40%
809/287	1/4	0.705	37	90	127
			9.16%	62.94%	23.22%
8809/287	1/5	0.564	94	60	154
			23.27%	41.96%	28.15%
809/287	1/6	0.470	99	55	154
			24.50%	38.46%	28.15%
809/287	1/7	0.403	102	48	150
			25.25%	33.57%	27.42%
809/287	1/8	0.352	119	39	158
			29.46%	27.27%	28.88%
809/287	1/9	0.313	119	39	158
			29.46%	27.27%	28.88%
809/287	1/10	0.282	137	29	166
			33.91%	20.28%	30.35%
809/287	1/15	0.188	194	22	216
			48.02%	15.38%	39.49%
809/287	1/20	0.141	227	9	236
			56.19%	6.20%	43.14%
809/287	1/30	0.094	296	4	300
			73.27%	2.80%	54.84%
809/287	1/50	0.056	404	0	404
			100.00%	0.00%	73.86%

Test Data Set
Number of Modules/Percent

404 *not fault-prone* modules (base of Type I %)

143 *fault-prone* modules (base of Type II %)

547 modules, total (base of Overall %) in the *test* data set

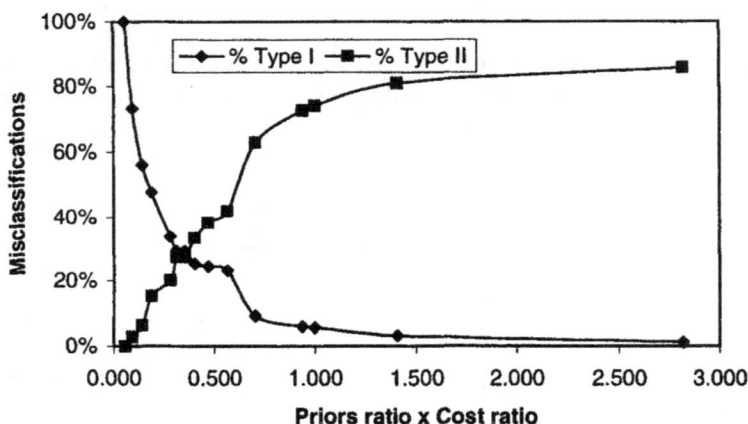

Fig. 1. The effects of costs on misclassification rates.

When we use nonuniform priors and let misclassification costs be equal, our classification rule minimized the overall misclassification rate for the fit data set. This model was also not useful due to a high Type II misclassification rate (86.01%).

The models on each end of the range of cost ratios shown in Table 2 were not useful to the project, because one type of misclassification rate or the other was so extreme. For a cost ratio of 1/2, the Type II misclassification rate was 81.12%, and for a cost ratio of 1/30, the Type I misclassification rate was 73.27%. The cost ratios in the middle are probably often more realistic. The cost of reliability enhancement early in the life cycle is on the order of man-hours, but late in the life cycle, is on the order of man-days per module. For example, a cost ratio of 1/8 has balanced misclassification rates of 29.46% for Type I and 27.27% for Type II, for an overall rate of 28.88%. Future work will seek to improve model performance by adding software product metrics and variable interactions to the list of candidate independent variables.

Another study with this data set found that a nonparametric discriminant analysis model had similar accuracy.[4] Similar to this study, the classification rule with $C_I/C_{II} = 1$ had poor accuracy. The rule with $C_I/C_{II} = 1/10$ had a Type I misclassification rate of 32.4%, a Type II rate of 26.6%, and an overall misclassification rate of 30.9%. In Table 2, the logistic regression model had a Type I misclassification rate of 33.9%, a Type II rate of 20.3%, and an overall misclassification rate of 30.4%. Even though nonparametric discriminant analysis and logistic regression balanced the misclassification rates differently for a given cost ratio, overall accuracy was similar. On the one hand, nonparametric discriminant analysis does not assume a functional form as logistic regression does, but on the other hand, logistic regression is more easily interpreted than nonparametric discriminant analysis.

Apply the model to a current project. Having validated the model, one can then use it to predict whether each module in a current similar project is *fault-prone* or not. Validation results indicate the level of accuracy one can expect. The predictions, in turn, may be the basis for extra reviews or other reliability enhancement measures.

Consider the accuracy of our model at $C_I/C_{II} = 1/8$. Suppose a current project actually has 738 *not fault-prone* modules and 262 *fault-prone* modules for a total of 1000 (i.e., similar proportions to π_{nfp} and π_{fp}). The expected cost of enhancement efforts would be $(738)(0.2946) = 217$ cost units wasted on *not fault-prone* modules, and $(262)(1 - 0.2727) = 191$ cost units invested in *fault-prone* modules, for a total investment of 408 cost units. The expected cost avoidance for the enhanced *fault-prone* modules would be $(191)(8) = 1528$ cost units for a profit of $1528 - 408 = 1120$ cost units. Thus, this level of accuracy could be useful to a similar current project.

4. Conclusions

Reliable software is mandatory for mission-critical software such as tactical military systems. During development, reliability enhancement processes try to find faults before the system becomes operational. Classifying modules as *fault-prone*, or not, is a valuable technique for guiding such development processes.

This research contributes an integrated method for using logistic regression in software quality modeling to predict whether each module will be *fault-prone* or not. A case study of a major subsystem of JSTARS, a tactical real-time system, illustrates the techniques. We saw that logistic regression parameters can be interpreted, and that using prior probabilities and costs of misclassification can improve model accuracy. This is the first research that we know of that uses prior probabilities and costs of misclassification in a logistic regression model of software quality.

The case study's model validation showed misclassification rates over a range of the ratios. Prior probabilities were set to the proportions of being *fault-prone*, or not, in the *fit* data set. At one end of the range, the costs of Type I and Type II misclassifications were equal, and the model was not useful due to excessive Type II misclassifications. At the other extreme, Type I misclassifications were excessive. We suspect that the typical relative costs of Type I to Type II misclassifications is on the order of 1:10, which, for this data, gave approximately balanced model accuracy. This case study gives empirical evidence that logistic regression, with thoughtful consideration to prior probabilities and costs of misclassification, can be used for software quality modeling. This case study also illustrates the utility of process measures alone in such models.

Future research will seek to improve model accuracy by combining software product metrics, process measures, and interaction terms in the same logistic regression models. A detailed comparison of classification modeling methods is also a research goal.

Acknowledgments

We thank Robert Halstead for his support and encouragement, and Ronald Flass and Gary P. Trio for helpful discussions regarding collected data and the software development process. We thank Lionel C. Briand for discussion regarding logistic regression. We thank the anonymous reviewers for their thoughtful comments. This work was supported in part by a grant from Northrop Grumman. The findings and opinions in this paper belong solely to the authors, and are not necessarily those of the sponsor. Moreover, our results do not in any way reflect the quality of the sponsor's software products.

References

1. T. M. Khoshgoftaar and E. B. Allen, "Classification techniques for predicting software quality: Lessons learned," *Proceedings of the Annual Oregon Workshop on Software Metrics* (Coeur d'Alene, Idaho, USA, 1997).
2. V. R. Basili, L. C. Briand, and W. Melo, "A validation of object-oriented design metrics as quality indicators," *IEEE Trans. Software Engineering* **22**(10) (1996), pp. 751–761.
3. P. G. Frankl and S. N. Weiss, "An experimental comparison of the effectiveness of branch testing and data flow testing," *IEEE Trans. Software Engineering* **19**(8) (1993), pp. 774–787.
4. T. M. Khoshgoftaar and E. B. Allen, "The impact of costs of misclassification on software quality modeling," *Proceedings of the Fourth International Software Metrics Symposium*, IEEE Computer Society (Albuquerque, New Mexico, USA, 1997), pp. 54–62.
5. T. M. Khoshgoftaar, E. B. Allen, R. Halstead, G. P. Trio, and R. Flass, "Process measures for predicting software quality," *Computer* **31**(4) (1998), pp. 66–72.
6. D. W. Hosmer, Jr. and S. Lemeshow, *Applied Logistic Regression* (John Wiley & Sons, New York, 1989).
7. T. M. Khoshgoftaar, E. B. Allen, K. S. Kalaichelvan, and N. Goel, "Early quality prediction: A case study in telecommunications," *IEEE Software* **13**(1) (1996), pp. 65–71.
8. R. H. Myers, *Classical and Modern Regression with Applications*, Duxbury Series (PWS-KENT, Boston, 1990).
9. M. E. Stokes, C. S. Davis, and G. G. Koch, *Categorical Data Analysis Using the SAS System* (SAS Institute, Cary, North Carolina, USA, 1995).
10. R. A. Johnson and D. W. Wichern, *Applied Multivariate Statistical Analysis* (2nd edition, Prentice Hall, Englewood Cliffs, NJ, 1992).
11. C. Ebert, "Classification techniques for metric-based software development," *Software Quality Journal* **5**(4) (1996), pp. 255–272.
12. T. M. Khoshgoftaar and D. L. Lanning, "A neural network approach for early detection of program modules having high risk in the maintenance phase," *J. Systems and Software* **29**(1) (1995), pp. 85–91.
13. N. F. Schneidewind, "Methodology for validating software metrics," *IEEE Trans. Software Engineering* **18**(5) (1992), pp. 410–422.
14. R. W. Selby and A. A. Porter, "Learning from examples: Generation and evaluation of decision trees for software resource analysis," *IEEE Trans. Software Engineering* **14**(12) (1998), pp. 1743–1756.
15. T. M. Khoshgoftaar, E. B. Allen, R. Halstead, and G. P. Trio, "Detection of fault-prone software modules during a spiral life cycle," *Proceedings of the International Conference on Software Maintenance*, IEEE Computer Society (Monterey, CA, 1996), pp. 69–76.
16. B. W. Boehm, "A spiral model of software development and enhancement," *Computer* **21**(5) (1988), pp. 61–72.

17. D. L. Lanning and T. M. Khoshgoftaar, "The impact of software enhancement on software reliability," *IEEE Trans. Reliab.* **44**(4) (1995), pp. 677–682.

About the Authors

Taghi M. Khoshgoftaar is a professor of the Department of Computer Science and Engineering, Florida Atlantic University. He is also the Director of the Empirical Software Engineering Laboratory, established through a grant from the National Science Foundation. His research interests are in software engineering, software complexity metrics and measurements, software reliability and quality engineering, computational intelligence, computer performance evaluation, multimedia systems, and statistical modeling. He has published more than 150 refereed papers in these areas. He has been a principal investigator and project leader in a number of projects with industry, government, and other research-sponsoring agencies. He is a member of the Association for Computing Machinery, the American Statistical Association, and the IEEE (Computer Society and Reliability Society). He was the General Chair of the 1999 International Symposium on Software Reliability Engineering (ISSRE'99). He has served on technical program committees of various international conferences, symposia, and workshops. He has served as North American editor of the *Software Quality Journal*, and is on the editorial board of the *Journal of Multimedia Tools and Applications*.

Edward B. Allen received the B.S. degree in Engineering from Brown University in 1971, the M.S. degree in Systems Engineering from the University of Pennsylvania in 1973, and the Ph.D. degree in Computer Science from Florida Atlantic University in 1995. He is currently a research associate in the Department of Computer Science and Engineering at Florida Atlantic University. He began his career as a programmer with the U.S. Army. From 1974 to 1983, he performed systems engineering and software engineering on military systems, first for Planning Research Corp. and then for Sperry Corp. From 1983 to 1992, he developed corporate data processing systems for Glenbeigh, Inc., a specialty health care company. His research interests include software measurement, software process modeling, software quality, and computer performance modeling. He has more than 50 refereed publications in these areas. He is a member of the IEEE Computer Society and the Association for Computing Machinery.

Chapter 7

A SURVEY ON DISCRETE LIFETIME DISTRIBUTIONS[#]

CYRIL BRACQUEMOND* and OLIVIER GAUDOIN[†]

Institut National Polytechnique de Grenoble, Laboratoire IMAG-LMC,
BP 53 - 38041 Grenoble Cedex 9, France
*Cyril.Bracquemond@imag.fr
†Olivier.Gaudoin@imag.fr

This paper presents a comprehensive survey of discrete probability distributions used in reliability for modeling discrete lifetimes of nonrepairable systems. The basic properties of each model are given. A classification into two families is proposed, highlighting the interest of using a Pólya urn scheme. The quality of the estimation of models parameters is numerically assessed. Some criteria are given in order to select among the presented distributions the most useful for applications.

Keywords: Discrete distributions; lifetime; reliability models; ageing; nonrepairable systems; failure rate; Pólya distributions.

Acronyms

CDF   Cumulative Distribution Function
CV   Coefficient of Variation
DFR   Decreasing Failure Rate
IFR   Increasing Failure Rate
MLE   Maximum Likelihood Estimator
MTTF   Mean Time To Failure
SB   Salvia and Bollinger

1. Introduction

Nearly all reliability studies assume that time is continuous. But sometimes, system lifetimes can not be measured with calendar time. This is the case, for example, when an equipment operates in cycles or on demands, and the number of cycles or demands prior to failure is observed. This is also the case when an equipment is monitored only once per period and the observation is the number of time periods successfully completed prior to failure. Moreover, reliability data are often grouped or truncated. In all these situations, system lifetime is a discrete random variable.

[#]This chapter appeared previously on the International Journal of Reliability, Quality and Safety Engineering. To cite this chapter, please cite the original article as the following: C. Bracquemond and O. Gaudoin, *Int. J. Reliab. Qual. Saf. Eng*, **10**, 69–98 (2003), doi:S0218539303001007.

Then, the usual reliability concepts for continuous lifetimes have to be defined again to be adapted to discrete time. In particular, discrete analogous of usual distributions for continuous lifetimes, such as the exponential or Weibull distributions, have to be defined. It is well known that the geometric distribution is the discrete counterpart of the exponential distribution, but it is not so easy for Weibull, since at least three distributions are known as "discrete Weibull distributions". Moreover, discrete lifetime distributions can be defined without any continuous counterpart.

Rather few work has been done in discrete reliability. Several discrete lifetime distributions have been proposed, but the links between them have not been studied. The aim of this paper is to provide a survey of discrete lifetime distributions, which can also be understood as discrete nonrepairable systems reliability models.

In 1975, Nakagawa and Osaki[16] were the first to propose a specific discrete lifetime distribution which is defined as the discrete counterpart of the usual continuous Weibull distribution. In spite of the extensive use of its continuous analogous, no further work on the discrete Weibull distribution has been done until Ali Khan, Khalique and Abouammoh[2] in 1984. In 1982, Salvia and Bollinger[20] presented basic results about discrete reliability and illustrated them with simple discrete lifetime distributions with only one parameter. Stein and Dattero[24] introduced in 1984 a second discrete Weibull distribution and a third one was proposed by Padgett and Spurrier[17] in 1985. These authors also generalized Salvia and Bollinger models by adding a second parameter in order to make them more flexible. In fact, these models appear to be inverse Pólya distributions (Refs. 11 and 12), also used in reliability by Clarotti, Lannoy and Procaccia[5] in 1997. In 1983, Xekalaki[27] characterized discrete lifetime distributions whose failure rate is inverse linear, such as the Waring distribution, which is in fact also an inverse Pólya distribution. A similar work on characterization of discrete distributions with linear mean residual life has been done by Roy and Gupta[19] in 1999. In 1997, Gupta, Gupta and Tripathi[9] proposed wide classes of discrete distributions with increasing failure rate, such as for example, extended Katz and Kemp families. Finally, in 1999, Klar[13] presented a complete statistical study of a family of discrete lifetime distributions proposed by Adams and Watson,[1] which can be understood as a generalized discrete truncated logistic distribution.

It appears that the above distributions can be classified into two families. The first class is constituted with discrete distributions derived from usual continuous lifetime distributions and the second class contains distributions based on a Pólya urn scheme. In our opinion, the interest of Pólya distributions in reliability has not been stressed enough.

It is noticeable that, even within the discrete reliability topic, very few papers deal with the study of parametric discrete lifetime distributions. Recent works in discrete reliability concern discrete models for repairable system reliability,[8,18] discrete competing risks,[6] multivariate discrete lifetime distributions,[25] or discrete ageing properties.[21,26]

This paper is organized as follows. In Sec. 2, the basic concepts of discrete reliability are briefly presented. In Secs. 3 and 4, discrete lifetime distributions are described according to the classification proposed. The basic properties of each distribution are given. Existing distributions are presented and new ones are proposed, such as the discrete "$\mathcal{S}$" distribution, derived from Soler.[23] Section 5 provides a numerical study of the quality of models parameters estimators. Finally, some criteria are given in order to select among the presented distributions the most useful for applications.

2. Basic Discrete Reliability Concepts

For the sake of simplicity, we will assume in this paper that a discrete lifetime is the number K of system demands until the first failure. Then, K is a random variable defined over the set N^* of positive integers. Note that several authors, such as Salvia and Bollinger[20] or Xekalaki[27] assumed that lifetimes are defined over N. So their results are slightly different from those presented here.

The probability function and cumulative distribution function (CDF) of K are respectively defined as $p(k) = P(K = k)$ and $F(k) = P(K \leq k) = \sum_{i=1}^{k} p(i)$, $\forall k \in N^*$. The basic discrete reliability concepts are defined hereafter.

- The reliability is:

$$\forall k \in N^*, \quad R(k) = P(K > k) = 1 - F(k) = 1 - \sum_{i=1}^{k} p(i).$$

- The mean residual life is:

$$\forall k \in N^*, \quad m(k) = E(K - k | K > k).$$

- The Mean Time To Failure is (if the series converges):

$$\text{MTTF} = E(K) = m(0) = \sum_{i=1}^{\infty} i p(i).$$

- The failure rate is:

$$\forall k \in N^*, \quad \lambda(k) = P(K = k | K \geq k) = \frac{P(K = k)}{P(K \geq k)} = \frac{p(k)}{R(k-1)}.$$

The failure rate (or hazard rate) in discrete time has been defined by Barlow, Marshall and Proschan.[3] It gives the conditional probability of failure of the system at time k, given that it did not fail before. It is interesting to note that in discrete time, $\lambda(k) \leq 1$, while the usual failure rate in continuous time is not bounded. Since a failure rate determines completely a lifetime distribution, Shaked, Shanthikumar and Valdez-Torres[22] gave necessary and sufficient conditions for a sequence $\{\lambda(k)\}_{k \geq 1}$ to be a failure rate:

(a) $\exists m \in N^*, \quad \forall i < m, \lambda(i) < 1$ and $\lambda(m) = 1$

or

(b) $\forall k \in N^*, \quad \lambda(k) \in [0, 1[$ and $\sum_{k=1}^{\infty} \lambda(k) = +\infty.$

The distribution is defined over $\{1, \ldots, m\}$ in case (a), and over N^* in case (b).

The sense of variation of the failure rate is of major concern since it indicates system wear-out (IFR: Increasing Failure Rate) or burn-in (DFR: Decreasing Failure Rate). It is often easy to determine the failure rate monotonicity given its expression. When it is not the case, it is usual in continuous time to look at the log-concavity or log-convexity of the distribution (Barlow and Proschan[4]). Gupta, Gupta and Tripathi[9] proposed analogous statements for discrete distributions with unbounded support ($\forall k \in N^*, p(k) \neq 0$):

- The distribution is log-concave if and only if $\left\{ \dfrac{p(k+1)}{p(k)} \right\}_{k \geq 1}$ is decreasing. Then the failure rate is increasing (IFR).

- The distribution is log-convex if and only if $\left\{ \dfrac{p(k+1)}{p(k)} \right\}_{k \geq 1}$ is increasing. Then the failure rate is decreasing (DFR).

- If the sequence $\left\{ \dfrac{p(k+1)}{p(k)} \right\}_{k \geq 1}$ is constant, the failure rate is constant and the distribution is geometric.

3. Discrete Lifetime Distributions Derived from Continuous Ones

There are several ways to derive discrete lifetime distributions from continuous ones. The first possibility is to consider a characteristic property of a continuous distribution and to build the similar property in discrete time. The second one is to consider discrete lifetime as the integer part of continuous lifetime.

In this section and in the following, the distributions are defined equivalently by functions p, R and λ. The MTTF and variance are mentioned only if they have a closed form expression, but this seldom happens. The interpretation of model parameters is given.

3.1. *Geometric distribution*

The geometric distribution is the analogous in discrete time of the exponential distribution, since it has the lack of memory property (no ageing, no burn-in): the system failure probabilities on each demand are independent and all equal to $p \in]0, 1[$. Equivalently, the failure rate is constant. This property can also be reformulated as:

$$\forall (i, k) \in N^{*2}, \qquad P(K > i + k | K > i) = P(K > k).$$

The geometric distribution $\mathcal{G}(p)$ is defined by:

- $p(k) = p(1-p)^{k-1}$
- $R(k) = (1-p)^k$
- $\lambda(k) = p.$

The Mean Time To Failure is MTTF $= \dfrac{1}{p}$.

3.2. *Shifted negative binomial distribution*

Assume that $K_1, \ldots, K_r$ are independent random variables from a geometric distribution with parameter p. Then $\sum_{i=1}^{r} K_i$ has a negative binomial distribution with parameters r and p, $\mathcal{BN}(r,p)$. So the negative binomial distribution is the analogous in discrete time of the Gamma distribution. In order to obtain a random variable defined over N^*, we have to shift the $\mathcal{BN}(r,p)$ distribution by setting $K = X - r + 1$ where X has the $\mathcal{BN}(r,p)$ distribution.

The shifted negative binomial distribution $\mathcal{BN}_s(r,p)$ is defined by:

- $p(k) = C_{k+r-2}^{r-1} p^r (1-p)^{k-1}$
- $R(k) = 1 - \displaystyle\sum_{i=1}^{k-1} C_{i+r-2}^{r-1} p^r (1-p)^{i-1}$
- $\lambda(k) = \dfrac{C_{k+r-2}^{r-1} p^r (1-p)^{k-1}}{R(k-1)}.$

$\dfrac{p(k+1)}{p(k)} = \left(1 + \dfrac{r-1}{k}\right)(1-p)$ is a decreasing function of k, so the distribution is log-concave and the failure rate is increasing. For $r = 1$, the distribution reduces to the geometric distribution.

The Mean Time To Failure is MTTF $= \dfrac{r(1-p)}{p} + 1$. Due to the shifting, parameters r and p have no practical interpretation.

3.3. *Type I discrete Weibull distribution*

The first discrete Weibull distribution has been defined by Nakagawa and Osaki[16] in order to obtain the analogous in discrete time of the continuous Weibull distribution. It was the first time a probability distribution was specifically defined to be a discrete lifetime distribution. The model is based on a similarity of expression of the reliability between discrete and continuous time. If T has a continuous Weibull distribution $\mathcal{W}(\eta, \beta)$, then $R(t) = e^{-(\frac{t}{\eta})^\beta}$. A similar expression for the reliability in discrete time is $R(k) = e^{-(\frac{k}{\eta})^\beta}$ or equivalently $R(k) = q^{k^\beta}$, where $\beta \in R^{+*}$ and $q \in]0, 1[$.

Thus, the type I discrete Weibull distribution $\mathcal{W}_1(q, \beta)$ is defined by:

- $p(k) = q^{(k-1)^\beta} - q^{k^\beta}$
- $R(k) = q^{k^\beta}$
- $\lambda(k) = 1 - q^{k^\beta - (k-1)^\beta}$

q is the probability of surviving the first demand. As for the continuous Weibull distribution, β is a shape parameter: the distribution is IFR for $\beta > 1$, DFR for $0 < \beta < 1$, and for $\beta = 1$, it reduces to the geometric distribution.

3.4. *Type II discrete Weibull distribution*

Stein and Dattero[24] defined another discrete Weibull distribution, whose construction is based on the preservation of the power function form of the failure rate. If T has a continuous Weibull distribution $\mathcal{W}(\eta, \beta)$, then $\lambda(t) = (\frac{t}{\eta})^{\beta-1}$. A similar expression in discrete time is $\lambda(k) = (\frac{k}{\eta})^{\beta-1}$ with $\eta \in R^{+*}$ and $\beta \in R^{+*}$.

But in discrete time, $\lambda(k) \le 1$, then k has to be less than η. So this distribution has a bounded support. In order to check the conditions presented in Sec. 2 for λ to be a failure rate, η has to be an integer. It is more usually denoted m.

Then, the type II discrete Weibull distribution $\mathcal{W}_2(m, \beta)$ with support in $\{1, 2, \cdots, m\}$ can be redefined as:

- $p(k) = \left(\dfrac{k}{m}\right)^{\beta-1} \prod\limits_{i=1}^{k-1} \left[1 - \left(\dfrac{i}{m}\right)^{\beta-1}\right]$ for $k \in \{1, 2, \ldots, m\}$

- $R(k) = \prod\limits_{i=1}^{\inf(k,m)} \left[1 - \left(\dfrac{i}{m}\right)^{\beta-1}\right]$

- $\lambda(k) = \left(\dfrac{k}{m}\right)^{\beta-1}$ for $k \in \{1, 2, \ldots, m\}$.

m is the maximal lifetime of the system and β is a shape parameter.

Another discrete lifetime distribution with bounded support has been proposed by Lai and Wang.[14] From the application point of view, the assumption of a bounded support seems not realistic: how can we be sure that a system will necessarily fail in less than m demands? Then, such models will not be practically very useful.

3.5. *Type III discrete Weibull distribution*

In Ref. 17, Padgett and Spurrier proposed a discrete model which is flexible with respect to the choice of a shape parameter, analogous to the Weibull distribution in the continuous case.

The type III discrete Weibull distribution $\mathcal{W}_3(c, \beta)$ is defined for $c \in R^{+*}$ and $\beta \in R$ by:

- $p(k) = (1 - e^{-ck^\beta})e^{-c\sum_{i=1}^{k-1} i^\beta}$
- $R(k) = e^{-c\sum_{j=1}^{k} j^\beta}$
- $\lambda(k) = 1 - e^{-ck^\beta}$.

The monotonicity of the failure rate depends on the value of the shape parameter β:

- For $\beta = 0$, the distribution reduces to the geometric distribution.
- For $\beta > 0$, the distribution is IFR.
- For $\beta < 0$, the distribution is DFR.

c is linked with the probability of failure at the first demand since $p(1) = 1 - e^{-c}$.

3.6. *"S" distribution*

Soler[23] proposed the so-called "S" distribution to describe continuous lifetimes of systems subjected to random stress. We propose here the analogous in discrete time. Let us consider a system such that, on each demand, a shock can occur with probability p, and not occur with probability $1 - p$. It is natural to assume that the failure rate at the kth demand, conditionally to the shock sequence, is an increasing function of the number N_k of shocks occurred at that time. One way of taking this assumption into account is to set:

$$\forall k \in N^*, \quad \lambda_N(k) = P\left(K = k | K \geq k, \{N_j\}_{j \geq 1}\right) = 1 - \pi^{N_k}, \quad \text{with } \pi \in]0,1].$$

Conditionnally to the shock sequence, the reliability is:

$$R(k) = P(K > k) = E\left[P(K > k | \{N_j\}_{j \geq 1})\right]$$

$$= E\left[\prod_{i=1}^{k} \left(1 - P(K = i | K \geq i, \{N_j\}_{j \geq 1})\right)\right] = E\left[\prod_{i=1}^{k} \pi^{N_i}\right] = E\left[\pi^{\sum_{i=1}^{k} N_i}\right].$$

The random variables N_i are not independent. $\forall i \geq 1$, let $U_i = N_i - N_{i-1}$. The U_i's are independent and have the Bernoulli distribution $\mathcal{B}(p)$, describing the occurrence of a shock at each demand. Then $\sum_{i=1}^{k} N_i = \sum_{i=1}^{k} (k - i + 1)U_i$. So the reliability becomes:

$$R(k) = E\left[\pi^{\sum_{i=1}^{k} N_i}\right] = E\left[\pi^{\sum_{i=1}^{k}(k-i+1)U_i}\right]$$

$$= \prod_{i=1}^{k} E\left[\pi^{(k-i+1)U_i}\right] = \prod_{i=1}^{k} G(\pi^{k-i+1})$$

where G is the probability generating function of the $\mathcal{B}(p)$ distribution: $G(u) = pu + 1 - p$.

Finally, The $S(p, \pi)$ distribution is defined by:

- $p(k) = p(1 - \pi^k) \prod_{i=1}^{k-1} (1 - p + p\pi^i)$

- $R(k) = \prod_{i=1}^{k} \left(1 - p + p\pi^i\right)$
- $\lambda(k) = p(1 - \pi^k)$.

p is the probability that a shock occurs on demand and π is the probability of surviving the first demand given that a shock has occurred.

If a shock occurs at each demand, then $p = 1$ and we obtain a very simple expression of the failure rate: $\lambda(k) = 1 - \pi^k$. This is a particular case of the type III discrete Weibull distribution with $\beta = 1$ and $c = -\ln \pi$.

3.7. *Discrete distributions derived from continuous ones by time discretization*

Let T be a real positive random variable describing a system lifetime in continuous time. Let $K = \lfloor T \rfloor + 1$ (where $\lfloor \; \rfloor$ is the integer part). K is a random variable defined over N^*. Let λ_K, F_K, R_K, and λ_T, F_T, R_T denote the failure rate, CDF and reliability related respectively to the random variables K and T.

The relationship between the probability function of K and the CDF of T is:

$$\forall k \in N^*, \quad p(k) = P(K = k) = P(k - 1 \leq T < k) = F_T(k) - F_T(k - 1).$$

Furthermore,

$$F_K(k) = P(K \leq k) = P(\lfloor T \rfloor + 1 \leq k) = P(T < k) = F_T(k).$$

Hence,

$$R_K(k) = R_T(k).$$

The failure rate of K can be written as follows:

$$\forall k \in N^*, \quad \lambda_K(k) = 1 - \frac{R_T(k)}{R_T(k-1)} = 1 - e^{-\int_{k-1}^{k} \lambda_T(u)du}$$

λ_T and λ_K have the same monotonicity property.

3.7.1. *Exponential distribution*

Let T have the exponential distribution $\mathcal{E}(\lambda)$, with CDF $F_T(t) = 1 - e^{-\lambda t}$. Using the above equations, we obtain:

$$R_K(k) = e^{-\lambda k} = (1 - (1 - e^{-\lambda}))^k.$$

This is the reliability function of the geometric distribution with parameter $1 - e^{-\lambda}$, whose (constant) failure rate is equal to $1 - e^{-\lambda}$. Consequently, the failure rate of a geometric distribution is not equal to the failure rate of the corresponding exponential distribution.

3.7.2. *Weibull distribution*

If T has the Weibull distribution $\mathcal{W}(\eta, \beta)$, then $R_K(k) = e^{-(\frac{k}{\eta})^\beta} = q^{k^\beta}$ with $q = e^{-\frac{1}{\eta^\beta}}$. This is the type I discrete Weibull distribution.

3.7.3. *Truncated logistic distribution*

Let T have the logistic distribution $\log(c, d)$, truncated on R^+ ($d \in R^{*+}$ and $c \in R$), with CDF:

$$F_T(t) = \frac{1 - e^{-\frac{t}{d}}}{1 + e^{-\frac{t-c}{d}}}.$$

Then, the discrete truncated logistic distribution is defined by:

- $p(k) = \dfrac{e^{-\frac{k-1}{d}}(1 - e^{-\frac{1}{d}})(1 + e^{\frac{c}{d}})}{(1 + e^{-\frac{k-c}{d}})(1 + e^{-\frac{k-1-c}{d}})}$
- $R(k) = \dfrac{e^{-\frac{k-c}{d}} + e^{-\frac{k}{d}}}{1 + e^{-\frac{k-c}{d}}}$
- $\lambda(k) = \dfrac{1 - e^{-1/d}}{1 + e^{-\frac{k-c}{d}}}.$

The failure rate is increasing. Parameters d and c have no simple practical interpretation.

Adams and Watson[1] proposed a general discrete lifetime distribution for which the failure rate can be written as $\lambda(k) = G(\sum_{j=0}^{m} \theta_j k^j)$, where G is the CDF of a continuous symmetric distribution. When G is chosen as the logistic function, as in Klar,[13] this model can be understood as a generalized discrete truncated logistic distribution.

3.7.4. *Geometric-Weibull distribution*

Industrial experts often consider that systems have two steps in their life. In the first step, the system has a constant failure rate until time τ, then, in the second step, the failure rate is larger than in step I and is increasing. The step I is the stable phase of the system while step II is its wear-out phase, and τ is a change-point.

In continuous time, if we consider that after time τ, the failure rate is increasing like one of a Weibull distribution, we obtain the exponential-Weibull distribution introduced by Zacks,[28] with CDF given by:

$$F_T(t) = 1 - e^{-\lambda t - [\lambda(t-\tau)^+]^\beta}$$

where $Y^+ = \max(0, Y)$, $\lambda \in R^{+*}$ is a scale parameter, $\beta \in R^+$ is a shape parameter and $\tau \in R^+$ is the change-point.

For the construction of the analogous of this distribution in discrete time, which will of course be called the geometric-Weibull distribution, τ takes values in N^*.

Then, the geometric-Weibull distribution is defined by:

- $p(k) = e^{-\lambda(k-1)-[\lambda(k-\tau-1)^+]^\beta} - e^{-\lambda k-[\lambda(k-\tau)^+]^\beta}$
- $R(k) = e^{-\lambda k-[\lambda(k-\tau)^+]^\beta}$
- $\lambda(k) = 1 - e^{-\lambda+\lambda^\beta[[(k-\tau-1)^+]^\beta-[(k-\tau)^+]^\beta]}$.

4. Pólya Urn Distributions

As stated by Johnson, Kotz and Kemp in Ref. 12, numerous discrete distributions can be built from urn representations. The urn scheme considered by Eggenberger and Pólya[7] is the following. An urn contains W white balls and R red balls. After each drawing of a ball, a replacement policy is chosen. Pólya distributions are the distributions of the number of times a red ball is drawn in N drawings. Inverse Pólya distributions(Ref. 11, p. 192) are the distributions of the number of drawings to obtain a specified number r of red balls. There are as many different distributions as possible replacement policies. For example, if after each drawing, only the chosen ball is returned in the urn, the Pólya distribution reduces to the binomial distribution, and the corresponding inverse Pólya distribution is the negative binomial distribution (geometric distribution for $r = 1$).

In the discrete reliability context, the drawing of a white ball corresponds to a demand successfully completed and the drawing of a red ball to a failure on demand. Thus, system lifetime is described by an inverse Pólya distribution with $r = 1$.

In this case, the failure rate at kth drawing can be understood as the probability of drawing for the first time a red ball at kth drawing, given that no red ball have been drawn during the first $k - 1$ drawings.

If the replacement policy consists in returning the drawn (white) ball, together with red balls, this scheme increases failure probability. So the corresponding lifetime will have an increasing failure rate. Conversely, if the drawn (white) ball is returned together with other white balls, the lifetime distribution has a decreasing failure rate.

It seems that only Clarotti, Lannoy and Procaccia[5] have explicitly used an inverse Pólya distribution for modelling a discrete lifetime. But it appears that the generalized Salvia and Bollinger distributions[17] are also members of this family. This section aims to show that inverse Pólya distributions can be very useful in reliability studies since they both have a practical interpretation and a simple expression of the failure rate. Moreover, most of these distributions have closed form expressions for the MTTF and variance, which provide a simple way to estimate parameters.

4.1. *IFR inverse Pólya distribution*

If the drawn white ball is returned in the urn with Δ red balls, the corresponding failure rate is increasing and given by:

$$\lambda(k) = \frac{R + (k-1)\Delta}{R + W + (k-1)\Delta}.$$

Let $\theta = \dfrac{R}{R + W} \in]0, 1[$ and $\delta = \dfrac{\Delta}{R + W} \in R^{+}$. From the lifetime point of view, θ is the probability of failure on the first demand and δ quantifies the importance of ageing. For $\delta = 0$, the distribution is geometric.

The IFR inverse Pólya distribution is defined by:

- $p(k) = \dfrac{(1 - \theta)^{k-1}[\theta + (k-1)\delta]}{\displaystyle\prod_{i=1}^{k}[1 + (i-1)\delta]} = \dfrac{(1 - \theta)^{k-1}[\theta + (k-1)\delta]}{\delta^{k}\left(\dfrac{1}{\delta}\right)_{(k)}}$

where $(a)_{(k)}$ is the Pochhammer symbol: $(a)_{(k)} = \begin{cases} \displaystyle\prod_{i=1}^{k}(a + i - 1) & \text{if } k \geq 1, \\ 1 & \text{if } k = 0. \end{cases}$

- $R(k) = \dfrac{(1 - \theta)^{k}}{\delta^{k}\left(\dfrac{1}{\delta}\right)_{(k)}}$

- $\lambda(k) = \dfrac{\theta + (k-1)\delta}{1 + (k-1)\delta} = 1 - \dfrac{1 - \theta}{1 + (k-1)\delta}.$

The Mean Time To Failure is MTTF $= \dfrac{(1 - \delta)\delta^{\frac{1}{\delta} - 2}}{(1 - \theta)^{\frac{1-\delta}{\delta}}} e^{\frac{1-\theta}{\delta}} \gamma\left(\dfrac{1 - \delta}{\delta}, \dfrac{1 - \theta}{\delta}\right)$, where $\gamma(b, x) = \displaystyle\int_{0}^{x} t^{b-1} e^{-t} dt$ is the incomplete Gamma function.

This distribution has been used by Clarotti, Lannoy and Procaccia[5] to describe ageing of components which fail on demands.

4.2. DFR inverse Pólya distribution

If the white ball drawn is replaced in the urn together with Δ white balls, the failure rate is decreasing and given by:

$$\lambda(k) = \frac{R}{R + W + (k-1)\Delta}.$$

With the same notations as before, the DFR inverse Pólya distribution is defined by:

- $p(k) = \dfrac{\theta \displaystyle\prod_{i=1}^{k-1}[1 - \theta + (i-1)\delta]}{\displaystyle\prod_{i=1}^{k}[1 + (i-1)\delta]} = \dfrac{\theta}{1 - \theta - \delta} \dfrac{\left(\dfrac{1 - \theta - \delta}{\delta}\right)_{(k)}}{\left(\dfrac{1}{\delta}\right)_{(k)}}$

- $R(k) = \dfrac{\displaystyle\prod_{i=1}^{k}[1 - \theta + (i-1)\delta]}{\displaystyle\prod_{i=1}^{k}[1 + (i-1)\delta]} = \dfrac{\left(\frac{1-\theta}{\delta}\right)_{(k)}}{\left(\frac{1}{\delta}\right)_{(k)}}$

- $\lambda(k) = \dfrac{\theta}{1 + (k-1)\delta}.$

The MTTF is only defined for $\theta > \delta$ and is equal to:

$$\text{MTTF} = \sum_{k=0}^{+\infty} \frac{\left(\frac{1-\theta}{\delta}\right)_{(k)}}{\left(\frac{1}{\delta}\right)_{(k)}} = {}_2F_1\left(\frac{1-\theta}{\delta}, 1, \frac{1}{\delta}, 1\right) = \frac{1-\delta}{\theta-\delta}$$

where $\ {}_2F_1(a, b, c, z) = \displaystyle\sum_{k=0}^{+\infty} \frac{(a)_{(k)}(b)_{(k)}}{(c)_{(k)}} \frac{z^k}{k!} = \frac{\Gamma(c)}{\Gamma(a)\Gamma(b)} \sum_{k=0}^{+\infty} \frac{\Gamma(a+k)\Gamma(b+k)}{\Gamma(c+k)} \frac{z^k}{k!}\ $ is a

Gauss hypergeometric series.

The distribution variance, only defined for $\theta > 2\delta$, is:

$$\text{Var}(K) = \frac{(1-\delta)\theta(1-\theta)}{(\theta-\delta)^2(\theta-2\delta)}.$$

The DFR Inverse Pólya distribution is also called the Waring distribution.[10] Its failure rate is inversely proportional to a linear function of time : $\lambda(k) = \dfrac{1}{a+bk}$. The studied case here corresponds to $b > 0$. Xekalaki[27] also considered the case $b < 0$. In this case, the random variable is defined on $\{1, \ldots, m\}$ and its failure rate is $\lambda(k) = \dfrac{1}{1 + c(m-k)}$, $c > 0$.

For the particular case where $\dfrac{1}{c}$ is an integer, the probability function is given by:

$$p(k) = \frac{C^{\frac{1}{c}-1}_{m+\frac{1}{c}-k-1}}{C^{\frac{1}{c}}_{m+\frac{1}{c}-1}}$$

and K has a shifted negative hypergeometric distribution.

It is easy to notice (see Roy and Gupta[19]) that an inverse linear failure rate is equivalent to a linear mean residual life. For the DFR inverse Pólya distribution, if $\theta > \delta$, we obtain:

$$m(k) = \frac{\delta(k-1)+1}{\theta-\delta}.$$

As a particular case, $\text{MTTF} = m(0) = \dfrac{1-\delta}{\theta-\delta}.$

At last, Gupta, Gupta and Tripathi[9] showed that the Waring distribution is a particular case of the extended Katz family, defined by the ratio of two consecutive probabilities given by:

$$\frac{p(k+1)}{p(k)} = \frac{\alpha + \beta k}{\gamma + k}, \quad \text{for } \alpha = \frac{1 - \theta - \delta}{\delta}, \quad \beta = 1 \text{ and } \gamma = \frac{1}{\delta}.$$

In a same way, the IFR inverse Pólya distribution can be defined by:

$$\frac{p(k+1)}{p(k)} = \frac{1 - \theta}{\delta} \frac{\frac{\theta}{\delta} + k}{\left(\frac{\theta}{\delta} - 1 + k\right)\left(\frac{1}{\delta} + k\right)}$$

which is a particular case of the Kemp family (Ref. 12, p. 86).

4.3. *Salvia and Bollinger distributions*

Salvia and Bollinger[20] proposed two distributions with only one parameter $c \in]0,1]$, based on a very simple expression of the failure rate. In fact, they are inverse Pólya distributions with $\delta = 1$.

4.3.1. *IFR SB distribution*

The IFR SB distribution is such that:

- $p(k) = (k - c)\dfrac{c^{k-1}}{k!}$
- $R(k) = \dfrac{c^k}{k!}$
- $\lambda(k) = 1 - \dfrac{c}{k}.$

MTTF $= e^c$. The distribution variance is $\text{Var}(K) = 2ce^c - e^c - 1$. This distribution has few practical interest because its mean is between 1 and e. So the values taken by K are mainly equal to 1.

4.3.2. *DFR SB distribution*

The DFR SB distribution is such that:

- $p(k) = \dfrac{c}{k!}(1 - c)_{(k-1)}$
- $R(k) = \dfrac{(1 - c)_{(k)}}{k!}$
- $\lambda(k) = \dfrac{c}{k}.$

The MTTF does not exist because the assumption $\theta > \delta$, here $c > 1$, is not verified.

4.4. *Generalized Salvia and Bollinger distributions*

The Salvia and Bollinger distributions are not flexible enough to fit a wide variety of situations. Padgett and Spurrier[17] proposed to generalize these distributions by adding a second parameter $\alpha \in R^+$. When $\alpha = 1$, the distributions reduce to Salvia and Bollinger ones. When $\alpha = 0$, the distribution is geometric.

The IFR generalized SB distribution is defined by its failure rate:

$$\lambda(k) = 1 - \frac{c}{(k-1)\alpha + 1}$$

and the DFR generalized SB distribution by:

$$\lambda(k) = \frac{c}{(k-1)\alpha + 1}.$$

It appears that these distributions are exactly inverse Pólya distributions. Padgett and Spurrier did not notice this fact.

4.5. *Eggenberger–Pólya distribution*

Le Breton and Martin[15] had the idea to use the Eggenberger–Pólya distribution[7] in the study of climatological sequences. This distribution can also be used in the reliability context because its properties in climatology deal with the persistence function, which is defined as the complementary to 1 of the failure rate. The Eggenberger–Pólya distribution can be understood as a limit of a Pólya distribution when the number of drawings goes to infinity (Ref. 11, pp 190–191).

The probability of system failure at demand k is:

$$p(k) = \frac{1}{(1+d)^{h/d}} \frac{(h/d)_{(k-1)}}{(k-1)!} \left(\frac{d}{d+1}\right)^{k-1}.$$

The failure rate is given by:

$$\frac{1}{\lambda(k)} = \sum_{j=0}^{+\infty} \frac{(h/d + k - 1)_{(j)}}{(k)_{(j)}} \left(\frac{d}{1+d}\right)^j \quad = \quad {}_2F_1\left(\frac{h}{d} + k - 1, 1, k, \frac{d}{1+d}\right).$$

The expression of the failure rate is very complex and the result on ratios of successive probabilities is required to study its monotonicity:

$$\frac{p(k+1)}{p(k)} = \frac{d}{1+d} + \frac{h-d}{1+d} \cdot \frac{1}{k}.$$

- If $h = d$, then the failure rate is constant and the model reduces to the geometric distribution with parameter $\frac{1}{1+d}$.
- If $h < d$, the distribution is log-convex and then DFR.
- If $h > d$, the distribution is log-concave and then IFR.

In the three cases, $\lim_{k \to \infty} \lambda(k) = \frac{1}{1+d}$.

Using the probability generating function of this distribution, the mean and variance can be derived:

- MTTF $= h + 1$
- $\text{Var}(K) = h(d + 1)$.

h represents the mean number of demands until the first failure minus one. There is no interpretation for parameter d.

5. Parameters Estimation

Let $K_1, \ldots, K_n$ be discrete system lifetimes, assumed to be independent and identically distributed, according to one of the previous distributions. For the analysis of these reliability data, it is necessary to estimate the distributions parameters. A reliability model will be practically useful only if its parameters can be efficiently estimated. At our knowledge, except for the geometric case, only Klar's paper[13] provides a detailed statistical study of a discrete lifetime distribution.

We present in this section some results for assessing the efficiency of these estimation procedures. Exact results are given for the geometric distributions. For all others, the estimators properties are assessed with Monte-Carlo simulations : for each sample size n and each chosen value of the parameters, 2000 samples have been simulated. For each sample, the parameters have been estimated. The results are given as plots of empirical bias and coefficient of variation of the estimators, as functions of n. The parameter values have been chosen in order to obtain data in the order of magnitude of 10^k, for $k \in \{1, 2, 3, 4\}$. The sample sizes are $n \in \{5, 10, 20, 40, 60, 100, 200\}$ in order to illustrate a large number of practical cases.

5.1. *Geometric distribution*

The maximum likelihood estimator (MLE) of p based on $K_1, \ldots, K_n$ is $\hat{p}_n = \dfrac{n}{\sum_{i=1}^{n} K_i}$.

The random variable $\sum_{i=1}^{n} K_i$ has a negative binomial distribution $\mathcal{BN}(n, p)$, so it is possible to compute the bias of $\hat{p}_n$:

$$E(\hat{p}_n) = p^n \, _2F_1(n, n, n+1, 1-p).$$

The unbiased minimal variance estimator of p is:

$$\tilde{p}_n = \frac{n-1}{\displaystyle\sum_{i=1}^{n} K_i - 1}.$$

The variance of $\tilde{p}_n$ is $\text{Var}(\tilde{p}_n) = p^n \, _2F_1(n-1, n-1, n, 1-p) - p^2$.

In order to appreciate the relative dispersion of $\tilde{p}_n$ rather than the absolute one, we prefer using the coefficient of variation (CV), defined as the ratio of the standard deviation to the mean:

$$\text{CV}(\tilde{p}_n) = \frac{\sqrt{\text{Var}(\tilde{p}_n)}}{E(\tilde{p}_n)} = \sqrt{p^{n-2} \, _2F_1(n-1, n-1, n, 1-p) - 1}.$$

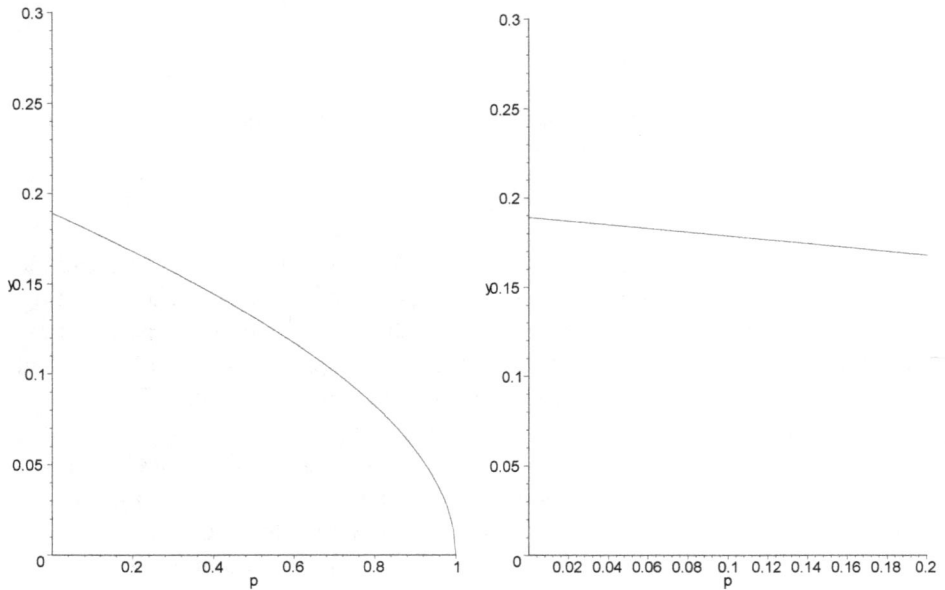

Fig. 1.   Geometric – Theoretical CV of $\tilde{p}_n$ for $n = 30$.

An empirical estimate of the coefficient of variation, denoted by CV_n, can be defined from a sample by replacing the mean and square deviation by their empirical estimates.

For assessing the quality of the estimators, we will use the same empirical criterion all along this section: the estimation is considered to be satisfactory when CV or CV_n are less than 15%.

For a fixed sample size, the theoretical coefficient of variation is a decreasing function of p (see the first plot of Fig. 1). Then p is well estimated when it is close to 1. But for small values of p, $\tilde{p}_n$ decreases slowly as we can see on the second plot of Fig. 1.

Figure 2 plots the theoretical coefficient of variation of $\tilde{p}_n$ as a function of the sample size n for $p = 0.1$. With the adopted criterion, the estimation of p is considered here to be satisfactory for $n \geq 40$.

5.2. *Type I discrete Weibull distribution*

In Ref. 2, Ali Khan, Khalique and Abouammoh proposed to estimate the parameters of the type I discrete Weibull distribution by the method of proportions. The idea is to remark that, since $P(K = 1) = 1 - q$, an estimator of q is the empirical frequency of observations strictly greater than one in the sample:

$$\hat{q} = \frac{1}{n} \sum_{i=1}^{n} \mathrm{II}_{\{K_i > 1\}} = R_n(1)$$

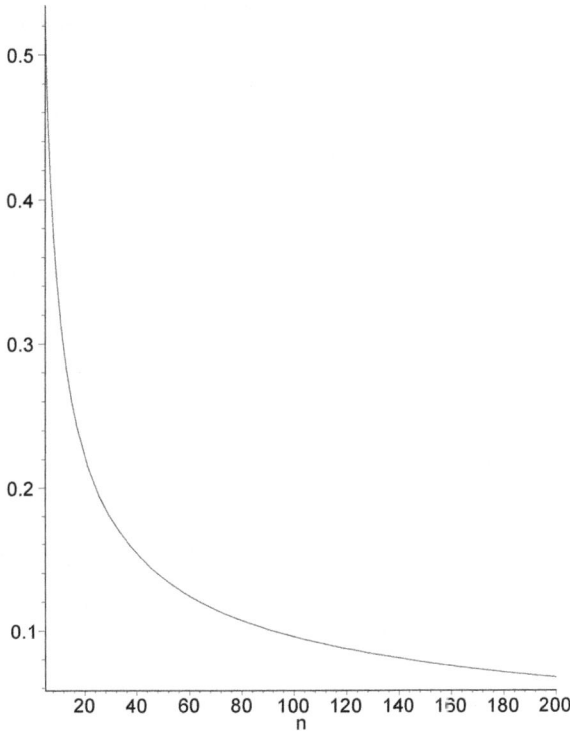

Fig. 2. Geometric – Theoretical CV of $\tilde{p}_n$ for $p = 0.1$.

where R_n denotes the empirical reliability function. In the same way $P(K = 2) = q - q^{2^\beta}$, so $\beta = \frac{1}{\ln 2} \{\ln \ln(q - p(2)) - \ln \ln q\}$ can be estimated by:

$$\hat{\beta} = \frac{1}{\ln 2} \{\ln \ln [R_n(1) - (R_n(1) - R_n(2))] - \ln \ln R_n(1)\} = \frac{1}{\ln 2} \ln \frac{\ln R_n(2)}{\ln R_n(1)}.$$

Of course, two other integers can be used instead of 1 and 2.

The method of proportions requires that lots of observations are equal to 1 and 2 (or two other values) in order that the empirical estimators of $R(1)$ and $R(2)$ provide good estimation of the true values. But this is very seldom the case in practice. Moreover, the main drawback of this method is that it does not use all the observations but only a few of them. For these reasons, only the maximum likelihood estimators have been considered in this paper.

Figures 3 to 5 show the empirical bias and CV of the MLE of q and β for several sets of parameters. The conclusions given have been confirmed by other simulations not presented here.

Simulation results indicate that the estimation of q is slightly biased: q is over-estimated when $q < 0.9$ and under-estimated when $q \geq 0.9$. Furthermore, other simulation results seem to show that the bias of $\hat{q}$ is independent of the value of β.

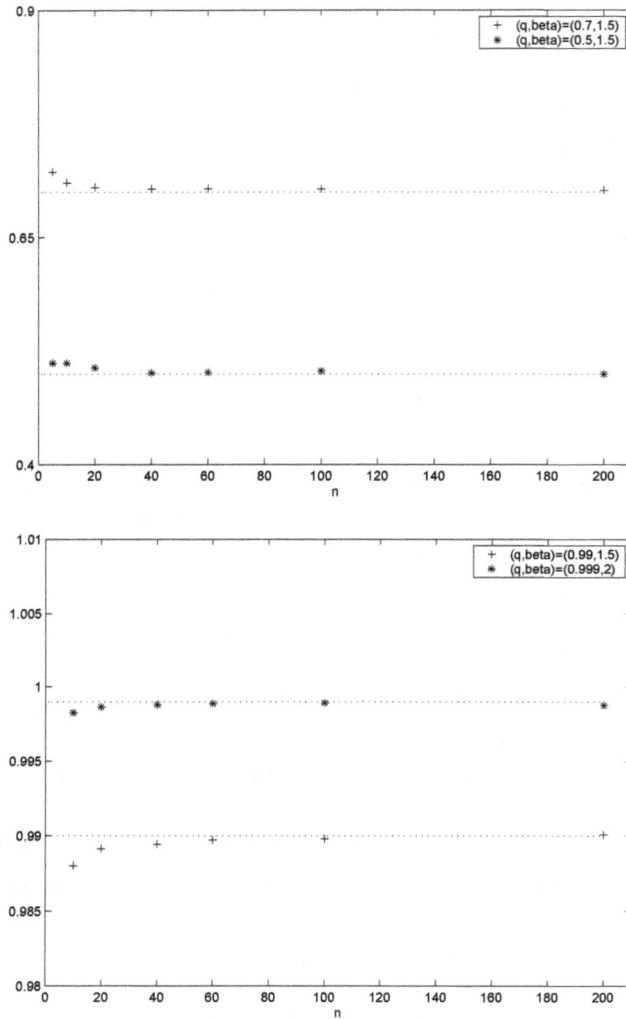

Fig. 3.   Weibull I–Empirical bias of the MLE of q.

The CV of the MLE of q is satisfactory even for rather small samples, especially when $q \geq 0.9$.

The estimation of β is slightly biased as we can see on the first plot of Fig. 5. The CV for β is larger than the CV for q.

Numerical estimation problems occurred for simulated samples with q close to zero, that is to say when the order of magnitude of the observations is high.

As a conclusion, the maximum likelihood estimation of the parameters of the type I discrete Weibull distribution is of good quality for large values of q but not for too small values of q.

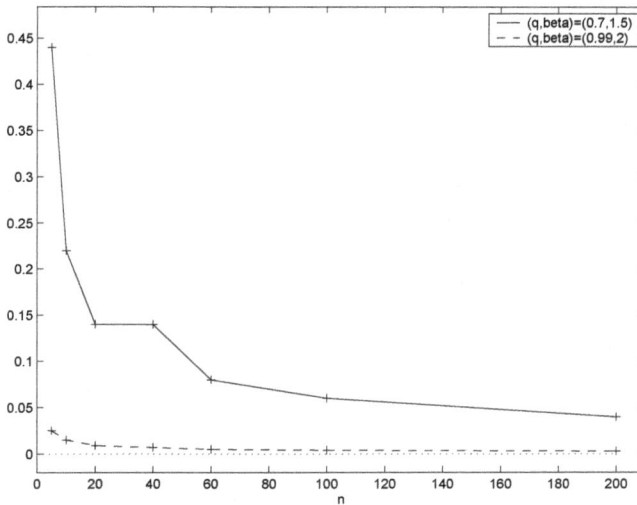

Fig. 4. Weibull I–Empirical CV of the MLE of q.

5.3. *Type III discrete Weibull distribution*

Compared to the type I Weibull distribution, the computer time needed for parameters estimation of the type III Weibull distribution is larger because of the complexity of its probability function. Simulation results indicate that the MLE of c is only a little biased (see Fig. 6). The estimation quality seems to increase with c. For $c = 0.001$, the smallest value used for simulations, the CV remains very large even for $n = 200$.

The bias of the MLE of β is low except for very small samples (see the first plot of Fig. 7). The CV is satisfactory only for large samples (see the second plot of Fig. 7). The simulations seem also to indicate that the estimation quality is independent of the value of c.

As for the type I Weibull distribution, numerical problems occurred for simulated samples with c close to zero (high order of magnitude of the observations).

5.4. *IFR inverse Pólya distribution*

Thanks to the simplicity of the first two moments of the DFR inverse Pólya distribution, its parameters can be easily estimated by the method of moments. But, for the purpose of modelling lifetimes of ageing systems, we are more interested by the IFR inverse Pólya distribution. Unfortunately, the complexity of the MTTF in this case is such that the method of moments is not applicable. So we used again the maximum likelihood method.

Problems appeared with the MLE for sample size less than 40 because the estimations were sometimes out of the parameters definition set (for example, θ was sometimes estimated larger than 1). That is the reason why the following plots

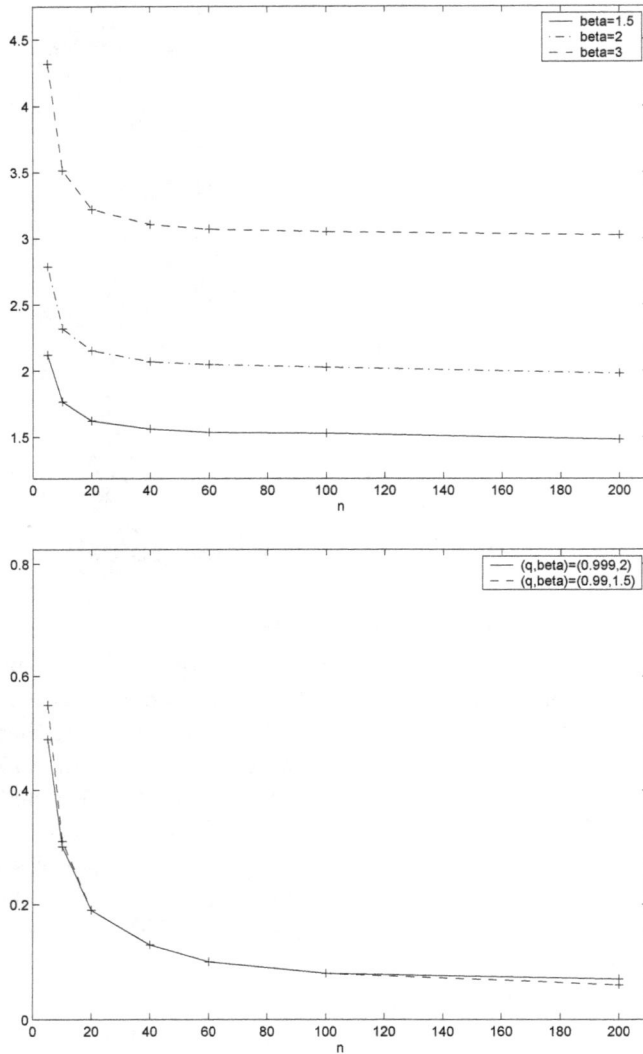

Fig. 5.   Weibull I–Empirical bias and CV of the MLE of β.

present only simulation results for $n \geq 40$. From Fig. 8, we can conclude that the estimation of θ is sligthly biased and the CV is satisfactory for $n \geq 40$. Plots of Fig. 9 indicate that the estimation of δ is highly biased and the CV is not satisfactory at all, even for large sample size.

Clarotti, Lannoy and Proccacia[5] used bayesian procedures to estimate the parameters. But the IFR inverse Pólya distribution does not belong to the exponential family. Therefore there is no natural conjugate distribution. The authors had to develop specific numerical integration techniques to obtain posterior distributions.

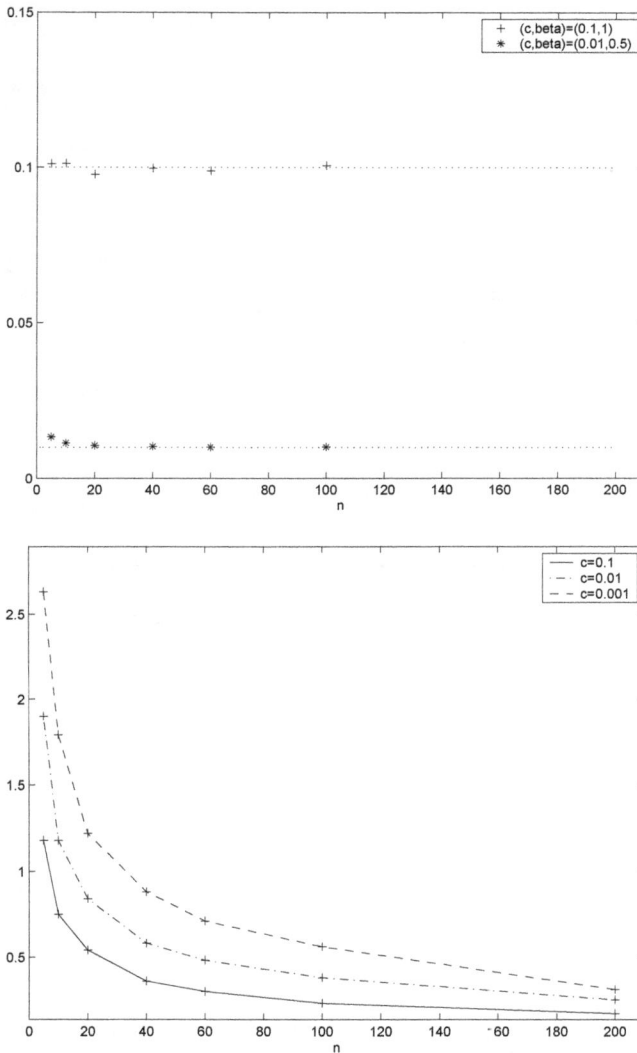

Fig. 6.   Weibull III–Empirical bias and CV of the MLE of c.

5.5. *Eggenberger–Pólya distribution*

Since the first two moments of the Eggenberger–Pólya distribution are very simple, it is easy to derive the moment estimators (ME) of parameters h and d:

$$\hat{h} = \bar{K}_n - 1 \quad \text{and} \quad \hat{d} = \frac{S_n^2}{\bar{K}_n - 1} - 1.$$

where $\bar{K}_n$ and S_n^2 are the sample mean and variance.

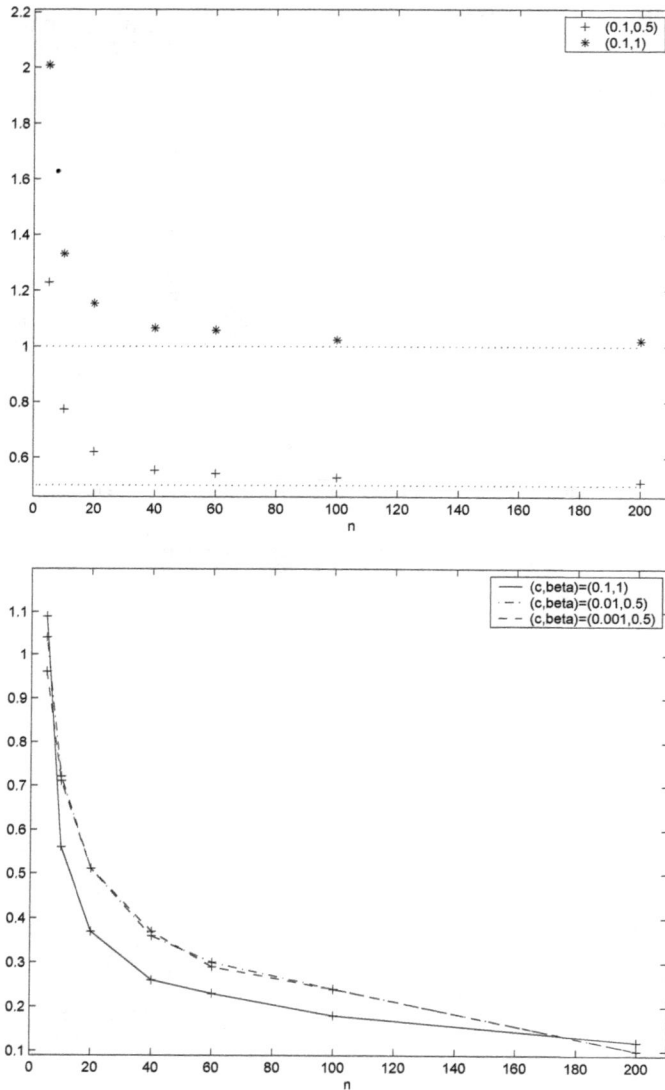

Fig. 7.   Weibull III–Empirical bias and CV of the MLE of β.

The estimator of h is unbiased and its CV is $\text{CV}(\hat{h}) = \sqrt{\dfrac{d+1}{hn}}$. On the examples of Fig. 10, it is satisfactory even for small samples.

Figure 11 shows that the estimator of d seems nearly unbiased but the CV is high even for large sample size. The difference in the estimation quality of parameters h and d lies probably in the fact that h is analogous to the mean of the distribution, and d to the variance.

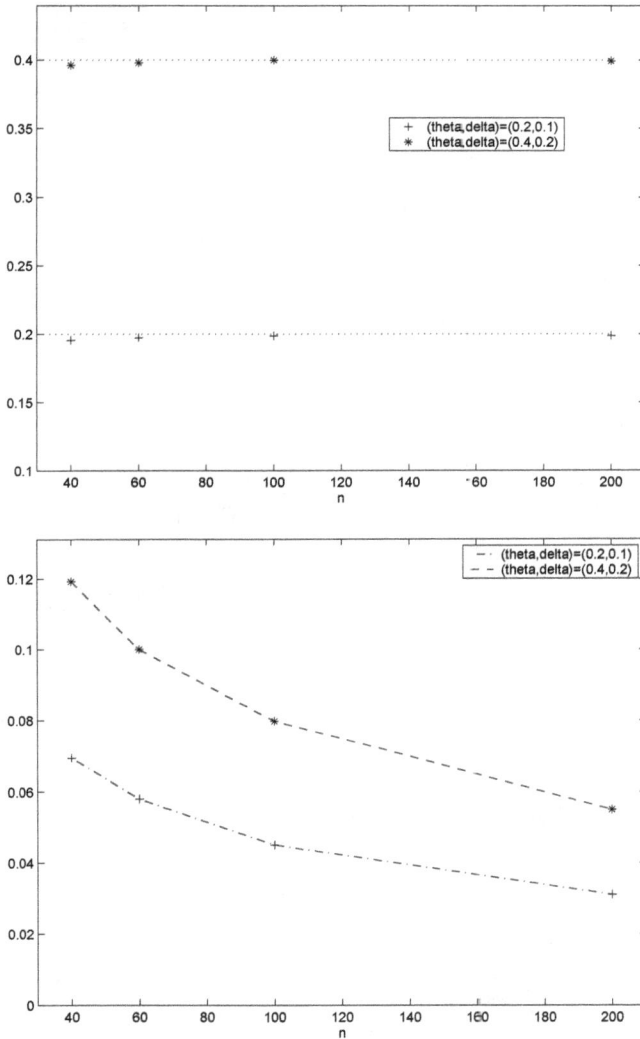

Fig. 8.   IFR Pólya–Empirical bias and CV of the MLE of θ.

5.6. *Discretized truncated logistic distribution*

The MLE of parameter c seems nearly unbiased and its CV is very satisfactory, whatever the values taken by c and d, as shown in Fig. 12.

For parameter d, we can see in Fig. 13 that the MLE of d is biased and its CV depends strongly on the values of c and d: for simulated values $(c, d) = (100, 5)$, the CV is low enough from $n = 40$, but for $(c, d) = (20000, 100)$, it remains large even for $n = 200$.

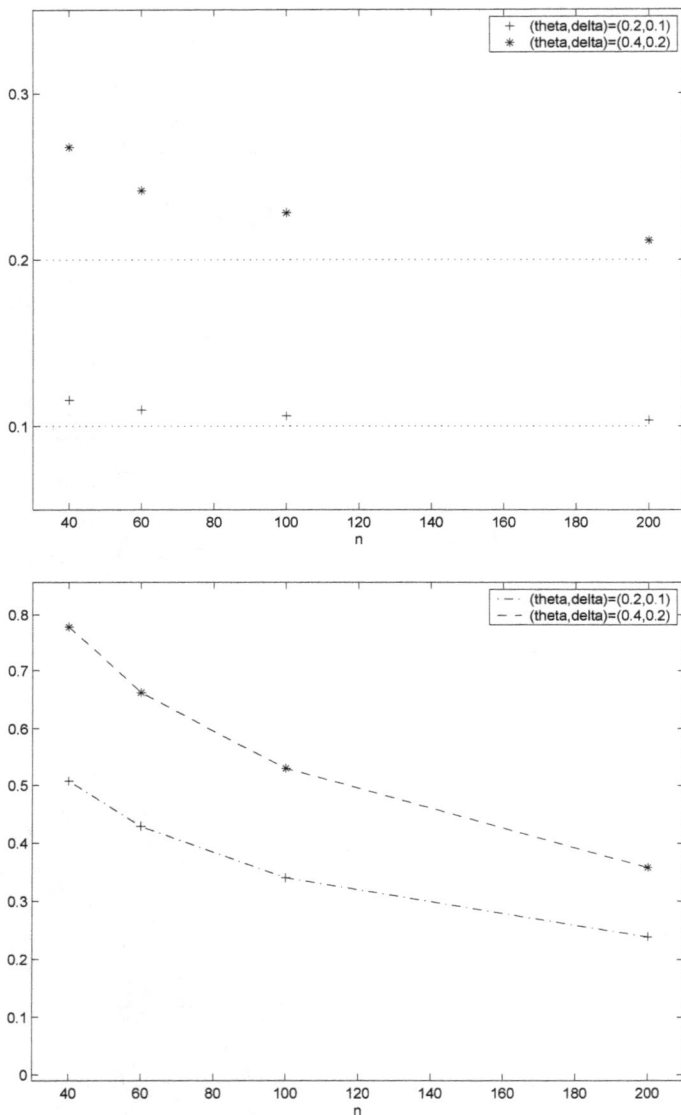

Fig. 9.   IFR Pólya–Empirical bias and CV of the MLE of δ.

6. Discussion and Conclusion

Faced to the observation of discrete reliability data, it is important to select an appropriate distribution. If the studied phenomenon is such that a constant failure rate is acceptable, the geometric distribution is appropriate. But if the observed systems are ageing, what distribution should be chosen? So it is interesting to provide criteria for selecting useful discrete reliability models in case of ageing.

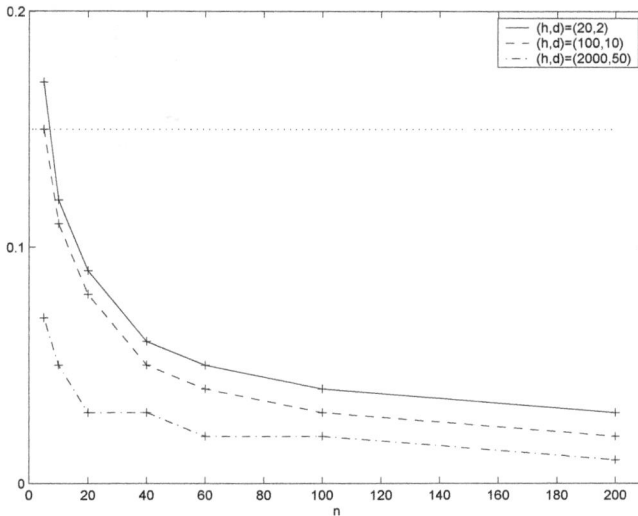

Fig. 10. Eggenberger–Pólya – $CV(\hat{h})$.

A first kind of criteria is based on the own distribution qualities, independently of observed data. Important points are:

- simplicity of expression of the reliability functions
- flexibility or ability to describe various situations
- physical basis of the distribution and interpretation of the parameters.

A second kind of criteria is the quality of parameter estimators, studied in the previous section.

Now we can compare the main presented distributions according to these criteria:

(1) The shifted negative binomial distribution is interesting only as the analogous of the gamma distribution. It is neither simple nor flexible (only IFR) and its parameters have no practical interpretation.
(2) The type I Weibull distribution is very simple, flexible according to the value of β, and its parameters have a physical meaning. The maximum likelihood estimation is satisfactory, except for data of too high order of magnitude.
(3) The type II Weibull distribution should not be used because of its bounded support.
(4) The type III Weibull distribution is simple, flexible, and its parameters have an interpretation (less obvious than for the type I distribution). The estimation quality is not very good.
(5) The "$\mathcal{S}$" distribution is simple and has an interesting physical meaning. But there are some numerical and identifiability problems in the parameter

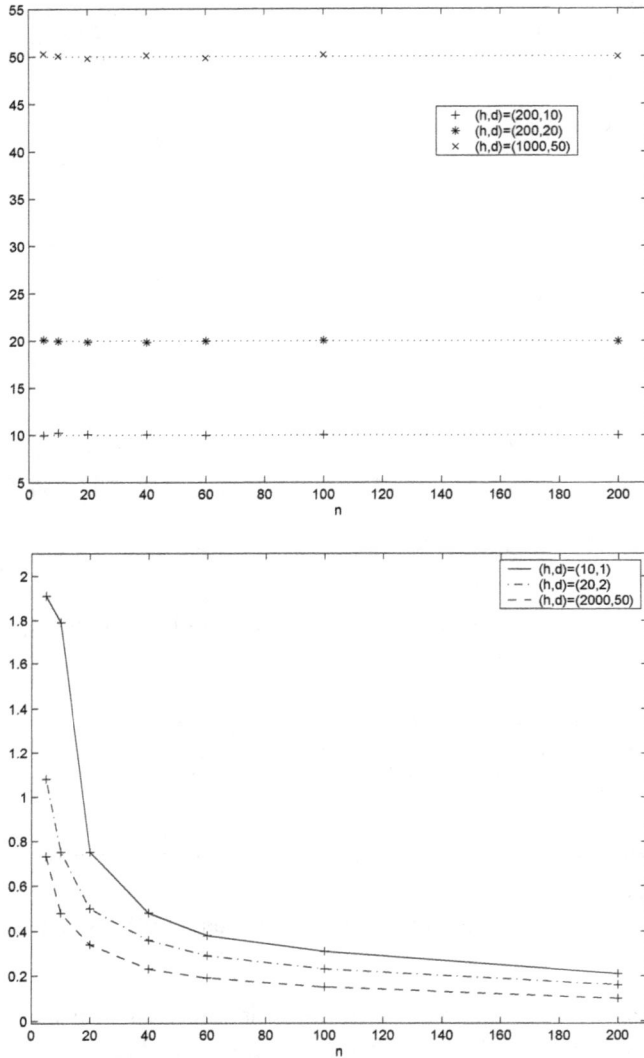

Fig. 11.   Eggenberger–Pólya–Empirical bias and CV of $\hat{d}$.

estimation which are not solved yet. That is why we have not presented statistical results for this distribution.

(6) The discrete truncated logistic distribution is slightly more complex than the Weibull distributions, is not flexible (only IFR), and its parameter have no practical interpretation. However, the parameter estimation is rather good and the sigmoidal shape of the failure rate seems to fit lots of failure data sets.

(7) The geometric-Weibull distribution is clearly of great practical interest and should be studied.

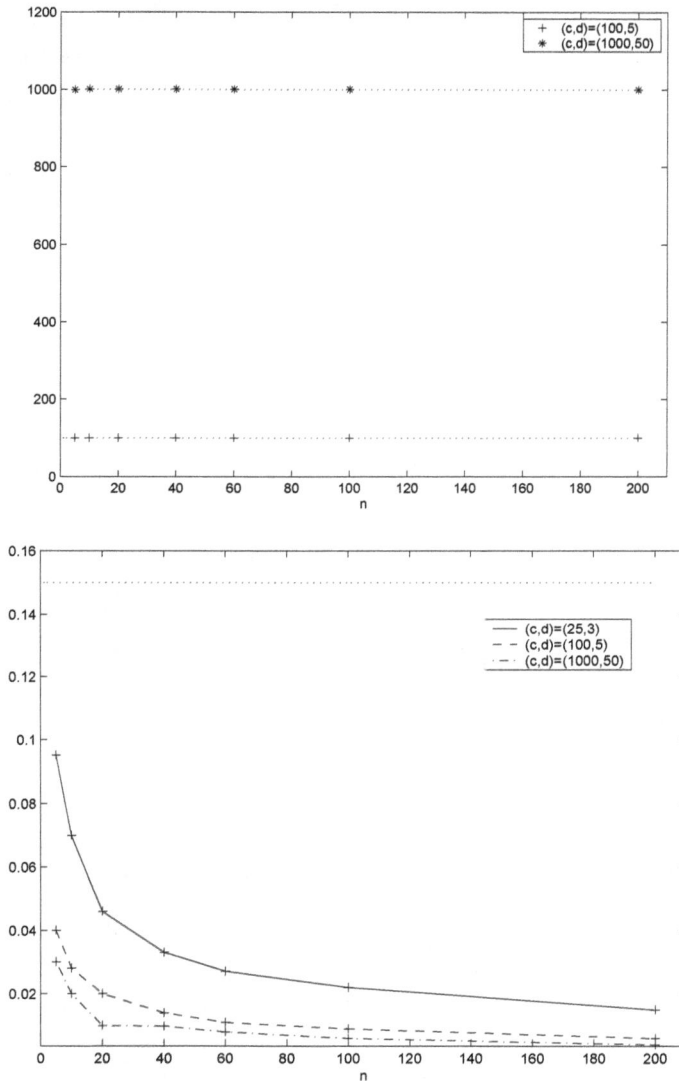

Fig. 12. Logistic–Empirical bias and CV of the MLE of c.

(8) The IFR Pólya distribution has a very interesting interpretation and its failure rate has a simple expression. However it is not flexible and the estimation is not very satisfactory. The derived models such as those of the Katz and Kemp family are much more complex.

(9) The Eggenberger–Pólya distribution has a very complex expression, but it is flexible and parameter h is easily interpreted. Its main advantage is that it is the only distribution for which the estimators have an explicit expression.

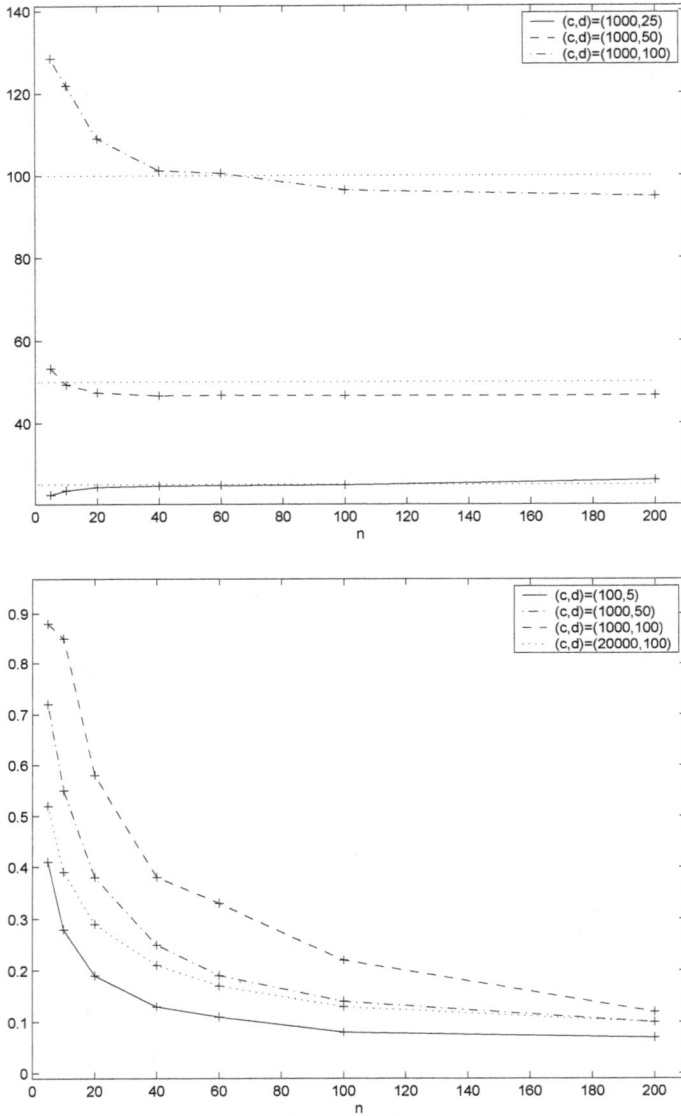

Fig. 13.   Logistic–Empirical bias and CV of the MLE of d.

On the basis of this study, we recommend to use the type I discrete Weibull and the Eggenberger–Pólya distributions for the modelling of discrete lifetimes of ageing nonrepairable systems. The discrete truncated logistic distribution can also be useful in some situations.

For nearly all distributions, numerical problems occur in the estimation when the order of magnitude of data is too large. The threshold at which these problems start is not the same for each distribution, but, roughly speaking, the sample mean should not be greater than 100. In fact, when the sample mean is very large, it is generally better to use continuous distributions.

Lot of work is needed in the statistical analysis of discrete lifetime distributions. First, it is certainly possible to find efficient estimation procedures for most of the presented distributions. Secondly, the selection of a model for some data should be done on the basis of a goodness-of-fit study. But there are very few studies on goodness-of-fit tests for discrete distributions, and, at our knowledge, only Klar's work[13] for discrete lifetime distributions. The study of some models (such as the "S" and geometric-Weibull) should be deepened and new ones will surely be proposed. Finally, data in practice are very often censored. So a study similar to that presented here should be done for censored data.

References

1. G. Adams and R. Watson, "A discrete time parametric model for the analysis of failure time data," *Australian Journal of Statistics* **31** (1989), pp. 365–384.
2. M. S. Ali Khan, A. Khalique and A. M. Abouammoh, "On estimating parameters in a discrete Weibull distribution," *IEEE Transactions on Reliability* **38** (3) (1989), pp. 348–350.
3. R. E. Barlow, A. W. Marshall and F. Proschan, "Properties of probability distributions with monotone hazard rate," *Annals of Statistics* **34** (1963), pp. 375–389.
4. R. E. Barlow and F. Proschan, *Statistical Theory of Reliability and Life Testing: Probability Models.* Second edition, (Silver Spring, MD: To begin with, 1975).
5. A. Clarotti, A. Lannoy and H. Procaccia, *Probability risk analysis of ageing components which fail on demand-a bayesian model: Application to maintenance optimization of diesel engine linings.* In *Ageing of Materials and Methods for the Assesment of Lifetimes of Engineering Plant* (Penny ed., Balkema, Rotterdam) (1997), pp. 85–93.
6. M. Crowder, "A test for independence of competing risks with discrete failure times," *Lifetime Data Analysis* **3** (1997), pp. 215–223.
7. F. Eggenberger and G. Pòlya, "Über die Statistik verketter Vorgänge," *Zeitschrift für angewandte Mathematik und Mechanik* **3** (1923), pp. 279–289 (in German).
8. A. Fries and A. Sen, "A survey of discrete reliability growth models," *IEEE Transactions on Reliability* **45** (4) (1996), pp. 582–604.
9. P. L. Gupta, R. C. Gupta and R. C. Tripathi, "On the monotonic properties of the discrete failure rates," *Journal of Statistical Planning and Inference* (65) (1997), pp. 255–268.
10. J. O. Irwin, "The generalized Waring distribution," *Journal of the Royal Statistical Society*, Series B, (**138**) pp. 18–31 (part I); pp. 204–227 (part II); pp. 374–384 (part III), (1975).
11. N. L. Johnson and S. Kotz, *Urn models and their application. An approach to modern discrete probability theory* (Wiley series in Probability and Mathematical Statistics), (1977).
12. N. L. Johnson, S. Kotz and A. W. Kemp, *Univariate discrete distributions* (Wiley series in Probability and Mathematical Statistics. Second edition, 1992).
13. B. Klar, "Model selection for a family of discrete failure time distributions," *Australia & New Zealand Journal of Statistics* **41** (3) (1999), pp. 337–352.

14. C. D. Lai and D. Q. Wang, "A finite range discrete life distribution," *International Journal of Reliability, Quality and Safety Engineering* **2** (2) (1995), pp. 147–160.
15. A. Le Breton and S. Martin, *Un modèle pour l'étude des séquences climatologiques,* Séminaire de l'équipe S.M.S, Université Grenoble I, 1979 (in French).
16. T. Nakagawa and S. Osaki, "The discrete Weibull distribution," *IEEE Transactions on Reliability* **24** (5) (1975), pp. 300–301.
17. W. J. Padgett and J. D. Spurrier, "Discrete failure models," *IEEE Transactions on Reliability* **34** (3) (1985), pp. 253–256.
18. J. M. Rocha-Martinez and M. Shaked, "A discrete-time model of failure and repair," *Applied Stochastic Models and Data Analysis* **11** (1995), pp. 167–180.
19. D. Roy and R. P. Gupta, "Characterizations and model selections through reliability measures in the discrete case," *Statistics & Probability Letters* **43** (1999), pp. 197–206.
20. A. A. Salvia and R. C. Bollinger, "On discrete hazard functions," *IEEE Transactions on Reliability* **31** (5) (1982), pp. 458–459.
21. D. Sengupta, A. Chatterjee and B. Chakraborty, "Reliability bounds and other inequalities for discrete life distributions," *Microelectronics and Reliability* **35** (12) (1995), pp. 1473–1478.
22. M. Shaked, J. G. Shanthikumar and J. B. Valdez-Torres, "Discrete hazard rate functions," *Computers and Operations Research* **22** (4) (1995), pp. 391–402.
23. J. L. Soler, "Croissance de fiabilité des versions d'un logiciel," *Revue de Statistique Appliquée* **XLIV** (1) (1996), pp. 5-20, (in French).
24. W. E. Stein and R. Dattero, "A new discrete Weibull distribution," *IEEE Transactions on Reliability* **33** (2) (1984), pp. 196–197.
25. N. Unnikrishnan Nair and G. Asha, "Some classes of multivariate life distributions in discrete time," *Journal of Multivariate Analysis* **62** (1997), pp. 181–189.
26. G. E. Willmot and J. Cai, "Aging and other distributional properties of discrete compound geometric distributions," *Insurance: Mathematics and Economics* **28** (2001), pp. 361–379.
27. E. Xekalaki, "Hazard functions and life distributions in discrete time," *Communications in Statistics, Theory and Methods* **12** (21) (1983), pp. 2503–2509.
28. S. Zacks, "Estimating the shift to wear-out of systems having exponential-Weibull life," *Operations Research* (32) (1984), pp. 741–749.

About the Authors

Cyril Bracquemond obtained his Ph.D. from the National Polytechnic Institute of Grenoble, France in 2001. His research interests are applied statistics and discrete reliability. He has now joined Pechiney as a Statistician, specialised in design of experiments.

Oliver Gaudoin is an Associate Professor at the National Polytechnic Institute of Grenoble, France. His research interests are stochastic modelling and statistical analysis for system reliability, software reliability, reliability of components subjected to random stresses, goodness-of-fit techniques and discrete reliability.

Chapter 8

CLUSTERING-BASED NETWORK INTRUSION DETECTION[#]

SHI ZHONG* and TAGHI M. KHOSHGOFTAAR[†]

Computer Science and Engineering
Florida Atlantic University
777 West Glades Road
Boca Raton, FL 33431, USA
*zhong@cse.fau.edu
†taghi@cse.fau.edu

NAEEM SELIYA

Computer and Information Science
University of Michigan – Dearborn
4901 Evergreen Road
Dearborn, MI 48128, USA
nseliya@umich.edu

Recently data mining methods have gained importance in addressing network security issues, including network intrusion detection — a challenging task in network security. Intrusion detection systems aim to identify attacks with a high detection rate and a low false alarm rate. Classification-based data mining models for intrusion detection are often ineffective in dealing with dynamic changes in intrusion patterns and characteristics. Consequently, unsupervised learning methods have been given a closer look for network intrusion detection. We investigate multiple centroid-based unsupervised clustering algorithms for intrusion detection, and propose a simple yet effective self-labeling heuristic for detecting attack and normal clusters of network traffic audit data. The clustering algorithms investigated include, k-means, Mixture-Of-Spherical Gaussians, Self-Organizing Map, and Neural-Gas. The network traffic datasets provided by the DARPA 1998 offline intrusion detection project are used in our empirical investigation, which demonstrates the feasibility and promise of unsupervised learning methods for network intrusion detection. In addition, a comparative analysis shows the advantage of clustering-based methods over supervised classification techniques in identifying new or unseen attack types.

Keywords: Network intrusion detection; clustering algorithms; classification techniques.

1. Introduction

Considerable attention has been given to data mining approaches for addressing network security issues.[1–3] This is particularly due to the increasing dependence on computer networks for personal, business, and government activities. An intrusion

---

[#] This chapter appeared previously on the International Journal of Reliability, Quality and Safety Engineering. To cite this chapter, please cite the original article as the following: S. Zhong, T. M. Khoshgoftaar and N. Seliya, *Int. J. Reliab. Qual. Saf. Eng*, **14**, 169–187 (2007), doi:10.1142/S0218539307002568.

attack can result in several severity levels of incapacity, from loss of personal privacy to an enormous loss of business capital. An intrusion is any use of the given network that compromises its stability and/or security of information stored across the network. Network intrusion detection models are used for detecting intrusions or anomalous behavior.[4–6] There are generally two types of approaches taken toward network intrusion detection: anomaly detection and misuse detection.

In misuse detection, each network traffic record is identified as either normal or one of many intrusion types.[7] A classifier is then trained to discriminate one category from the other, based on network traffic attributes. Lee *et al.*[8] proposed a data mining framework for such a system. They used a series of data mining techniques, such as frequent episodes and association rules, to help extract discriminative features, which include various network traffic statistics. Obtaining correctly-labeled data instances, however, is difficult and time intensive, especially for new attack types and patterns. On the other hand, anomaly detection amounts to training models for normal traffic behavior and then classifying as intrusions any network behavior that significantly deviates from the known normal patterns.[9] Traditional anomaly detection algorithms often require a set of purely normal traffic data from which models can be trained to represent normal traffic patterns.

Clustering algorithms have recently gained attention[10–15] in related literature, since they can help current intrusion detection systems in several respects. An important advantage of using clustering or unsupervised learning to detect network attacks is the ability to find new attacks not seen before. This implies that attack types with unknown pattern signatures can be detected. A main objective of this study is to demonstrate this advantage of unsupervised learning using a simple (clustering-based attack detection) method. The proposed method is also shown to improve (assist) traditional classification-based intrusion detection models that often have difficulty in classifying new or unseen attacks correctly.

Clustering results can also assist the network security expert with labeling network traffic records as normal or intrusive. The amount of available network traffic audit data is usually large, making the expert-based labeling process of all records very tedious, time-consuming, and expensive. Additionally, labeling a large number of network traffic records can lead to errors being incorporated during the process. Grouping similar data together eases the task of labeling by experts. Instead of evaluating each data instance one by one, the expert can simultaneously label all (tens or hundreds) data instances in a cluster by observing the common characteristics of the cluster, with possibly very few mistakes, provided the clusters obtained are relatively "pure". A completely (100%) pure cluster is one that contains data instances only from one category (normal or a specific attack type or category). We investigate the performance of several clustering algorithms in terms of cluster purity as well as other performance criteria.

Recent works on clustering-based intrusion detection focus on constructing a set of clusters (based on unlabeled[11,13] or labeled[12] training data) to classify (future) test data instances. Such approaches do not completely exploit the advantages of clustering-based methods. In contrast, we advocate clustering the (future) test data instances to be classified, and then use heuristics or existing labeled data to help label the clustered instances.

Portnoy *et al.*[11] presented a clustering method for detecting intrusions from unlabeled data. Unlike traditional anomaly detection methods, they cluster data instances that contain both normal behaviors and attacks, using a modified incremental k-means algorithm. After clustering, each cluster is labeled (as normal or attacks) based on the number of instances in the cluster. The heuristic is that very small clusters tend to be attacks. The self-labeled clusters are then used to detect attacks in a separate test dataset. Guan *et al.*[13] worked on the same idea of detecting intrusions but with a different clustering algorithm, namely an improved k-means algorithm that addresses the selection of number of clusters and the elimination of empty clusters.

Although unsupervised intrusion detection in general looks promising, we feel the approach used by Portnoy *et al.*[11] and Guan *et al.*[13] has a few problems. First, each cluster is self-labeled as attacks or normal, based purely on the number of instances in it. This is not reliable since we have seen in our experiments many relatively large attack clusters as well as small normal clusters. Secondly, the idea of detecting intrusions in a new dataset using the self-labeled clusters of the training dataset seems misguided. We feel that the purpose of unsupervised intrusion detection is to discover new attacks in a new dataset. It is more practical to run clustering algorithms on the new dataset and identify attacks by self-labeling.

Ye and Li[12] proposed a supervised clustering technique that constructs clusters from labeled training data and uses them to score the possibility of being attacks for test data instances. Performance better than decision tree classification models was reported. Lee *et al.*[8] emphasized the data-flow environment of network intrusion detection, aiming at real-time feature extraction and classification from network traffic data. We agree that an online, real-time, and adaptive intrusion detection system is the ultimate goal, towards which our online clustering-based approaches have provided a promising tool.

This paper extends our previous study[14,15] to demonstrate that our clustering-based method can outperform and enhance the state-of-the-art support vector machine algorithm in detecting unseen intrusion (attack) types. The primary contributions of this paper are:

1. A comprehensive comparative study of multiple clustering algorithms for analyzing large intrusion detection datasets was conducted. We analyzed the suitability of different clustering algorithms and empirically compared several centroid-based algorithms — k-means,[16] Mixture-Of-Spherical Gaussians,[17] Self-Organizing Map,[18] and Neural-Gas,[19] in terms of both clustering quality and run-time efficiency. To our knowledge the different clustering algorithms (in the paper) have not been presented in a relative comparative study for the network intrusion detection problem. Such a study holds practical merit and will benefit

future studies in this area, especially given the existence of many clustering methods.

2. A simple and effective self-labeling heuristic is proposed to detect and label attack clusters. The detection performance is evaluated by detection accuracies and ROC (Receiver's Operating Characteristics) curves for each of the aforementioned clustering algorithms.

3. A set of empirical investigations are designed to show the main advantage of our clustering-based intrusion detection method in identifying new attack instances. Results also demonstrate that our method can be used to help classification methods achieve better overall intrusion detection accuracies. More specifically, useful intrusion detection performance are obtained using clustering for classification of network traffic data.

This paper is structured as follows. The clustering algorithms investigated in this paper are introduced in the next section. We then present the proposed heuristic for self-labeling of clusters, followed by a description of the intrusion detection dataset and empirical settings, and a discussion of our empirical results. Finally, we present concluding remarks and some suggestions for future work. Some key abbreviations/notations used in this paper are summarized in Table 1.

2. Clustering Techniques

Clustering techniques can generally be divided into two categories: pairwise clustering and central clustering. Pairwise clustering algorithms are based on the pairwise

Table 1.   Key notations.

Notation	Description
adr	attack detection rate
avg-pur	average purity metric of cluster
dos	denial of service attack type
EM	expectation-maximization algorithm
fpr	false positive rate
kmo	online k-means clustering algorithm
MOSG	mixture-of-spherical gaussian clustering algorithm
mse	mean squared error metric
N	number of instances in dataset
Neural-Gas	online Neural-Gas clustering algorithm
probe	reconnaissance attack type
r2l	remote-to-local attack type
ROC	Receiver's Operating Characteristic curve
SOM	self-organizing maps clustering algorithm
SVM	support vector machine classification algorithm
u2r	user-to-root attack type
x_i	the ith data vector
y_k	the kth cluster
η	percentage of normal instances
μ_k	centroid of the kth cluster

proximity affinities between the instances of a dataset. As pairwise distances need to be computed for all pairs in the dataset, such algorithms are based on efficient Eigen vector calculations and implement a combinatorial optimization method for data grouping based on proximity data.[20] Graphical clustering and spectral clustering methods fall into this category.[21,22] The computational complexity of pairwise clustering increases quickly with the number of instances, i.e., dataset size.

Central or centroid-based clustering refers to algorithms such as k-means where a cluster is defined by its centroid values for the different attributes. The cluster centroid is a model for a given cluster, and hence; such clustering methods are also known as model-based clustering. Data points are clustered based on closest distance to the centroids of the different clusters. Such clustering algorithms generally need an initialization parameter where the initial value of the centroid is specified. Central clustering algorithms are often more efficient than pairwise clustering algorithms.[23] Regular k-means [16] and EM (Expectation Maximization) clustering algorithms [17] fall into this category.

We choose centroid-based clustering over pairwise-based clustering due to the large size of the network traffic audit dataset. This was decided after initial studies on a variety of methods, including the CLUTO toolkit,[22] which is based on graph partitioning techniques for clustering. Preliminary experiments using CLUTO show that it takes several hours to cluster about 110K network traffic instances and the algorithm generated $500 \sim 700$ clusters due to required (for efficiency) sparsity for the constructed graph.[a] We could not efficiently get a desired (for comprehensibility) number of clusters, e.g., 100 or 200 as set by users.

Pairwise-based algorithms usually have a complexity of at least $O(N^2)$ (for computing the data-pairwise proximity measures), where N is the number of data instances. In contrast, centroid-based algorithms are more scalable, with a complexity of $O(NKM)$, where K is the number of clusters and M the number of batch iterations. In addition, all centroid-based clustering techniques have an online version which can be suitably used for adaptive attack detection in a data-flow environment.[8] An online version, compared to batch version, generally refers to how the centroid values are updated — either after each data point cluster assignment or after cluster assignments for a group of data points.

2.1. *K-means*

The widely used standard k-means algorithm[16] can be found in many other papers and its details are omitted here. It minimizes the mean-squared error (mse) objective function

$$E = \frac{1}{N} \sum_n \|x_n - \mu_{y_n}\|^2 \,, \tag{1}$$

---

[a]See the CLUTO toolkit manual for more details.

Algorithm: online k-means (kmo)

Input: A set of N data vectors $X = \{x_1, \ldots, x_N\}$ in $I\!\!R^d$ and number of clusters K.

Output: A partition of the data vectors given by the cluster identity vector
$Y = \{y_1, \ldots, y_N\}$, $y_n \in \{1, \ldots, K\}$.

Steps:

1. Initialization: initialize the cluster centroid vectors $\{\mu_1, \ldots, \mu_K\}$;

2. Loop for M iterations

   For each data vector x_n, set $y_n = \arg\min_k \|x_n - \mu_k\|^2$, and update the centroid μ_{y_n} as

   $$\mu_{y_n}^{(new)} = \mu_{y_n} - \frac{\partial E}{\partial \mu_{y_n}} = \mu_{y_n} + \xi(x_n - \mu_{y_n}),$$

   where ξ is a learning rate usually set to be a small positive number (e.g., 0.05). The number can also gradually decrease in the learning process.

Fig. 1.   Online k-means algorithm.

where $y_n = \arg\min_k \|x_n - \mu_k\|^2$ is the cluster identity of data vector x_n and μ_{y_n} is the centroid of cluster y_n. Unless specified otherwise, $\|\cdot\|$ represents L_2 norm. The popularity of the k-means algorithm is largely due to its simplicity, low time complexity, and fast convergence. Since batch k-means has been described in many papers,[24] here we instead present an online k-means algorithm in Fig. 1. Both are studied in Sec. 4.3, and they are named batch k-means and kmo, respectively.

2.2. *Mixture-Of-Spherical Gaussians (MOSG)*

The MOSG clustering using the EM algorithm[25] is shown in Fig. 2. It is a special case of Mixture-Of-Gaussians clustering, with an identity covariance matrix used for the Gaussian distribution of a cluster. Using an identity covariance matrix makes the algorithm more scalable to higher data dimensionality. A detailed derivation of the parameter estimation for the Mixture-Of-Gaussians and an excellent tutorial on the EM algorithm can be found in Blimes.[26] The MOSG algorithm is a "soft" clustering technique because each data vector is fractionally assigned to multiple clusters. In contrast, the data assignment in the k-means algorithm is considered "hard" (i.e., each data vector is assigned to only one cluster).

2.3. *Self-Organizing Map (SOM)*

Self-organizing map[18] is a competitive learning technique that can extract structural information from data and provide low-dimensional (1, 2, or 3-D) visualization through a topological map. Each cluster has a fixed coordinate in the topological map.

Algorithm: Mixture-Of-Spherical Gaussians clustering

Input: A set of N data vectors $X = \{x_1, \ldots, x_N\}$, model structure
$\Lambda = \{\mu_k, \sigma_k, \alpha_k\}_{k=1,\ldots,K}$, where μ's and σ's are the parameters for Gaussian models and α's are prior parameters that are subject to $\alpha_k \geq 0, \forall k$ and $\sum_k \alpha_k = 1$.

Output: Trained model parameters Λ that maximizes the data likelihood $P(X|\Lambda) = \prod_n \sum_k \alpha_k p(x_n|\lambda_k)$, and a partition of the data vectors given by the cluster identity vector $Y = \{y_1, \ldots, y_N\}$, $y_n \in \{1, \ldots, K\}$.

Steps:

1. Initialization: initialize the model parameters Λ;

2. E-step: the posterior probability of model k, given a data vector x_n and current model parameters Λ, is estimated as

$$P(k|x_n, \Lambda) = \frac{\alpha_k p(x_n|\lambda_k)}{\sum_j \alpha_j p(x_n|\lambda_j)},$$

   where the pdf $p(x|\lambda)$ is given by

$$p(x_n|\lambda_k) = \frac{1}{(\sqrt{2\pi}\sigma)^d} \exp\left(-\frac{\|x_n - \mu_k\|^2}{c_k^2}\right);$$

3. M-step: the maximum likelihood re-estimation of model parameters Λ is given by

$$\mu_k^{(new)} = \frac{\sum_n P(k|x_n, \Lambda)x_n}{\sum_n P(k|x_n, \Lambda)},$$

$$\sigma_k^{(new)} = \frac{1}{d}\frac{\sum_n P(k|x_n, \Lambda)\|x_n - \mu_k\|^2}{\sum_n P(k|x_n, \Lambda)},$$

   and

$$\alpha_k^{(new)} = \frac{1}{N}\sum_n P(k|x_n, \Lambda);$$

4. Stop if $P(X|\Lambda)$ converges, otherwise go back to Step 2;

5. For each data vector x_n, set $y_n = \arg\max_k (\alpha_k p(x_n|\lambda_k))$.

Fig. 2. Mixture-Of-Spherical Gaussians clustering algorithm.

Let the map location of cluster k be ϕ_k; $K_\alpha(\phi_1, \phi_2) = \exp\left(-\frac{\|\phi_1 - \phi_2\|^2}{2\alpha^2}\right)$ be a neighborhood function; and $y_n = \arg\min_y \|x_n - \mu_y\|^2$. The batch SOM algorithm amounts to iterating between the following two steps:

$$P(y|x_n) = \frac{K_\alpha(\phi_y, \phi_{y_n})}{\sum_{y'} K_\alpha(\phi_{y'}, \phi_{y_n})},$$

and

$$\mu_y = \frac{\sum_x P(y|x)x}{\sum_x P(y|x)},$$

where α is a parameter controlling the width of the neighborhood function and decreases gradually during the clustering process. Note that the α parameter has the same functionality of a temperature parameter in an annealing process.

A unique feature of SOM is that the calculation of $P(y|x)$ is constrained by a topological map structure, which gives SOM the advantage that all resulting clusters are structurally related according to the user-specified topological map (which is good for visualization). However, SOM is usually not as good as k-means or Neural-Gas (discussed next) in terms of minimizing the mse function (1).

2.4. *Neural-Gas*

The (batch-version) Neural-Gas algorithm differs from the SOM clustering, only in how $P(y|x)$ is computed, i.e.,

$$P(y|x) = \frac{e^{-r(x,y)/\beta}}{\sum_{y'} e^{-r(x,y')/\beta}},$$

where β is an equivalent temperature parameter and $r(x,y)$ is a rank function that takes the value $k-1$ if y is the kth closest cluster centroid to data vector x.

The original Neural-Gas algorithm[19] was proposed as an online algorithm, in which all cluster centroids are updated each time a data instance x_n is given, according to

$$\mu_y = \mu_y + \xi e^{-r(x_n,y)/\beta}(x_n - \mu_y).$$

In the online learning process, both the learning rate ξ and the temperature β gradually decrease, simulating an annealing procedure. It has been shown that the online version can converge faster and find better local solutions than clustering with SOM for certain problems.[19]

3. Intrusion Detection with Clustering

In addition to cluster sizes, inter-cluster distances are used in our clustering-based detection process. For the k-means, SOM, and Neural-Gas algorithms, the inter-cluster distance is defined as the Euclidean distance between two cluster centroids. For the MOSG algorithm, it is defined as the Mahalanobis distance between two cluster centroids,[27] calculated as $D(y_1, y_2) = (\frac{1}{\sigma_1^2} + \frac{1}{\sigma_2^2})\|\mu_1 - \mu_2\|^2$.

We use a practical assumption about the intrusion data — the number of normal instances is much larger than that of attack instances. This is usually true in reality. However, unlike in Portnoy et al.[11], we do not make the strict hypothetical requirement that the percentage of attacks has to be less than a certain threshold (e.g., $\sim 1.5\%$).

Based on the assumption that normal instances dominate attack instances, our simple self-labeling heuristic for unsupervised intrusion detection consists of the following steps:

- Find the largest cluster, i.e., the one with the most number of instances, and label it *normal*. Assume its centroid is μ_0.
- Sort the remaining clusters in ascending order of the distance from each cluster centroid to μ_0. Within a cluster, sort the data instances in the same way (i.e., ascending order of distance from each data instance to μ_0).
- Select the first $N_1 = \eta N$ instances, and label them as *normal*, where η is the percentage of normal instances. The parameter n is the (given or estimated) fraction of all data instances as normal ones. By varying η, we can get a series of accuracy numbers can be used to draw ROC curves.
- Label all the other instances as *attacks*.

4. Empirical Investigation

4.1. *Case study description*

The network traffic audit data used in our case study comes from the data supplied as part of the 1998 DARPA off-line intrusion detection project. The original dataset of about 5 million records was collected from a controlled experiment in which a real-life military network was intentionally subjected to various attacks at specified time periods.

Each record, representing a connection between two network hosts according to some well defined network protocol, is described by 41 attributes (38 continuous or discrete numerical features and three categorical features) such as, duration of connection, number of bytes transferred, number of failed logic attempts, etc. The attributes are of three types[28]: intrinsic, content-based, and traffic-based. A record was labeled as either normal or one of four intrusion categories: denial of service (*dos*), reconnaissance (*probe*), remote-to-local (*r2l*), and user-to-root (*u2r*). A 10% sample consisting of about 500,000 records obtained from the UCI machine learning data repository was used in our study.

The *dos* attack type accounted for about 80% of the data, and consisted mostly of redundant *Neptune* and *Smurf* attacks. To reduce such a redundancy and the computational complexity of the problem (i.e., a dataset with mostly *dos* instances), the dataset was reduced to *109,910* records by randomly selecting only 1% of *Neptune* and 1% of *Smurf* attack types from the *dos* category. This is possible since the attack types of all instances in the dataset is known. For example, among 1000 *Neptune* attack instances, any 10 are randomly selected. Similarly, among 1000 *Smurf* attack instances, any 10 are randomly selected. This process is commonly adopted by other researchers. The *u2r* type was almost negligible.

We are interested in anomaly detection via unsupervised clustering algorithms; hence, all records labeled as attacks were considered as intrusion, while the remaining were considered as normal. All clustering algorithms were implemented to cluster the network records without their intrusion/normal labels, thereby making it an unsupervised approach for intrusion detection. Once the set of clusters are formed we use a self-labeling heuristic to assign intrusion/normal labels to the traffic records. The actual labels are not used during the clustering process, but are only used for evaluating the detection performance of the algorithms.

The clustering algorithms used in our comparative study do not directly handle categorical data. The 3 categorical features, i.e., protocol type, service type, and connection flag, in the dataset were converted using the 1-of-N encoding scheme, which expanded the categorical features into 67 dimensions based on the counts for each categorical feature. Thus, the dataset used in our study consisted of 109,910 records, each having 105 feature dimensions. All feature dimensions were scaled to be within $[0, 1]$ to avoid potential scale problems.

It is worth mentioning here that feature construction (data preprocessing) is an important step in intrusion detection[8,29] and involves tremendous efforts. However, in this paper we simply use the same set of features that recent papers[11,30] have used, since our focus is on unsupervised intrusion detection algorithms.

4.2. *Empirical setting*

The k-means and Neural-Gas algorithms are written in C and compiled into mex files which can be run from Matlab. We implemented the MOSG algorithm in Matlab and used the SOM algorithm from the SOM Toolbox for Matlab provided by the Helsinki University of Technology.[b] Note that the Matlab-coded MOSG and SOM algorithms are relatively efficient due to vectorized programming and active optimization. All experiments are run on a PC with a 3.06GHz Pentium-4 CPU with 1GB DRAM and running Windows XP.

For the kmo algorithm, we use a learning rate that follows $\xi_m = 1.0(0.01/1.0)^{\frac{m}{NM}}$, where m is the online iteration number (from 0 to NM, N is the number of instances and M the number of batch iterations). For the online Neural-Gas algorithm, the learning rate follows $\xi_m = 0.5(0.005/0.5)^{\frac{m}{NM}}$, and the temperature follows $\beta_m = \beta_0(\beta_f/\beta_0)^{\frac{m}{NM}}$, where $\beta_0 = K/2$ (K is the number of clusters) and $\beta_f = 0.01/M$.

In order to study the effect of the total number of clusters on the intrusion detection results, we performed empirical studies with 100 and 200 total number of clusters. For clustering quality, we use the mean squared error (*mse*) objective (1) and average purity (*ave-pur*). The purity of a cluster is defined as the percentage of the most dominated instance category in the cluster, and average purity is the mean over all clusters. Its value can range from 0 to 1, with higher values representing better average purity. The run time of each algorithm is also recorded and compared.

---

[b]Available from http://www.cis.hut.fi/projects/somtoolbox/.

Each experiment is run ten times and the averages and standard deviations of the above measures are reported.

For evaluating intrusion detection results, we report false positive rate (*fpr*), attack detection rate (*adr*), and overall accuracy. The false positive rate is the percentage of normal instances that are labeled as attacks. The attack detection rate represents the percentage of all attack instances that are detected, i.e., labeled as attacks. The overall accuracy measures the percentage of all instances that are correctly labeled. We also report *ROC* curves, by varying the parameter η used in the detection method, to show the tradeoff between the false positive rate and the detection rate.

In the next two sections, we present two sets of experiments, each designed to demonstrate a different point. The first set is used to compare the algorithms presented in Sec. 2 whereas the second set is used to show the situations in which clustering-based intrusion detection methods outperform classification-based techniques (in terms of detection unseen attacks). We also show that clustering methods can be employed to help the performance of classification-based intrusion detection techniques.

4.3. *Experimental results — I*

In this section, we compare the aforementioned clustering algorithms on the whole data set (with 109,910 instances). Our results are promising based on a network traffic audit dataset with approximately 110K-instances, of which around 11.5% were attack instances. The proposed self-labeling heuristic (with $\eta = 11.5\%$) achieved an overall accuracy of 93.6%, a false positive rate of 3.3%, and a detection rate of 72%. It is logical that these numbers are not comparable to the accuracy that can be achieved by a supervised classifier since unsupervised intrusion detection tries to learn attacks without any labeled data. Detailed comparisons are presented below.

The *mse*, *average purity*, and *run time* results for the clustering algorithms with 100 clusters and 200 clusters, are shown in Tables 2 and 3, respectively. A standard deviation of 0.0 in the table signifies a very small (< 0.001) number. The *mse* results are not indicated for the MOSG algorithm, because it optimizes a different objective (maximizing data likelihood as seen in Fig. 2).

The kmo and Neural-Gas algorithms perform significantly better ($p < 5\%$) than the others in terms of *mse* and average purity, but kmo achieves the same high

Table 2.   Summary of clustering results with 100 clusters.

	mse	*ave-pur*	*time* (seconds)
k-means	0.15 ± 0.03	0.93 ± 0.01	78.6 ± 14.6
kmo	0.06 ± 0.0	0.964 ± 0.0	249.7 ± 9.6
SOM	0.28 ± 0.0	0.932 ± 0.0	124.8 ± 1.3
Neural-Gas	0.05 ± 0.0	0.962 ± 0.003	1481.9 ± 204.5
MOSG		0.958 ± 0.006	366.7 ± 8.0

Table 3.   Summary of clustering results with 200 clusters.

	mse	ave-pur	time (seconds)
k-means	0.11 ± 0.01	0.932 ± 0.01	155.1 ± 22.8
kmo	0.04 ± 0.0	0.977 ± 0.004	363.6 ± 19.0
SOM	0.14 ± 0.0	0.956 ± 0.0	147.5 ± 1.7
Neural-Gas	0.03 ± 0.0	0.970 ± 0.003	2524.9 ± 214.2
MOSG		0.963 ± 0.003	923.5 ± 31.8

quality with much less run time. Comparing the results for 100 clusters and those for 200 clusters, we observe that the k-means and Neural-Gas algorithms scale linearly with the number of clusters whereas the SOM algorithm does sub-linearly and the MOSG algorithm super-linearly. Although the batch k-means and SOM algorithms are computationally efficient, they do not generate coherent clusters like the other algorithms, as indicated by mse and purity measures. The kmo algorithm seems to be a desirable choice, with high clustering quality and relatively low time complexity.

Since our aim is to detect network intrusion using clustering algorithms, we now analyze the unsupervised intrusion detection accuracies. Using the simple self-labeling heuristic presented in Sec. 3, we sort clusters according to their possibility of being normal in decreasing order and arrange data instances in a cluster in the same way. The possibility of being normal is measured by the distance to the centroid of the largest cluster. *ROC* curves can be constructed by dividing the sorted data instances into normal and intrusive categories at a series of cutting points.

Figures 3(a) and 3(b) show the *ROC* curves for the kmo, SOM, and Neural-Gas algorithms, with 100 clusters and 200 clusters, respectively. The curve for the batch k-means and MOSG algorithms are omitted for comprehensibility and better visualization, particularly because they are visibly worse than the other three algorithms. It can be seen that for 100 clusters, the SOM algorithm is the worst and the k-means and Neural-Gas algorithms work equally well for most of the false positive rates. In contrast, for 200 clusters, all three algorithms perform comparably most of the time and the SOM algorithm clusters performs extremely well at very low false positive rates, e.g., it can detect more than 50% attacks with a false positive rate of almost 0. Overall, the kmo and Neural-Gas algorithms seem to be the better ones and stable across different number of clusters.

We now discuss the accuracy results at one cutting point from the sorted list of clusters. Suppose the approximate percentage of attack instances is known *a priori* or from heuristics, we split the sorted cluster list at a point that generates the desired percentage. In this paper, we group the clusters as normal or intrusive in such a way that the number of data instances in attack clusters account for about 11.5% of the total population, reflecting the assumed distribution of the training data. We run each experiment 10 times and report the average and standard

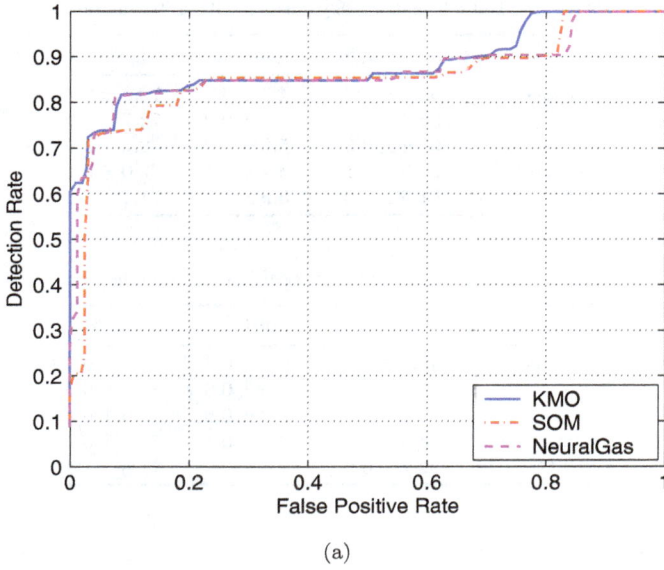

(a)

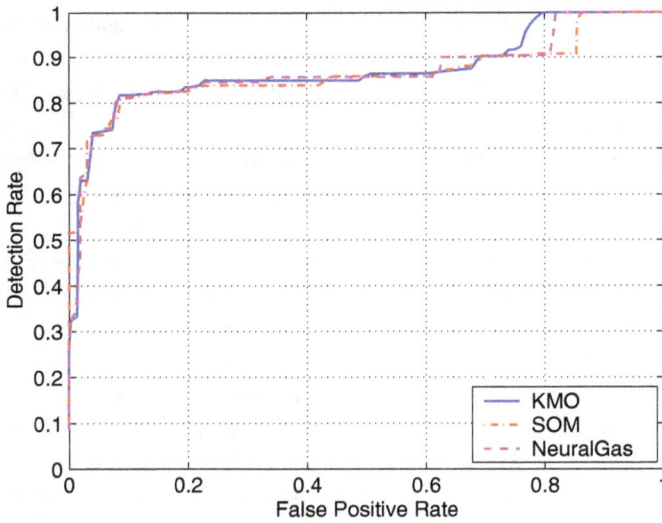

(b)

Fig. 3.   ROC curves for kmo, SOM, and Neural-Gas, with (a) 100 clusters and (b) 200 clusters.

deviation for overall accuracy, false positive rate, and attack detection rate, in Tables 4 and 5. The kmo and Neural-Gas algorithms again perform better than others, generating low *fpr* values and high overall accuracies; the performance difference to other algorithms are significant ($p < 5\%$) according to paired *t*-tests.

Table 4.   Detection accuracy results with 100 clusters.

	accuracy(%)	fpr(%)	adr(%)
k-means	92.6 ± 0.7	4.7 ± 0.7	72.1 ± 2.7
kmo	93.7 ± 0.0	3.6 ± 0.0	72.7 ± 0.0
SOM	92.4 ± 0.0	5.0 ± 0.0	72.2 ± 0.0
Neural-Gas	93.5 ± 0.1	3.7 ± 0.1	72.0 ± 0.1
MOSG	91.5 ± 1.0	5.9 ± 1.0	71.4 ± 3.0

Table 5.   Detection accuracy results with 200 clusters.

	accuracy(%)	fpr(%)	adr(%)
k-means	93.1 ± 0.3	4.2 ± 0.3	72.8 ± 0.7
kmo	93.6 ± 0.0	3.6 ± 0.0	72.1 ± 0.0
SOM	93.1 ± 0.0	4.3 ± 0.0	72.9 ± 0.0
Neural-Gas	93.6 ± 0.1	3.6 ± 0.1	72.5 ± 0.1
MOSG	92.7 ± 0.9	4.6 ± 0.9	71.8 ± 1.4

4.4.   *Experimental results — II*

Now we present results to verify our hypothesis that clustering-based methods have an edge over classification techniques in identifying new or unseen attack types. The classification model to be compared is the state-of-the-art support vector machine method. We used the svm-light software package available online.[c] Default settings in the package are used.

We divided the 109,190 data instances into two parts — a training set and a test set. The training set is formed by randomly picking half of all *dos* attack instances and half of all normal instances. The rest goes into the test set. Therefore, the test set contains new attack types (*probe*, *r2l*, and *u2r*) that are not seen in the training set. The SVM algorithm is trained using the training set and then evaluated on the test set. An ROC curve can be obtained since the SVM algorithm outputs a continuous numerical score that indicates the probability of being normal (or intrusive) and can be used to order data instances. The kmo algorithm (picked according to our studies in the previous section) is applied directly to the test set.

In addition to the SVM and kmo-based intrusion detection algorithms, we also compare a combined method, which orders data instances using both SVM and kmo results. Here we only present a simple method and show the benefit of combining classification and clustering for intrusion detection. More sophisticated combining methods can be studied in our future work. The simple combining strategy works like this: first we use the order in SVM results for those instances that are classified as attacks; then we use the order in kmo results for the remaining instances. The idea is to trust SVM for identifying seen attacks in the training data but kmo for identifying unseen attacks.

---

[c]http://www.cs.cornell.edu/People/tj/svm_light/

Figure 4 presents the ROC curves for the kmo, SVM, and combined methods. It can be seen that, for very small *fpr* values, SVM delivers better detection rates, but for *fpr* > 5%, kmo works better. The combined method always generates better performance than both SVM and kmo methods.

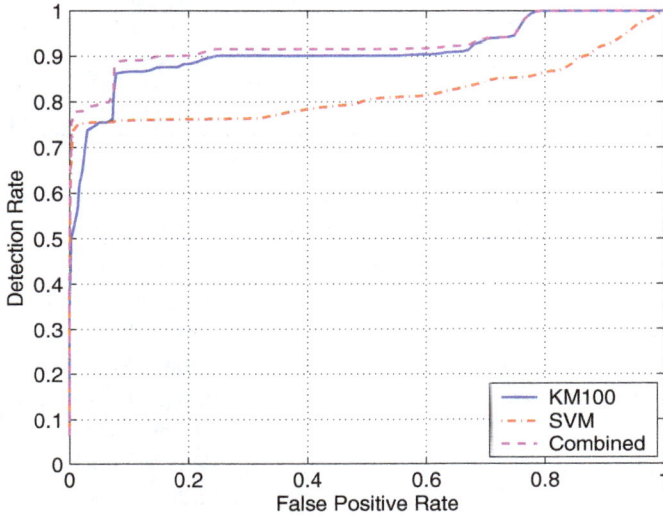

(a)

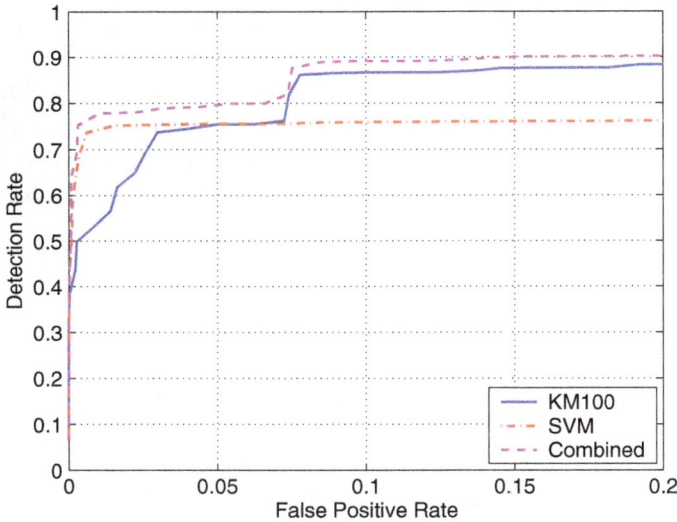

(b)

Fig. 4. (a) ROC curves for kmo (100 clusters), SVM, and the combined method; and (b) a close-up look at low fpr area.

Table 6. Detection accuracy results with 100 clusters using k-means.

		Detection accuracy (%)				
		dos	*probe*	*r2l*	*u2r*	*overall*
fpr = 5%	kmo	69.81	99.62	6.48	49.45	75.39
	SVM	99.9	67.31	29.09	0	75.47
	combined	75.76	99.61	22.24	49.45	79.80
fpr = 10%	kmo	69.81	99.92	92.27	100	86.61
	SVM	99.96	68.1	29.16	0	75.86
	combined	75.76	99.93	92.53	100	89.08

To inspect the details of detection accuracies for individual attack types, we present the results for two fixed *fpr* values (5% and 10%, respectively) in Table 6. A quick glance at the table would suggest the overall accuracy measure does not provide a good insight into relative performances (especially at *fpr* = 10%) of the clustering methods, and may lead to an erred conclusion that there is little significant difference in the relative performances. While this is reflective of the skewed distribution of the dataset with respect to the five categories (*dos, probe, r2l, u2r,* and *normal*), the classification accuracy results of the individual attack categories is of more interest, e.g., for the *dos* attack category SVM is clearly better than kmo.

The kmo-based method detects more *probe* and *u2r* attacks. In fact, the SVM method detects zero *u2r* attacks for the two given *fpr* levels, which indicates that *u2r* attacks are quite different from *dos* attacks in the current representation. As stated earlier, the SVM method always detects more *dos* attacks than the kmo-based method due to its exposure to *dos* attacks in the training data. For *r2l* type attacks, SVM works better at *fpr* = 5% but kmo is better at *fpr* = 10%. The combined method catches more other-than-*dos* attacks than SVM and produces higher overall detection accuracies, demonstrating the potential of combining classification and clustering for intrusion detection. There is also room for further improvements, as indicated by the observation that the combined method still cannot compare with SVM in detecting *dos* attacks.

5. Conclusion

The feasibility of unsupervised intrusion detection using multiple centroid-based clustering algorithms is investigated in this study. Considering the dynamic nature of network traffic intrusions, unsupervised intrusion detection is more appropriate for anomaly detection than classification-based intrusion detection methods. Since the attack labels are not used during unsupervised intrusion detection, changes in the behavior and characteristics of network attacks can be accommodated with relative ease.

An empirical study consisting of different clustering algorithms is performed with a case study of network traffic data obtained from the DARPA 1998 offline intrusion detection project. A comparative analysis and evaluation of the clustering algorithms yielded reasonable intrusion detection rates.

In addition, a simple yet effective self-labeling heuristic for labeling clusters as normal or intrusive (anomalous) is proposed and evaluated. Comparisons with classification-based intrusion detection methods on a completely new network traffic dataset demonstrated the usefulness and feasibility of clustering-based methods in both detecting new attack types and assisting classification techniques in achieving better intrusion detection performance.

Promising clustering and detection results encourage us to proceed our future work in several directions. A further detailed analysis of individual clusters can be done by identifying the precise attack category associated with a cluster and the discriminating features that are unique to a given cluster. In addition, feature selection/weighting for clustering will be investigated. This will eventually enhance our understanding and detection of new attack categories. Sophisticated self-labeling techniques, taking into consideration of additional network security domain knowledge, can be developed to improve the performance of clustering-based intrusion detection. Finally, incremental algorithms can be built by extending the competitive learning versions of k-means (e.g., according to[31,32]), making them more suitable for adaptive intrusion detection systems.

Acknowledgments

We thank Dr. Hoang Pham, Editor-in-chief, and the anonymous referees for their comments and critique which contributed in improving this paper. We are also grateful for getting assistance with reviews from the current and former members of the Empirical Software Engineering Laboratory and the Data Mining and Machine Learning Laboratory, both located at Florida Atlantic University, Boca Raton, FL.

References

1. V. Kumar, Parallel and distributed computing for cybersecurity. *IEEE Distributed Systems* **6**(10) 2005.
2. Z.-X. Yu, J.-R. Chen and T.-Q. Zhu, A novel adaptive intrusion detection system based on data mining, in *Proceedings IEEE International Conference on Machine Learning and Cybernetics* **4** (IEEE Computer Society) (2005), pp. 2390–2395.
3. X. Zhu, Z. Huang and H. Zhou, Design of a multi-agent based intelligent intrusion detection system, in *Proceedings 1st International Symposium on Pervasive Computing and Applications* (Urumqi, China, 2006) (IEEE Computer Society), pp. 290–295.
4. W. Li, K. Zhang, B. Li and B. Yang, An efficient framework for intrusion detection based on data mining, in *Proceedings 2005 ICSC Congress on Computational Intelligence Methods and Applications* (IEEE Computer Society) (2005), p. 4.
5. C.-T. Lu, A. P. Boedihardjo and P. Manalwar, Exploiting efficient data mining techniques to enhance intrusion detection systems, in *Proceedings IEEE International Conference on Information Reuse and Integration* (Las Vegas, NV) (IEEE Computer Society, 2005), pp. 512–517.
6. T. M. Khoshgoftaar, C. Seiffert and N. Seliya, Labeling network event records for intrusion detection in a wireless lan, in *Proceedings IEEE International Conference on*

Information Reuse and Integration (Waikoloa Village, HI) (IEEE Computer Society) (2006), pp. 200–206.

7. J. Zhang and M. Zulkernine, A hybrid network intrusion detection technique using random forests, in *Proceedings 1st International Conference on Availability, Reliability and Security* (IEEE Computer Society) (2006), p. 8.

8. W. Lee, S. Stolfo and K. Mok, Mining in a data-flow environment: Experience in network intrusion detection, in *Proc. 5th ACM SIGKDD Int. Conf. Knowledge Discovery Data Mining* (San Diego, CA) (1999), pp. 114–124.

9. D. E. Denning, An intrusion detection model, *IEEE Trans. Software Engineering* **13** (1987) 222–232.

10. E. Eskin, Anomaly detection over noisy data using learned probability distributions, in *Proc. 17th Int. Conf. Machine Learning* (San Francisco, CA) (2000), pp. 255–262.

11. L. Portnoy, E. Eskin and S. Stolfo, Intrusion detection with unlabeled data using clustering, in *ACM Workshop on Data Mining Applied to Security* (Philadelphia, PA) (2001).

12. N. Ye and X. Li, A scalable clustering technique for intrusion signature recognition, in *Proc. 2nd IEEE SMC Information Assurance Workshop* (2001), pp. 1–4.

13. Y. Guan, A. A. Ghorbani and N. Belacel, Y-means: A clustering method for intrusion detection, in *Canadian Conference on Electrical and Computer Engineering*, Montral, Qubec, Canada (2003), pp. 1–4.

14. T. M. Khoshgoftaar, S. V. Nath, S. Zhong and N. Seliya, Intrusion detection in wireless networks using clustering techniques with expert analysis, in *Proceedings 4th International Conference on Machine Learning and Applications* (Los Angeles, CA) (2005), p. 6.

15. S. Zhong, T. M. Khoshgoftaar and N. Seliya, Evaluating clustering techniques for network intrusion detection, in *10th ISSAT Int. Conf. on Reliability and Quality Design* (Las Vegas, Nevada, USA) (2004), pp. 173–177.

16. J. MacQueen, Some methods for classification and analysis of multivariate observations. In *Proc. 5th Berkeley Symp. Math. Statistics and Probability* (1967), pp. 281–297.

17. J. D. Banfield and A. E. Raftery, Model-based Gaussian and non-Gaussian clustering, *Biometrics* **49**(3) (1993) 803–821.

18. T. Kohonen, *Self-Organizing Map* (Springer-Verlag, New York, 1997).

19. T. M. Martinetz, S. G. Berkovich and K. J. Schulten, Neural-Gas network for vector quantization and its application to time-series prediction, *IEEE Trans. Neural Networks* **4**(4) (1993) 558–569.

20. B. Fischer, T. Zoller and J. M. Buhmann, Path based pairwise data clustering with application to texture segmentation, *Lecture Notes in Computer Science* **2134** (2001) 235–250.

21. G. Karypis, E.-H. Han and V. Kumar, Chameleon: Hierarchical clustering using dynamic modeling, *IEEE Computer* **32**(8) (1999) 68–75.

22. G. Karypis, *CLUTO — A Clustering Toolkit*, Dept. of Computer Science, University of Minnesota, May 2002. http://www-users.cs.umn.edu/~karypis/cluto/.

23. S. Zhong and J. Ghosh, A unified framework for model-based clustering, *Journal of Machine Learning Research* **4** (2003) 1001–1037.

24. S. Zhong, T. M. Khoshgoftaar and N. Seliya, Analyzing software measurement data with clustering techniques, *IEEE Intelligent Systems* **19**(2) (2004) 20–27.

25. A. P. Dempster, N. M. Laird and D. B. Rubin, Maximum-likelihood from incomplete data via the EM algorithm, *Journal of the Royal Statistical Society B* **39**(1) (1977) 1–38.

26. J. A. Blimes, A gentle tutorial of the EM algorithm and its application to parameter estimation for Gaussian mixture and hidden Markov models, Technical report,
University of California at Berkeley (1998).

27. C. M. Bishop, *Neural Networks for Pattern Recognition* (Oxford University Press,
1995).

28. T. M. Khoshgoftaar and M. E. Abushadi, Resource-sensitive intrusion detection models for network traffic, in *Proceedings of the 8th IEEE International Symposium on
High Assurance Systems Engineering* (Tampa, Florida) (2004), pp. 249–258.

29. W. Lee, S. Stolfo and K. Mok, A data mining framework for building intrusion detection models, in *Proc. IEEE Symposium on Security and Privacy* (1999).

30. A. H. Sung and S. Mukkamala, Identifying important features for intrusion detection
using support vector machines and neural networks, in *Proc. IEEE Symposium on
Applications and the Internet* (Orlando, FL) (2003), pp. 209–216.

31. P. S. Bradley, U. M. Fayyad and C. Reina, Scaling clustering algorithms to large
databases, in *Proc. 4th ACM SIGKDD Int. Conf. Knowledge Discovery Data Mining*
(1998), pp. 9–15.

32. F. Farnstrom, J. Lewis and C. Elkan, Scalability for clustering algorithms revisited,
SIGKDD Explorations **2**(1) (2000) 51–57.

About the Authors

Shi Zhong was an Assistant Professor in Florida Atlantic University. Currently,
he works at Yahoo! Data Mining and Research Group in Sunnyvale, California. He
holds a Ph.D. in Computer Engineering from The University of Texas at Austin. His
research interest includes data mining, machine learning, and intelligent information
retrieval. He is a member of IEEE and IEEE computer society.

Taghi M. Khoshgoftaar is a Professor of the Department of Computer Science and
Engineering, Florida Atlantic University and the Director of the Empirical Software
Engineering and Data Mining and Machine Learning Laboratories. His research
interests are in software engineering, software metrics, software reliability and quality engineering, computational intelligence, computer performance evaluation, data
mining, machine learning, and statistical modeling. He has published more than
300 refereed papers in these areas. He is a member of the IEEE, IEEE Computer
Society, and IEEE Reliability Society. He was the program chair and General Chair
of the IEEE International Conference on Tools with Artificial Intelligence in 2004
and 2005 respectively. He has served on technical program committees of various
international conferences, symposia, and workshops. Also, he has served as North
American Editor of the Software Quality Journal, and is on the editorial boards of
the journals Software Quality and Fuzzy systems.

Naeem Seliya is an Assistant Professor of Computer and Information Science at the
University of Michigan — Dearborn. He received his Ph.D. in Computer Engineering from Florida Atlantic University, Boca Raton, FL in 2005. His research interests
include software engineering, data mining and machine learning, computer data and
network security, and computational intelligence. He is a member of IEEE, IEEE
Computer Society, and ACM.

© 2025 World Scientific Publishing Company
https://doi.org/10.1142/9789819812547_0009

Chapter 9

RANDOM AND AGE REPLACEMENT POLICIES*

MINGCHIH CHEN

*Department of Industrial Engineering and Management
Chaoyang University of Technology
168 Jifong East Road, Wufong, 41349, Taiwan, R.O.C.
mchen@cyut.edu.tw*

SATOSHI MIZUTANI

*Department of Media Information, Aichi University of Technology
502 Manori, Nishihazama-cho, Gamagori, 443-0047, Japan
mizutani@aut.ac.jp*

TOSHIO NAKAGAWA

*Department of Business Administration, Aichi Institute of Technology
1247 Yachigusa, Yakusa-cho, Toyota, 470-0392, Japan
toshi-nakagawa@aitech.ac.jp*

This paper proposes a random and age replacement policy for an operating unit which works at random times. First, the unit is replaced before failure at a planned time T or at the completion of a working time, whichever occurs first. The expected cost rate is obtained. Next, as one extended model, the unit is replaced before failure at a number N of working times or at a planned time T. An optimal policy which minimizes the expected cost rate is discussed analytically, and its numerical example is given. Two modified models, where the unit is replaced at the first completion of the working time over time T or at number N, and it is replaced at time T or number N, whichever occurs last, are considered. Furthermore, we show that this corresponds to a cumulative damage model by replacing shock with work. Finally, one optimization problem of how much time to preset for scheduling the completion of N works is proposed.

Keywords: Random maintenance; working time; age replacement; damage model; scheduling.

1. Introduction

Most units deteriorate with age and use, and eventually, fail from either or both causes in random environment. If their failure rates increase with age and use, it may be wise to make some maintenance at periodic times or at a certain number of failures. The recent published books[1–4] collected many reliability and maintenance

---

*This chapter appeared previously on the International Journal of Reliability, Quality and Safety Engineering. To cite this chapter, please cite the original article as the following: M. Chen, S. Mizutani and T. Nakagawa, *Int. J. Reliab. Qual. Saf. Eng*, **17**, 29–39 (2010), doi:10.1142/S0218539310003652.

models and their optimal policies. However, some systems in offices and industry successively execute jobs and computer processes. For such systems, it would be impossible or impractical to maintain them in a strictly periodic fashion. For example, when a job has a variable working cycle and processing time, it would be better to do some maintenance after it has completed its work and process.[5] The various schedules of jobs that have random working and processing times were summarized.[6]

First, it was proved that the optimal age replacement is nonrandom.[7] When a unit is replaced only at random times, the properties of replacement times between two successive failed units were investigated.[8] Random replacement models in which a unit is replaced at planned times or at the same random times as its working times were proposed, and their optimal policies with random life cycle were discussed analytically using a discount rate.[9] Recently, multiple modular systems of error detection for tasks with random processing times were considered, and their optimal checkpoint numbers were obtained.[10]

This paper takes up an operating unit which works at successive random times and its age replacement policies. First, it is assumed that the unit is replaced before failure at a planned time T $(0 < T \leq \infty)$ or a working time, whichever occurs first. The expected cost rate is obtained, and the optimal time T^* which minimizes it is discussed when the failure rate increases strictly. Next, as an extended model, the unit is replaced before failure at a planned time T or at the number N $(N = 1, 2, \ldots)$ of working times, whichever occurs first. When $N = \infty$, this corresponds to a standard age replacement policy.[3,7] Introducing two replacement costs at time T and number N, the expected cost rate is obtained, and optimal T^* and N^* which minimize it are discussed analytically. This is one kind of optimization problems with two variables.[4] When the working time of the unit is exponential and its time to failure has a Weibull distribution, optimal time T^* and number N^* are computed numerically.

It might be useless to replace an operating unit even when a planned time T comes and be wise to replace it at the first completion of the working time over time T. We sometimes want to use the unit as long as possible. From such viewpoints, we propose two modified replacement models where the unit is replaced before failure at the first comletion over time T, or at the number N and it is replaced at time T or at number N, whichever occurs last. We obtain explicitly the expected cost rates of these modified models.

Suppose that each work causes some damage to the unit and it fails when the total damage has exceeded a specified failure level. Then, this corresponds to a cumulative damage model,[11] replacing *shock* with *work*. Thus, a variety of optimal policies studied in Ref. 11 can be applied easily to this model.

Finally, as applied problems with random working times, we take up the number N of works that needs to be scheduled and achieved by the unit.[4] We discuss how much time we should preset for the scheduling the completion of N works by introducing the shortage and excess costs.

2. Age Replacement

2.1. *Random replacement*

It is assumed that Y is the working time of the unit and has a general distribution $G(t) \equiv \Pr\{Y \leq t\}$ with finite mean $1/\theta$. Suppose that the unit with a failure time X deteriorates increasingly with its age, and fails according a general distribution $F(t) \equiv \Pr\{X \leq t\}$ with finite mean $\mu \equiv \int_0^\infty \overline{F}(t)dt$, where $\overline{\Phi}(t) \equiv 1 - \Phi(t)$ for any function $\Phi(t)$. When $F(t)$ has a density function $f(t)$, the failure rate $h(t) \equiv f(t)/\overline{F}(t)$ for $F(t) < 1$ is assumed to increase to $h(\infty) \equiv \lim_{t\to\infty} h(t)$.

The unit is replaced at time T, Y or at failure, whichever occurs first, where T $(0 < T \leq \infty)$ is a constant and Y is a random variable with distribution $G(t)$.[3] The probability that the unit is replaced at time T is

$$\Pr\{X > T, Y > T\} = \overline{G}(T)\overline{F}(T), \tag{1}$$

the probability that it is replaced at random time Y is

$$\Pr\{Y \leq T, Y \leq X\} = \int_0^T \overline{F}(t)dG(t), \tag{2}$$

and the probability that it is replaced at failure is

$$\Pr\{X \leq T, X \leq Y\} = \int_0^T \overline{G}(t)dF(t), \tag{3}$$

where $(1) + (2) + (3) = 1$. Thus, the mean time to replacement is

$$T\overline{G}(T)\overline{F}(T) + \int_0^T t\overline{F}(t)dG(t) + \int_0^T t\overline{G}(t)dF(t) = \int_0^T \overline{G}(t)\overline{F}(t)dt. \tag{4}$$

Next, introduce the following maintenance costs: Cost c_F is the replacement cost at failure, and c_T and c_N are the respective replacement costs at time T and at random time Y, where $c_F > c_T$ and $c_F > c_N$. Then, the expected cost rate is, from (1)–(4),

$$C_1(T) = \frac{c_T + (c_F - c_T)\int_0^T \overline{G}(t)dF(t) + (c_N - c_T)\int_0^T \overline{F}(t)dG(t)}{\int_0^T \overline{G}(t)\overline{F}(t)dt}. \tag{5}$$

Clearly,

$$C_1(0) \equiv \lim_{T\to 0} C_1(T) = \infty,$$

$$C_1(\infty) \equiv \lim_{T\to\infty} C_1(T) = \frac{c_F - (c_F - c_N)\int_0^\infty \overline{F}(t)dG(t)}{\int_0^\infty \overline{G}(t)\overline{F}(t)dt}. \tag{6}$$

We find an optimal replacement time T^* which minimizes $C_1(T)$ in (5). Differentiating $C_1(T)$ with respect to T and setting it equal to zero,

$$(c_F - c_T)\left[h(T)\int_0^T \overline{G}(t)\overline{F}(t)dt - \int_0^T \overline{G}(t)dF(t)\right]$$

$$- (c_T - c_N)\left[r(T)\int_0^T \overline{G}(t)\overline{F}(t)dt - \int_0^T \overline{F}(t)dG(t)\right] = c_T, \qquad (7)$$

where $r(t) \equiv g(t)/\overline{G}(t)$ and $r(\infty) \equiv \lim_{t\to\infty} r(t)$, and the minimum expected cost rate is

$$C_1(T^*) = (c_F - c_T)h(T^*) - (c_T - c_N)r(T^*). \qquad (8)$$

It is assumed that $c_T \geq c_N$, $h(t)$ increases strictly and $r(t)$ decreases with t. Then, it can be easily seen that the left-hand side of (7) strictly increases with T to

$$(c_F - c_T)\int_0^\infty \overline{G}(t)\overline{F}(t)[h(\infty) - h(t)]dt$$

$$- (c_T - c_N)\int_0^\infty \overline{G}(t)\overline{F}(t)[r(\infty) - r(t)]dt. \qquad (9)$$

Thus, if the function (9) is greater than c_T, then there exists a finite and unique T^* ($0 < T^* < \infty$) which satisfy (7), and the resulting cost rate is given in (8). It is clear that if $h(\infty) = \infty$ then (9) goes to infinity.

In particular, when $G(t) = 1 - e^{-\theta t}$, i.e. $r(t) = \theta$, (7) becomes

$$h(T)\int_0^T e^{-\theta t}\overline{F}(t)dt - \int_0^T e^{-\theta t}dF(t) = \frac{c_T}{c_F - c_T}, \qquad (10)$$

which corresponds to the age replacement policy.[3]

2.2. *Replacement at time T or number N*

It is assumed that Y_j ($j = 1, 2, \ldots$) is the jth working time of the unit (Fig. 1), and is independent and has an identical distribution $G(t) \equiv \Pr\{Y_j \leq t\}$ with finite mean $1/\theta$. That is, the unit works at a renewal process with its distribution $G(t)$. Then, the probability that the unit works exactly j times in the interval $[0, t]$ is $G^{(j)}(t) - G^{(j+1)}(t)$, where $\Phi^{(j)}(t)$ ($j = 1, 2, \ldots$) denote the j-fold Stieltjes convolution of $\Phi(t)$ with itself and $\Phi^{(0)}(t) \equiv 1$ for $t \geq 0$ for any function $\Phi(t)$.

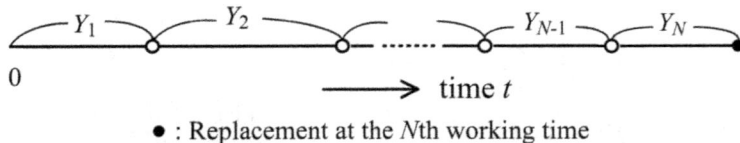

$\bullet$: Replacement at the Nth working time

Fig. 1.   Process of the working times.

Suppose that the unit with a failure time X deteriorates increasingly with its age, i.e. the total working time, irrespective of the number of working times, and fails according a general distribution $F(t) \equiv \Pr\{X \leq t\}$ with finite mean $\mu \equiv \int_0^\infty \overline{F}(t)dt$. When $F(t)$ has a density function $f(t)$, the failure rate $h(t) \equiv f(t)/\overline{F}(t)$ for $F(t) < 1$ is assumed to increase to $h(\infty)$. Similarly, when $G^{(j)}(t)$ has a density function $g^{(j)}(t)$, i.e. $g^{(j)}(t) \equiv dG^{(j)}(t)/dt$, $r_j(t) \equiv g^{(j)}(t)/[1 - G^{(j)}(t)]$. Note that $r_j(t)dt$ represents the probability that the jth work of the unit terminates in the interval $[t, t + dt]$, given that it operates at the jth number of working times.

As the preventive replacement, the unit is replaced before failure at a planned time T $(0 < T < \infty)$ of age or at a planned number N $(N = 1, 2, \ldots)$ of working times, whichever occurs first. Then, the probability that the unit is replaced at time T is

$$\overline{F}(T)[1 - G^{(N)}(T)], \tag{11}$$

the probability that it is replaced at number N is

$$\int_0^T \overline{F}(t)dG^{(N)}(t), \tag{12}$$

and the probability that it is replaced at failure is

$$\int_0^T [1 - G^{(N)}(t)]dF(t). \tag{13}$$

Thus, the mean time to replacement is

$$T\overline{F}(T)[1 - G^{(N)}(T)] + \int_0^T t\,\overline{F}(t)dG^{(N)}(t) + \int_0^T t\,[1 - G^{(N)}(t)]dF(t)$$

$$= \int_0^T [1 - G^{(N)}(t)]\overline{F}(t)dt. \tag{14}$$

Next, introduce the following costs: Cost c_N is the replacement cost at number N, and c_F and c_T are given in (5), where $c_F > c_T$ and $c_F > c_N$. Then, the expected cost rate is, from (11)–(14),

$$C_2(T, N) = \frac{c_T + (c_F - c_T)\int_0^T [1 - G^{(N)}(t)]dF(t) + (c_N - c_T)\int_0^T \overline{F}(t)dG^{(N)}(t)}{\int_0^T [1 - G^{(N)}(t)]\overline{F}(t)dt}. \tag{15}$$

In particular, when the unit is replaced only at time T,

$$C_2(T) \equiv \lim_{N \to \infty} C_2(T, N) = \frac{c_F - (c_F - c_T)\overline{F}(T)}{\int_0^T \overline{F}(t)dt}, \tag{16}$$

which agrees with the result for the standard age replacement.[3,7] When the unit is replaced only at number N,

$$C_2(N) \equiv \lim_{T \to \infty} C_2(T, N) = \frac{c_F - (c_F - c_N)\int_0^\infty \overline{F}(t)dG^{(N)}(t)}{\int_0^\infty [1 - G^{(N)}(t)]\overline{F}(t)dt}. \tag{17}$$

Further, when $N = 1$, $C_2(T, 1)$ agrees with (5).

We discuss optimal policies which minimize the expected cost rates: First, we derive an optimal number N^* which minimizes $C_2(N)$ in (17). From the inequality $C_2(N+1) - C_2(N) \geq 0$,

$$Q_2(N) \int_0^\infty [1 - G^{(N)}(t)]\overline{F}(t)dt - \int_0^\infty [1 - G^{(N)}(t)]dF(t) \geq \frac{c_N}{c_F - c_N}, \tag{18}$$

where

$$Q_2(T, N) \equiv \frac{\int_0^T [G^{(N)}(t) - G^{(N+1)}(t)]dF(t)}{\int_0^T [G^{(N)}(t) - G^{(N+1)}(t)]\overline{F}(t)dt}, \tag{19}$$

and $Q_2(N) \equiv \lim_{T\to\infty} Q_2(T, N)$. It is easily proved that if $Q_2(N)$ increases strictly, then the left-hand side of (18) also increases strictly. Thus, if there exist some N such that (18), an optimal number N^* is given by a finite and unique minimum which satisfies (18).

In particular, when $G(t) = 1 - e^{-\theta t}$, i.e. $G^{(N)}(t) = 1 - \sum_{j=0}^{N-1}[(\theta t)^j/j!]\,e^{-\theta t}$,

$$Q_2(N) \equiv \frac{\int_0^\infty (\theta t)^N e^{-\theta t}dF(t)}{\int_0^\infty (\theta t)^N e^{-\theta t}\overline{F}(t)dt} \tag{20}$$

increases strictly with N to $h(\infty)$.[4] Therefore, if $h(\infty) > c_F/[\mu(c_F - c_N)]$, then there exists a finite and unique minimum N^* $(1 \leq N^* < \infty)$ which satisfies (18).

Second, we discuss optimal T^* and N^* which minimize $C_2(T, N)$ in (15). Differentiating $C_2(T, N)$ with respect to T and setting it equal to zero,

$$(c_F - c_T)\left\{h(T)\int_0^T [1 - G^{(N)}(t)]\overline{F}(t)dt - \int_0^T [1 - G^{(N)}(t)]dF(t)\right\}$$
$$- (c_T - c_N)\left\{r_N(T)\int_0^T [1 - G^{(N)}(t)]\overline{F}(t)dt - \int_0^T \overline{F}(t)dG^{(N)}(t)\right\} = c_T. \tag{21}$$

From the inequality $C_2(T, N+1) - C_2(T, N) \geq 0$,

$$(c_F - c_T)\left\{Q_2(T, N)\int_0^T [1 - G^{(N)}(t)]\overline{F}(t)dt - \int_0^T [1 - G^{(N)}(t)]dF(t)\right\}$$
$$+ (c_T - c_N)\left\{\frac{\int_0^T \overline{F}(t)d[G^{(N)}(t) - G^{(N+1)}(t)]}{\int_0^T [G^{(N)}(t) - G^{(N+1)}(t)]\overline{F}(t)dt} \times \int_0^T [1 - G^{(N)}(t)]\overline{F}(t)dt\right.$$
$$\left. + \int_0^T \overline{F}(t)dG^{(N)}(t)\right\} \geq c_T. \tag{22}$$

In addition, the inequality (22) is, from (21),

$$(c_F - c_T)[Q_2(T, N) - h(T)] + (c_T - c_N)$$
$$\times \left\{\frac{\int_0^T \overline{F}(t)d[G^{(N)}(t) - G^{(N+1)}(t)]}{\int_0^T [G^{(N)}(t) - G^{(N+1)}(t)]\overline{F}(t)dt} + r_N(T)\right\} \geq 0. \tag{23}$$

Thus, when $c_T \le c_N$, there does not exists any finite optimal N^* for any $T > 0$ because $Q_2(T, N) \le h(T)$, i.e. $N^* = \infty$.

Next, we derive optimal T^* and N^* when $G(t) = 1 - e^{-\theta t}$ and $c_T > c_N$. Then, it is easily proved that

$$r_N(T) = \frac{\theta(\theta T)^{N-1}/(N-1)!}{\sum_{j=0}^{N-1}[(\theta T)^j/j!]} \tag{24}$$

decreases strictly with N to 0 and increases strictly with T from 0 to θ for $N \ge 2$. In addition,

$$\frac{\int_0^T [\theta(\theta t)^{N-1}/(N-1)!]e^{-\theta t}\overline{F}(t)dt}{\int_0^T [(\theta t)^N/N!]e^{-\theta t}\overline{F}(t)dt} > \theta \tag{25}$$

for $N > \theta T$. Thus, from (24) and (25), there exists a finite N^* which satisfies (23) for any $T > 0$. Further, because $h(t)$ increases strictly and $r_N(t)$ decreases with t, the left-hand side of (21) increases with T to

$$(c_F - c_T) \int_0^\infty [h(\infty) - h(t)][1 - G^{(N)}(t)]\overline{F}(t)dt$$

$$+ (c_T - c_N) \int_0^\infty \overline{F}(t)dG^{(N)}(t). \tag{26}$$

Thus, if the function (26) is greater than c_T, then there exists a finite and unique T^* $(0 < T^* < \infty)$ which satisfies (21) for any $N \ge 1$. It can be clearly seen that if $h(\infty) = \infty$, then (26) becomes infinity.

Table 1 presents the optimal number N^* which minimize $C_2(N)$ in (17), and T^* and N^* which minimize $C_2(T, N)$ in (15) when $G(t) = 1 - e^{-1}$ and $F(t) = 1 - e^{-(t/10)^2}$. This indicates that optimal N^* for (17) is a constant for different c_T. The optimal T^* for (21) increase, and N^* for (23) decrease with c_T, however are almost constant except for $N^* = \infty$, and become equal to N^* for (17) as c_T becomes larger.

Table 1. Optimal T^* and N^* when $G(t) = 1 - e^{-1}$, $F(t) = 1 - e^{-(t/10)^2}$ and $c_F = 100$.

c_T	$c_N = 10$			$c_N = 15$		
	N^*	T^*	N^*	N^*	T^*	N^*
10.0	4	3.36	∞	5	3.36	∞
11.0	4	3.64	6	5	3.55	∞
12.0	4	4.05	5	5	3.74	∞
13.0	4	4.68	4	5	3.91	∞
14.0	4	5.14	4	5	4.09	∞
15.0	4	5.64	4	5	4.26	∞
16.0	4	6.17	4	5	4.49	8
17.0	4	6.72	4	5	4.92	6
18.0	4	7.32	4	5	5.27	6
19.0	4	7.94	4	5	5.96	5
20.0	4	∞	4	5	∞	5

2.3. *Replacement over time T*

It might be wise to replace practically the unit at the completion of its working time even if T comes because it continues to work for some job. Suppose in the age replacement that the unit is replaced before failure at the Nth ($N = 1, 2, \ldots$) completion of working times or at the first completion over time T, whichever occurs first. Then, the probability that the unit is replaced at number N before time T,

$$\int_0^T \overline{F}(t)dG^{(N)}(t), \tag{27}$$

and the probability that it is replaced at the first completion of working times over time T is

$$\sum_{j=0}^{N-1} \int_0^T \left[\int_{T-t}^\infty \overline{F}(t+u)dG(u) \right] dG^{(j)}(t). \tag{28}$$

Furthermore, the probability that the unit is replaced at failure before time T is

$$\sum_{j=0}^{N-1} \int_0^T [G^{(j)}(t) - G^{(j+1)}(t)]dF(t), \tag{29}$$

and the probability that it is replaced at failure after time T is

$$\sum_{j=0}^{N-1} \int_0^T \left\{ \int_{T-t}^\infty [F(t+u) - F(T)]dG(u) \right\} dG^{(j)}(t). \tag{30}$$

Thus, the mean time to replacement is

$$\int_0^T t\,\overline{F}(t)dG^{(N)}(t) + \int_0^T t\,[1 - G^{(N)}(t)]dF(t)$$

$$+ \sum_{j=0}^{N-1} \int_0^T \left\{ \int_{T-t}^\infty \left[\int_T^{T+u} v\,dF(v) \right] dG(u) \right\} dG^{(j)}(t)$$

$$+ \sum_{j=0}^{N-1} \int_0^T \left[\int_{T-t}^\infty (t+u)\overline{F}(t+u)dG(u) \right] dG^{(j)}(t)$$

$$= \int_0^T [1 - G^{(N)}(t)]\overline{F}(t)dt + \sum_{j=0}^{N-1} \int_0^T \left[\int_T^\infty \overline{G}(u-t)\overline{F}(u)du \right] dG^{(j)}(t). \tag{31}$$

Therefore, the expected cost rate is, from (27)–(31),

$$C_3(T, N) = \frac{\begin{aligned} &c_F - (c_F - c_T)\sum_{j=0}^{N-1}\int_0^T \left[\int_{T-t}^\infty \overline{F}(t+u)dG(u)\right]dG^{(j)}(t) \\ &\quad - (c_F - c_N)\int_0^T \overline{F}(t)dG^{(N)}(t)\end{aligned}}{\begin{aligned}&\int_0^T [1 - G^{(N)}(t)]\overline{F}(t)dt \\ &\quad + \sum_{j=0}^{N-1}\int_0^T \left[\int_T^\infty \overline{G}(u-t)\overline{F}(u)du\right]dG^{(j)}(t)\end{aligned}}, \tag{32}$$

where c_T is the replacement cost over time T, and the other costs are the same ones as in (15).

It is noted that if $N \to \infty$, this corresponds to the replacement model where the unit is replaced at failure or is always replaced at the first completion of working times over time T.

2.4. *Replacement whichever occurs last*

The unit might be working as long as possible in the case where the replacement cost after failure is not so high. In this case, the unit should be replaced at T or N, whichever occurs last.

Suppose that the unit is replaced before failure at a planned time T or at a planned number N, whichever occurs last. Then, the probability that the unit is replaced at number N is

$$\int_T^\infty \overline{F}(t)dG^{(N)}(t), \tag{33}$$

the probability that it is replaced at time T is

$$\overline{F}(T)G^{(N)}(T), \tag{34}$$

and the probability that it is replaced at failure is

$$F(T) + \int_T^\infty [1 - G^{(N)}(t)]dF(t). \tag{35}$$

Thus, the mean time to replacement is, from (33)–(35),

$$\int_T^\infty t\overline{F}(t)dG^{(N)}(t) + T\overline{F}(T)G^{(N)}(T) + \int_0^T tdF(t) + \int_T^\infty t[1 - G^{(N)}(t)]dF(t)$$

$$= \int_0^T \overline{F}(t)dt + \int_T^\infty \overline{F}(t)[1 - G^{(N)}(t)]dt.$$

Therefore, the expected cost rate is

$$C_4(T, N) = \frac{c_F - (c_F - c_T)\overline{F}(T)G^{(N)}(T) - (c_F - c_N)\int_T^\infty \overline{F}(t)dG^{(N)}(t)}{\int_0^T \overline{F}(t)dt + \int_T^\infty \overline{F}(t)[1 - G^{(N)}(t)]dt}. \tag{36}$$

Compared with the expected cost rate $C_2(T, N)$ in (15) when $c_T = c_N$, both numerator and denominator are larger than those in (15). In particular, when $N = 0$, i.e. $G^{(0)}(T) \equiv 1$ for $T \geq 0$, $C_4(T, 0)$ agrees with (16).

3. Other Reliability Models

3.1. *Cumulative damage model*

Suppose that each work causes some damage to the unit in the amount W_j with distribution $K(x) \equiv \Pr\{W_j \leq x\}$. This corresponds to a cumulative damage model,[11] by replacing *shock* with *work*, and the total damage $Z(t)$ at time t is additive. Then, the distribution of $Z(t)$ is, from Ref. 11,

$$\Pr\{Z(t) \le x\} = \sum_{j=0}^{\infty} K^{(j)}(x)[G^{(j)}(t) - G^{(j+1)}(t)]. \tag{37}$$

In addition, it is assumed that the unit fails when the total damage has exceeded a failure level Z.

As the preventive replacement, the unit is replaced before failure at time T and at number N, whichever occurs first. Then, the expected cost rate is, from Ref. 11

$$C_5(T, N) = \frac{c_F - (c_F - c_T)\sum_{j=0}^{N-1}[G^{(j)}(T) - G^{(j+1)}(T)]K^{(j)}(Z) - (c_F - c_N)G^{(N)}(T)K^{(N)}(Z)}{\sum_{j=0}^{N-1} K^{(j)}(Z)\int_0^T [G^{(j)}(t) - G^{(j+1)}(t)]dt}, \tag{38}$$

where c_F is the replacement cost when the total damage exceeds Z. A variety of optimal policies for cumulative damage models were discussed extensively.[11]

3.2. *Scheduling time*

Suppose that we have to finish a specified number N of works and set up on scheduling time:[6] If the Nth work is not accomplished up to scheduling time, its time is prolonged, and this causes a great loss to scheduling. Conversely, if the work is accomplished too early before the scheduling time, this involves a waste of time or cost. The problem is how to determine the scheduling time of N works in advance.[4]

Suppose that the scheduling time for N works is L $(0 \le L < \infty)$ whose cost is $c_0(L)$. If the work is accomplished up to time L, i.e. $L \ge S_N \equiv \sum_{j=1}^{N} Y_j$, it needs the excess cost $c_E(L - S_N)$, and if it is completed after time L, i.e. $L < S_N$, it needs the shortage cost $c_S(S_N - L)$ (Fig. 2). Then, the total expected cost until the Nth work completion is

$$C_6(L, N) = \int_L^{\infty} c_S(t - L)dG^{(N)}(t) + \int_0^L c_E(L - t)dG^{(N)}(t) + c_0(L). \tag{39}$$

When $c_0(L) = c_0 L$, $c_S(t) = c_S t > c_0 t$ and $c_E(t) = c_E t$, (39) is

$$C_6(L, N) = c_S \int_L^{\infty} [1 - G^{(N)}(t)]dt + c_E \int_0^L G^{(N)}(t)dt + c_0 L. \tag{40}$$

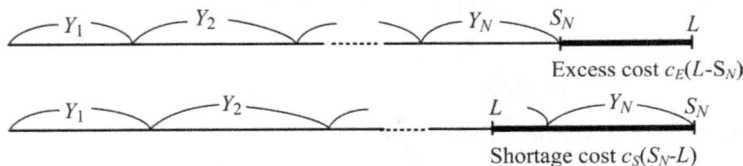

Fig. 2.   Excess and shortage costs of scheduling.

We find an optimal time L^* which minimizes $C_6(L, N)$. Differentiating $C_6(L, N)$ with respect to L and setting it equal to zero,

$$G^{(N)}(L) = \frac{c_S - c_0}{c_S + c_E}. \tag{41}$$

Because $G^{(N)}(t)$ increases with t from 0 to 1 for any $N \geq 1$, there exists a finite and unique L^* $(0 < L^* < \infty)$ which satisfies (41). It can be clearly seen that L^* increase strictly with N. This corresponds to the spare part problem.[7]

For example, when $G(t) = 1 - e^{-\theta t}$, (41) is

$$\sum_{j=N}^{\infty} \frac{(\theta L)^j}{j!} e^{-\theta L} = \frac{c_S - c_0}{c_S + c_E}. \tag{42}$$

Conversely, we consider another problem of presetting the number N of works for a specified time $L > 0$. Then, from the inequality $C_6(L, N+1) - C_6(L, N) \geq 0$,

$$\frac{\int_L^{\infty} [G^{(N)}(t) - G^{(N+1)}(t)] dt}{\int_0^L [G^{(N)}(t) - G^{(N+1)}(t)] dt} \geq \frac{c_E}{c_S}. \tag{43}$$

In particular, when $G(t) = 1 - e^{-\theta t}$, (43) is

$$\sum_{j=0}^{N} \frac{(\theta L)^j}{j!} e^{-\theta L} \geq \frac{c_E}{c_S + c_E}, \tag{44}$$

whose left-hand side increases strictly from $e^{-\theta L}$ to 1. If $e^{-\theta L} \geq c_E/(c_S + c_E)$, then $N^* = 0$, i.e. we should do no work for the interval L.

4. Conclusions

We have considered the age replacement policy where the unit is replaced before failure at time T or at a working time, whichever occurs first. The expected cost rate has been obtained. The extended model in which the unit is replaced at time T or at number N of working times, whichever occurs first, has been proposed and the optimal policy has been analytically discussed. It has been shown that if $c_F > c_T > c_N$, then there exist finite T^* and N^* when the working time is exponential. Furthermore, a numerical example has been given when the time to failure has a Weibull distribution and this result has been verified.

Subsequently, we have proposed two modified age replacement models where the unit is replaced at the first completion of the working time over time T or number N, and it is replaced at time T or number N, whichever occurs last. The expected cost rates for the two models are also derived. Furthermore, we have shown that these models correspond to a cumulative damage model by replacing work with shock. The methods demonstrated in this paper would be applied to other replacement and preventive maintenance policies.[3]

Finally, one optimization problem of how much time we should preset for scheduling the completion of N works has been given. There exist such optimization problems in practical fields, which could be solved by reliability techniques suggested in this paper.

References

1. S. Osaki (ed.), *Stochastic Models in Reliability and Maintenance* (Springer, Berlin, 2002).
2. H. Pham (ed.), *Handbook of Reliability Engineering* (Springer, London, 2003).
3. T. Nakagawa, *Maintenance Theory of Reliability* (Springer, London, 2005).
4. T. Nakagawa, *Advanced Reliability Models and Maintenance Policies* (Springer, London, 2008).
5. T. Sugiura, S. Mizutani and T. Nakagawa, Optimal random replacement policies, in *Proceedings of the Tenth ISSAT International Conference on Reliability and Quality Design* (Las Vegas, 2004), 99–103.
6. M. Pinedo, *Scheduling Theory, Algorithms and Systems* (Prentice Hall, NJ, 2002).
7. R. E. Barlow and F. Proschan, *Mathematical Theory of Reliability* (John Wiley & Sons, New York, 1965).
8. W. Stadje, Renewal analysis of a replacement process, *Operations Research Letters* **31** (2003) 1–6.
9. W. Y. Yun and C. H. Choi, Optimum replacement intervals with random time horizon, *Journal of Quality in Maintenance Engineering* **6** (2000) 269–274.
10. K. Naruse, T. Nakagawa and S. Maeji, Optimal checking intervals for error detection by multiple modular redundancies, in *Proceeding of the Third Asia International Workshop on Advanced Reliability Modeling*, eds. S. H. Sheu and T. Dohi (Taichung, Taiwan) (MC-Graw-Hill, 2008), 670–677.
11. T. Nakagawa, *Shock and Damage Models in Reliability Theory* (Springer, London, 2007).

About the Authors

Mingchih Chen received B.S. degree in Industrial Engineering from Chung-Yuan Christian University in 1988, M.S. and Ph.D. degrees both in Industrial Engineering from Texas A&M University in 1990 and 1993, respectively. She is now with the Industrial Engineering and Management Department, Chaoyang University of Technology as an associated professor. Her research interests include the reliability and maintainability, optimization and operation research.

Satoshi Mizutani was born in Yokkaichi City, Mie, Japan on 1976 June 10. He received Ph.D. degree from Aichi Institute of Technology in 2004. He worked as a Visiting Researcher Staff of Institute of Consumer Sciences and Human Life at Kinjyo Gakuin University in Nagoya City from 2004 to 2006. He is now Assistant Professor of Department of Media Informations at Aichi University of Technology in Gamagori City, Aichi, Japan. His research interests are analysis of optimum inspection policies for Information Systems.

Toshio Nakagawa received BSE and MS degrees from Nagoya Institute of Technology in 1965 and 1967, respectively, and Doctor of Engineering degree from Kyoto

University in 1977. He worked as a Research Associate at Syracuse University from 1972 to 1973. He is now a Professor of Business Administration Department at Aichi Institute of Technology in Toyota City. He has published 3 books from Springer, 6 books chapters and about 200 journal papers. His research interests are optimization problems in Operations Research and Management Science, and analysis for stochastic and computer systems in reliability theory.

© 2025 World Scientific Publishing Company
https://doi.org/10.1142/9789819812547_0010

Chapter 10

ANALYSES OF ACCELERATED LIFE TEST DATA UNDER TWO FAILURE MODES*

C. M. KIM and D. S. BAI

*Department of Industrial Engineering
Korea Advanced Institute of Science and Technology
Gusong-dong 373-1, Yusong-gu, Taejon 3C5-701, Korea*

This paper proposes a method of estimating the lifetime distribution at use condition for constant stress accelerated life tests when an extrinsic failure mode as well as intrinsic one exists. A mixture of two distributions is introduced to describe these failure modes. It is assumed that the log lifetime of each failure mode follows a location-scale distribution and a linear relation exists between the location parameter and the stress. An estimation procedure using the expectation and maximization algorithm is proposed and specific formulas for Weibull distribution are obtained. Simulation studies are performed to investigate the properties of the estimates and the effects of stress level. Numerical comparisons with the masked data model are also performed.

Keywords: Accelerated life test; expectation and maximization algorithm; intrinsic and extrinsic failures; missing variable; maximum likelihood.

1. Introduction

Accelerated life tests (ALTs) of units under higher-than-usual levels of stress involving high temperature, voltage, pressure, vibration, cycle rate or load, are commonly used to quickly obtain information on the lifetime distribution of durable products at use condition stress. Data collected at such accelerated conditions are extrapolated by means of a model to estimate the lifetime distribution at use condition stress. The stress can be applied in various ways; constant, step, and progressive stress loading. Most common method in practice is the constant stress test in which the stress applied to each unit is constant throughout the test.

The inference on ALTs usually assumes that the lifetime distribution at each stress comes from a prespecified parametric family of distributions such as exponential, Weibull or lognormal. See, for instance, Nelson and Meeker[20] for Weibull distribution and Nelson and Kielpinski[19] for lognormal distribution. See Nelson[18] for detailed treatments of ALTs.

Most of previous works assume that there exists only intrinsic failure. However, this assumption may not be appropriate in some populations of electronic

---

*This chapter appeared previously on the International Journal of Reliability, Quality and Safety Engineering. To cite this chapter, please cite the original article as the following: C. M. Kim and D. S. Bai, *Int. J. Reliab. Qual. Saf. Eng*, **9**, 111–125 (2002), doi:S0218539302000706.

devices or other system components since extrinsic failures also exist. Wear-out causes intrinsic failures, while extrinsic failures are attributed to the presence of randomly occurring defects in the manufacturing process. For example, failures due to pinholes, particulates and contaminants in the capacitors are extrinsic failures. Environmental stress screening and burn-in reduce the defect-related failures. However, they cannot eliminate all of them as Sichart and Vollertsen[21] pointed out. Consequently, extrinsic failures have an important effect upon the lifetime distribution at use condition. Therefore, it is necessary to estimate the parameters of extrinsic failure distribution as well as intrinsic one. Mori *et al.*[15] gave an example of capacitor lifetime data which include extrinsic failures. Martin *et al.*[11] monitored intrinsic and extrinsic breakdown properties of capacitors, and Croes *et al.*[4] fitted resistor failure data on log-normal probability paper and found strong indications of two failure modes.

This paper proposes a method of estimating the lifetime distribution at use condition for constant stress ALTs when an extrinsic failure mode as well as intrinsic one exists. A mixture of two distributions is used to describe the two failure modes. A mixed distribution is commonly used for modeling the lives of electrical and mechanical components when failures are caused by more than one failure mode. Assuming that the log lifetime of each failure mode follows a location-scale distribution and its location parameter is a linear function of stress, the maximum likelihood estimates (MLEs) of the distribution parameters and the mixing proportion are obtained by the expectation and maximization (EM) algorithm. The properties of the MLEs are investigated by Monte Carlo simulations. Section 2 describes an ALT model under the situation where intrinsic and extrinsic failure modes coexist. EM algorithm and estimators of the lifetime distribution are presented in Sec. 3. Simulation results on the properties of the estimators are given in Sec. 4. The following notations will be used in this paper.

Notation

h	Number of stress levels.
k	Failure mode index; 1 (intrinsic), 2 (extrinsic).
s_j	jth stress level, $j = 1, \ldots, h$.
s_d	Design stress level.
n_j	Number of test units at stress s_j, $j = 1, \ldots, h$.
n	$\sum_{j=1}^{h} n_j$.
ξ_j	Standardized stress, $\xi_j = \frac{s_j - s_d}{s_h - s_d}$, $j = 1, \ldots, h$.
α_{0k}, α_{1k}	Parameters of linear relation.
β_{0k}, β_{1k}	Parameters of standardized linear relation.
μ_{jk}	Location parameter at stress s_j, $j = 1, \ldots, h$.
σ_k	Scale parameter.
η_j	Censoring time at stress s_j, $j = 1, \ldots, h$.
$f_k(\cdot)$	Location-scale probability density function (pdf).

$F_k(\cdot)$ Location-scale cumulative distribution function (cdf).
$S_k(\cdot)$ Location-scale survival functions.
$S(\cdot)$ Mixture of location-scale survival functions
Θ_k $\{\beta_{0k}, \beta_{1k}, \sigma_k\}$.
Θ $\{\pi_1, \Theta_1, \Theta_2\}$.
Y_{ij} Log-lifetime of unit i under stress s_j, $i = 1, 2, \ldots, n_j$, $j = 1, 2, \ldots, h$.
π_1, π_2 Mixing proportions of intrinsic and extrinsic failure modes, $\pi_1 + \pi_2 = 1$.

2. The Model

2.1. *Assumptions*

(1) At any stress s_j, the log-lifetime of a test unit follows a mixture of two distributions with location and scale parameters, μ_{jk} and σ_k, $k = 1$ (intrinsic), 2 (extrinsic).
(2) μ_{jk} is a linear function of a (possibly transformed) stress s_j; that is, $\mu_{jk} = \alpha_{0k} + \alpha_{1k}s_j$.
(3) σ_k is constant and is independent of the stress.
(4) The lifetimes of test units are independent and identically distributed.

2.2. *Standardized model*

The location parameters of log-lifetime distributions of test units at stress s_j can be rewritten in terms of ξ_j as $\mu_{jk} = \beta_{0k} + \beta_{1k}\xi_j$ where $\beta_{0k} = \alpha_{0k} + \alpha_{1k}s_d$ and $\beta_{1k} = \alpha_{1k}(s_h - s_d)$. We note that for $s = s_d$, $\xi_d = 0$ and $\mu_{0k} = \beta_{0k}$ and for $s = s_h$, $\xi_h = 1$ and $\mu_{1k} = \beta_{0k} + \beta_{1k}$.

2.3. *Lifetime distribution*

From the assumption of the mixture of two distributions, the pdf of Y_{ij}, $i = 1, 2, \ldots, n_j$, $j = 1, 2, \ldots, h$, is:

$$f(y_{ij}; \Theta) = \pi_1 f_1(y_{ij}; \Theta_1) + \pi_2 f_2(y_{ij}; \Theta_2). \tag{2.1}$$

where y_{ij} is the realization of Y_{ij}.

3. Estimation with EM Algorithm

When more than one failure mode exist, the pdf of the time to failure can be multimodal, and a mixed distribution is a good candidate to describe the times to failure. See, for instance, Mann *et al.* (Ref. 10, Chap. 4), Lawless (Ref. 9, Chap. 5) and Moosa *et al.*[14] In the mixed distribution, however, one cannot obtain an explicit solution to the maximum likelihood equation for the mixing proportion since the number of intrinsic failures is unobservable. It can be regarded as a missing variable and the EM algorithm can be utilized.

3.1. *EM algorithm*

The EM algorithm is a powerful tool for computing MLEs with missing data. On each iteration of the EM algorithm there are two steps; the expectation step (E-step) and the maximization step (M-step). In the E-step, log-likelihood including missing data is replaced by its conditional expectation given the observed data. In the M-step, MLEs of the parameters are computed which maximize the conditional expectation of the log-likelihood calculated in the expectation step. See Dempster *et al.*[5] and Mclachlan and Krishnan[12] for details.

Several authors have considered the problem of estimating MLEs of the mixed distribution with the EM algorithm. See, for instance, Aitkin and Wilson[1] for normal or lognormal distribution and Jiang and Kececioglu[7] for Weibull distribution. However, all these works considered the situation where the data are obtained at use condition and no ALT model was used.

In this section, an estimation procedure using the EM algorithm is proposed for the data obtained at higher-than-use condition when an extrinsic failure mode as well as intrinsic one exists.

3.2. *Estimation procedure for ALT data*

When the data at stress j consist of r_j failure times and $(n_j - r_j)$ censoring times out of n_j units tested, the log-likelihood becomes

$$\log L = \sum_{j=1}^{h} \left[\sum_{i=1}^{r_j} \log f(y_{ij}; \Theta) + (n_j - r_j) \log S(\eta_j; \Theta) \right]$$

$$= \sum_{j=1}^{h} \left[\sum_{i=1}^{r_j} \log(\pi_1 f_1(y_{ij}; \Theta) + \pi_2 f_2(y_{ij}; \Theta_2)) \right.$$

$$\left. + (n_j - r_j) \log(\pi_1 S_1(\eta_j; \Theta_1) + \pi_2 S_2(\eta_j; \Theta_2)) \right]. \tag{3.1}$$

To estimate π_1, we need to know the number of data from intrinsic failure mode. It is, however, unobservable or missing, since we only get to observe the sum of intrinsic and extrinsic failures.

Let I_{ij1} and $I_{ij2}(= 1 - I_{ij1})$ be the indicator variables denoting whether unit i at stress j follows intrinsic or extrinsic failure mode, respectively. If these I_{ijk}'s were observable, then the log-likelihood of a complete data set would become

$$\log L_c = \sum_{j=1}^{h} \sum_{k=1}^{2} \left\{ \sum_{i=1}^{r_j} I_{ijk}[\log f_k(y_{ij}; \Theta_k) + \log \pi_k] \right.$$

$$\left. + (n_j - r_j) I_{ijk}[\log S_k(\eta_j; \Theta_k) + \log \pi_k] \right\}. \tag{3.2}$$

However, here, I_{ijk}'s are the missing variables.

E-step: As (3.2) is linear in I_{ijk}'s, the expectation step simply requires the calculation of the conditional expectation of I_{ijk} given the observed data y_{ij}. We have

$$\Pr\{I_{ijk} = 1|y_{ij}\} = \frac{\pi_k f_k(y_{ij}; \Theta_k)}{f(y_{ij}; \Theta)} \equiv \tau_k(y_{ij}; \Theta), \qquad (3.3)$$

for $i = 1, 2, \ldots, r_j$, $j = 1, 2, \ldots, h$ and $k = 1, 2$. The quantity $\tau_1(y_{ij}; \Theta) = 1 - \tau_2(y_{ij}; \Theta)$ is the posterior probability that the failure is intrinsic given y_{ij}. Similarly, for $(n_j - r_j)$ the censored units, the conditional probability of a unit's belonging to intrinsic failure mode, given the censoring time η_j, is

$$\Pr\{I_{ijk} = 1|\eta_j\} = \frac{\pi_k S_k(\eta_j; \Theta_k)}{S(\eta_j; \Theta)} \equiv \tau_k(\eta_j; \Theta). \qquad (3.4)$$

Thus the conditional expectation of I_{ijk} given y_{ij} on the pth iteration is

$$E_{\Theta^{(p-1)}}(I_{ijk}|y_{ij}) = \begin{cases} \tau_k(y_{ij}; \Theta^{(p-1)}) & \text{if a unit fails at } y_{ij} \\ \tau_k(\eta_j; \Theta^{(p-1)}) & \text{if a unit is censored at } \eta_j, \end{cases}$$

where $\Theta^{(p-1)}$ is the parameter set obtained on the $(p-1)$th iteration. Thus the conditional expectation of log-likelihood is

$$Q(\Theta; \Theta^{(p-1)}) = \sum_{j=1}^{h} \sum_{k=1}^{2} \left\{ \sum_{i=1}^{r_j} \tau_k(y_{ij}; \Theta^{(p-1)})[\log f_k(y_{ij}; \Theta_k) + \log \pi_k] \right.$$

$$\left. + (n_j - r_j)\tau_k(\eta_j; \Theta^{(p-1)})[\log S_k(\eta_j; \Theta_k) + \log \pi_k] \right\}. \qquad (3.5)$$

M-step: At the M-step of the pth iteration, the intent is to maximize $Q(\Theta; \Theta^{(p-1)})$ with respect to Θ to produce a new estimate $\hat{\Theta}^{(p)}$ of Θ. In particular,

$$\hat{\pi}_1^{(p)} = \frac{\sum_{j=1}^{h} \left\{ \sum_{i=1}^{r_j} \tau_1(y_{ij}; \hat{\Theta}^{(p-1)}) + (n_j - r_j)\tau_1(\eta_j; \hat{\Theta}^{(p-1)}) \right\}}{n}. \qquad (3.6)$$

The values $(\hat{\Theta}_1^{(p)}, \hat{\Theta}_2^{(p)})$ can be obtained by solving maximum likelihood equations obtained from (3.5).

As the iteration of the expectation and the maximization steps progresses, $\hat{\Theta}$ converges to the stationary solution. Wu[23] showed that if the likelihood function is unimodal, the stationary solution of the algorithm is the unique MLE. Even if the likelihood function is not unimodal, MLE could still be obtained by choosing the solution with the largest maximum among the local maximums (Jiang and Kececioglu).[7]

The Fisher information matrix $I(\Theta)$ is

$$I(\Theta) = E\left[-\frac{\partial^2 \log L}{\partial \Theta \partial \Theta'} \right], \qquad (3.7)$$

where Θ' is the transpose of parameter vector Θ, and asymptotic variance and covariance matrix is the inverse of $I(\Theta)$.

The qth quantile of the lifetime distribution, t_q, is of great interest in many applications of ALTs and it satisfies the following equation.

$$F(t_q) = \pi_1 F_1(t_q) + \pi_2 F_2(t_q) = q \qquad (3.8)$$

In the mixture of two distributions, qth quantile can only be obtained numerically. Therefore, the variance and covariance matrix of $\hat{\Theta}$ cannot be used to compute the confidence interval of t_q.

One way of solving this problem is the random walk approximation of confidence interval suggested by Murdoch.[17] The basic idea is to approximate the target confidence region by generating many uniformly distributed points within confidence region. He proposed two methods of calculating the confidence interval with the approximate confidence region obtained from the random walk approximation. One is a simple approach that takes the observed maximum and minimum values in this region as the limits of the confidence interval. The other method uses an extrapolated quantile-quantile (Q-Q) plot.

Although the extrapolated Q-Q plot gives more accurate endpoints of the confidence interval when the number of unknown parameter is large, we found that in our case the confidence intervals of t_q from the maximum-minimum approach are fairly close to the ones obtained by the extrapolated Q-Q plot. Therefore we take this maximum-minimum approach to obtain the confidence interval of t_q. See Murdoch[17] and Chan and Meeker[3] for details.

3.3. Example

If the lifetime T follows a Weibull distribution with scale parameter λ and shape parameter δ, then the log lifetime $Y = \log T$ has an extreme value distribution with location parameter $\mu = \log \lambda$ and scale parameter $\sigma = 1/\delta$. Under ALT situation, a mixture of two extreme value distributions has a parameter set $\Theta = \{\pi_1, \beta_{01}, \beta_{11}, \sigma_1, \beta_{02}, \beta_{12}, \sigma_2\}$.

The proposed method is illustrated with the ALT data generated from a mixture of two extreme value distributions with parameters;

$$\pi_1 = 0.8\,;$$

$$\beta_{01} = 16\,, \quad \beta_{11} = -6\,, \quad \sigma_1 = 1.0\,;$$

$$\beta_{02} = 12\,, \quad \beta_{12} = -8\,, \quad \sigma_2 = 0.5\,;$$

$$\eta_1 = 13\,, \quad \eta_2 = 10\,;$$

In this example, the simple stress ALT is considered. Given $\xi_1 = 0.5$, $\xi_2 = 1.0$, $n_1 = 40$ and $n_2 = 20$, Table 1 contains the failure and censoring times in minutes, rounded to one decimal point, under each stress level.

Table 1.   Failure times with censoring: Weibull case.

Low stress				High stress	
8.6	12.0	9.0	7.8	4.4	9.7
12.8	10.4	11.7	*13.0	9.0	*10.0
*13.0	7.8	*13.0	7.8	3.6	9.4
12.1	12.1	10.9	*13.0	3.9	4.6
10.3	12.0	12.8	8.5	2.9	9.6
12.6	7.3	12.7	13.0	*10.0	3.4
*13.0	*13.0	*13.0	10.6	9.6	7.8
12.5	*13.0	*13.0	12.7	8.2	8.2
*13.0	11.7	10.8	8.5	8.2	*10.0
*13.0	8.0	8.5	11.7	9.9	*10.0

'*' denotes censored observation.

Initial step: Initial estimates $\pi_1^{(0)} = 0.5$, $\beta_{01}^{(0)} = 20$, $\beta_{11}^{(0)} = -4$, $\sigma_1^{(0)} = 1.0$, $\beta_{02}^{(0)} = 11$, $\beta_{12}^{(0)} = -9$ and $\sigma_2^{(0)} = 1.0$ are chosen.

E-step: With the initial estimates, $\tau_k(y_{ij}; \Theta^{(0)})$ can be computed from (3.3) for $i = 1, 2, \ldots, n_j$, $j = 1, 2, \ldots, h$ and $k = 1, 2$. For example,

$$\tau_1(y_{11}; \Theta^{(0)}) = \frac{\pi_1^{(0)} f_1(y_{11}; \Theta_1^{(0)})}{f(y_{11}; \Theta^{(0)})} = 0.0363.$$

M-step: The first partial derivatives of (3.5) for the extreme value distribution can be found in Kim and Bai.[8] With $\tau_1(y_{ij}; \Theta^{(0)})$ and $\tau_2(y_{ij}; \Theta^{(0)})$, $\Theta_1^{(1)}$ and $\Theta_2^{(1)}$ can be obtained by solving maximum likelihood equations in Kim and Bai[8] using a numerical method such as Newton–Rapson algorithm. In particular, $\pi_1^{(1)}$ can be obtained from formula (3.6).

Computations are iterated until the differences between $(p-1)$th and pth value of parameters are smaller than 10^{-5}.

The stationary solutions

$$\hat{\pi}_1 = 0.7575,$$

$$\hat{\beta}_{01} = 16.1600, \quad \hat{\beta}_{11} = -6.3916, \quad \hat{\sigma}_1 = 0.8913,$$

$$\hat{\beta}_{02} = 12.6479, \quad \hat{\beta}_{12} = -8.5137, \quad \hat{\sigma}_2 = 0.4664,$$

are obtained after 88 iterations and 0.7 seconds on a Pentium III PC with 500 Mhz and the estimates of qth quantile of the lifetime distribution at use condition are

$$\hat{t}_{0.01} = 11.0390, \quad \hat{t}_{0.05} = 11.8942.$$

For a mixture of two extreme value distributions, see Kim and Bai[8] for the elements of the Fisher information matrix. Using the formulas in Kim and Bai,[8] we obtain the following Fisher information matrix with $n_1 = 40$ and $n_2 = 20$.

$$I(\pi_1, \beta_{01}, \beta_{11}, \sigma_1, \beta_{02}, \beta_{12}, \sigma_2)$$

$$= \begin{bmatrix} 312.2 & 3.0 & 1.7 & -12.9 & 6.2 & 3.7 & 9.1 \\ & 37.9 & 25.8 & -4.2 & -1.3 & -0.8 & -1.9 \\ & & 19.8 & -2.4 & -0.8 & -0.5 & -1.1 \\ & & & 37.7 & 3.9 & 2.5 & 7.6 \\ & & & & 55.2 & 37.6 & 13.3 \\ & & & & & 28.7 & 9.6 \\ & & & & & & 93.3 \end{bmatrix}.$$

The asymptotic variances are

$$As\,\mathrm{varr}(\hat{\pi}_1) = 0.0033\,,$$

$$As\,\mathrm{varr}(\hat{\beta}_{01}) = 0.2401\,, \quad As\,\mathrm{varr}(\hat{\beta}_{11}) = 0.4575\,, \quad As\,\mathrm{varr}(\hat{\sigma}_1) = 0.0279\,,$$

$$As\,\mathrm{varr}(\hat{\beta}_{02}) = 0.1642\,, \quad As\,\mathrm{varr}(\hat{\beta}_{12}) = 0.3151\,, \quad As\,\mathrm{varr}(\hat{\sigma}_2) = 0.0113\,,$$

and 95% confidence intervals of the parameters are

$$0.6449 \le \pi_1 \le 0.8701\,,$$

$$15.1996 \le \beta_{01} \le 17.1204\,, \quad -7.7173 \le \beta_{11} \le -5.0659\,, \quad 0.5639 \le \sigma_1 \le 1.2187\,,$$

$$11.8537 \le \beta_{02} \le 13.4421\,, \quad -9.6139 \le \beta_{12} \le -7.4135\,, \quad 0.2580 \le \sigma_2 \le 0.6748\,.$$

To obtain confidence intervals for t_q with the random walk approximation, many uniformly distributed points are generated. Murdoch[17] lists the required number of points depending on the number of unknown parameters to attain 2-digit accuracy. Our model is a 7-parameter model and the required number of simulated trials is 2×10^6. The confidence intervals of t_q calculated by the maximum-minimum approach are

$$9.8572 \le t_{0.01} \le 11.8042\,, \quad 10.9344 \le t_{0.05} \le 12.6186\,.$$

4. Simulation Study

In this section, finite sample properties of the estimators of parameters and qth quantile are investigated by Monte Carlo simulation. The estimators obtained by the proposed method are also compared with ones by assuming that the causes of failure are masked. To obtain mixture data, we generate Weibull random variates with location parameters $\mu_{jk} = \beta_{0k} + \beta_{1k}\xi_j$ and scale parameters σ_k, $k = 1, 2$, for intrinsic and extrinsic failure modes and then mix them in the ratio of π_1 to π_2.

4.1. *Properties of the estimators*

In practice, the degree of separation between intrinsic and extrinsic pdfs is very important in estimating the parameters (Jiang and Kececioglu).[8] Thus we investigate the effect of the distance (D) between intrinsic and extrinsic location parameters at use condition.

Given two levels of stress, $h = 2$, and with $\xi_1 = 0.5$ and $\xi_2 = 1.0$, simulation studies are performed using the iterative equations in Sec. 3. 10,000 estimates were computed with censoring times $\eta_1 = 13$ and $\eta_2 = 10$, and deviations and squared deviations of the estimates from true values were averaged to obtain (estimated) biases and MSEs for the following parameters:

$$n_1 = 400, \quad n_2 = 200;$$

$$\beta_{01} = 16.0, \quad \beta_{11} = -6.0, \quad \sigma_1 = 1.0, \quad \beta_{12} = -8.0, \quad \sigma_2 = 0.5;$$

$$\pi_1 = 0.7, 0.8, 0.9, \quad D = 2.0, 3.0, 4.0;$$

where $D = \beta_{01} - \beta_{02}$.

Since the likelihood in the proposed model is not unimodal, maximizing the log-likelihood may result in multiple solutions depending on the starting points. Thus the iteration began from 4 different initial points for each set of data and we choose the largest local maximum among the solutions by using the method of Jiang and Kececioglu.[7] The results, rounded to 4 decimal points, are shown in Table 2.

One can see from the table that:

(1) The MSEs of parameters for intrinsic (extrinsic) failure distribution decrease as π_1 (π_2) increases.
(2) The average MSEs become smaller as D increases, implying that the estimators are more accurate when the discriminations between intrinsic and extrinsic failures are clearer.

We also investigated the properties of qth quantile estimator of the lifetime distribution which is of interest in many applications of ALTs. Table 3 shows biases and MSEs, rounded to 3 decimal points, of $\hat{t}_{0.01}$ and $\hat{t}_{0.05}$, from which we note that:

(1) MSEs of $\hat{t}_{0.01}$ and $\hat{t}_{0.05}$ decrease as π_1 (π_2) decreases (increases) since early failures are mostly extrinsic ones.
(2) In some cases, MSEs of $\hat{t}_q$ do not decrease as D increases. The reason is that the variance of $\hat{t}_q$ is affected by covariances as well as variances of the parameter estimates.
(3) Under given parameter sets, low stress levels that minimize the MSEs of $\hat{t}_q$ are 0.1 when $D = 4$ and 0.3 when $D = 2$. That is, as the probability that a unit fails at use condition gets larger, the optimum low stress decreases. This results agree with those of Meeker and Nelson[13] and Bai and Kim.[2]

4.2. *Comparisons with the masked data model*

The problem of estimating component reliability from system-life data has usually been addressed by making a series-system assumption and applying a

Table 2.   Performance of the estimators.

D	π_1	Biases							MSEs						
		π_1	β_{01}	β_{11}	σ_1	β_{02}	β_{12}	σ_2	π_1	β_{01}	β_{11}	σ_1	β_{02}	β_{12}	σ_2
2	0.7	0.0011	0.0015	−0.0013	0.0051	0.0020	−0.0036	0.0045	0.0007	0.0377	0.0709	0.0079	0.0221	0.0397	0.0018
	0.8	0.0008	0.0022	−0.0033	0.0049	0.0012	−0.0029	0.0065	0.0006	0.0321	0.0604	0.0067	0.0407	0.0718	0.0034
	0.9	0.0026	−0.0007	−0.0014	0.0081	−0.0111	0.0031	0.0239	0.0006	0.0297	0.0564	0.0066	0.1520	0.2628	0.0139
3	0.7	0.0003	0.0023	−0.0025	0.0012	0.0036	−0.0033	0.0042	0.0004	0.0353	0.0683	0.0055	0.0173	0.0319	0.0012
	0.8	−0.0003	0.0005	−0.0072	0.0011	0.0056	−0.0053	0.0048	0.0003	0.0294	0.0563	0.0045	0.0291	0.0529	0.0020
	0.9	0.0006	0.0015	−0.0027	0.0034	0.0004	0.0005	0.0159	0.0003	0.0283	0.0550	0.0045	0.0885	0.1621	0.0068
4	0.7	0.0005	0.0023	−0.0028	0.0012	0.0034	−0.0028	0.0031	0.0003	0.0336	0.0652	0.0041	0.0151	0.0286	0.0010
	0.8	−0.0005	0.0051	−0.0079	0.0003	0.0057	−0.0051	0.0031	0.0003	0.0282	0.0543	0.0034	0.0241	0.0449	0.0015
	0.9	0.0002	0.0016	−0.0025	0.0021	0.0047	−0.0019	0.0123	0.0002	0.0276	0.0541	0.0034	0.0631	0.1200	0.0045

Table 3. Biases and MSEs of t_q.

D	π	Low stress	$t_{0.01}$ Biases	$t_{0.01}$ MSEs	$t_{0.05}$ Biases	$t_{0.05}$ MSEs
2	0.7	0.1	0.008	0.149	0.037	0.036
		0.3	0.019	0.087	−0.001	0.018
		0.5	0.033	0.093	0.011	0.026
		0.7	0.051	0.122	0.036	0.061
		0.9	0.293	0.973	0.271	0.764
	0.8	0.1	−0.006	0.152	0.039	0.039
		0.3	0.014	0.097	−0.001	0.023
		0.5	0.030	0.101	0.014	0.033
		0.7	0.046	0.146	0.040	0.084
		0.9	0.313	1.113	0.334	1.021
	0.9	0.1	−0.013	0.133	0.031	0.040
		0.3	0.011	0.110	0.000	0.035
		0.5	0.019	0.122	0.015	0.053
		0.7	0.052	0.177	0.068	0.139
		0.9	0.449	1.701	0.538	1.905
3	0.7	0.1	0.009	0.042	−0.004	0.012
		0.3	0.010	0.029	−0.004	0.011
		0.5	0.015	0.032	0.003	0.018
		0.7	0.033	0.058	0.018	0.047
		0.9	0.240	0.657	0.159	0.550
	0.8	0.1	0.006	0.052	−0.007	0.016
		0.3	0.012	0.036	−0.005	0.015
		0.5	0.021	0.040	0.006	0.024
		0.7	0.037	0.081	0.018	0.071
		0.9	0.297	0.859	0.220	0.801
	0.9	0.1	0.006	0.069	−0.010	0.025
		0.3	0.010	0.053	−0.010	0.026
		0.5	0.018	0.066	0.002	0.047
		0.7	0.057	0.135	0.038	0.140
		0.9	0.464	1.565	0.413	1.660
4	0.7	0.1	−0.007	0.017	−0.006	0.008
		0.3	−0.005	0.019	−0.005	0.011
		0.5	0.001	0.024	0.000	0.017
		0.7	0.018	0.048	0.014	0.045
		0.9	0.138	0.495	0.080	0.483
	0.8	0.1	−0.008	0.022	−0.007	0.011
		0.3	−0.005	0.024	−0.005	0.014
		0.5	0.006	0.031	0.003	0.024
		0.7	0.012	0.075	0.007	0.073
		0.9	0.191	0.733	0.116	0.754
	0.9	0.1	−0.019	0.035	−0.011	0.019
		0.3	−0.014	0.039	−0.008	0.026
		0.5	−0.005	0.058	0.000	0.050
		0.7	0.023	0.142	0.016	0.154
		0.9	0.353	1.466	0.257	1.574

competing-risk model. The observable quantities of interest are system-life and the exact component causing failure. In practice, however, this type of analysis is often confounded by the problem of masking. That is, the exact cause of system failure is unknown. See Usher[22] and Mukhopadhyay and Basu[16] for detailed treatment of the masked data. All these works on the masked data problem considered the situation where the data are obtained at use condition.

When there exist two causes of failure and the data are masked, the likelihood at higher-than-use condition becomes

$$
\log L = \sum_{j=1}^{h} \left[\sum_{i=1}^{r_j} \log(f_1(y_{ij}; \Theta) S_2(y_{ij}; \Theta) + f_2(y_{ij}; \Theta) S_1(y_{ij}; \Theta)) \right.
$$

$$
\left. + (n_j - r_j) \log S_1(\eta_j; \Theta) S_2(\eta_j; \Theta) \right].
\tag{4.1}
$$

For the method of obtaining the MLEs by EM algorithm, see Kim and Bai.[8]

In this section, with the data generated from the mixture of two Weibull distributions, the qth quantile estimators obtained by the proposed method in Sec. 3 is compared, with respect to biases and MLEs, with ones by assuming that the causes of failure are masked. Table 4 shows biases and MSEs, rounded to 3 decimal points, of qth quantile estimator.

It can be seen from the table that:

(1) The biases and MSEs of $\hat{t}_q$ obtained under the masked data model are found to be very large when the data are generated from the mixed distribution.

Table 4.   Comparisons with the masked data model.

(1) Biases and MSEs of $\hat{t}_{0.01}$

D	π_1	Mixed distribution Biases	MSEs	Masked data model Biases	MSEs
2	0.7	0.033	0.093	3.501	12.358
	0.8	0.030	0.101	2.553	6.647
	0.9	0.019	0.122	1.418	2.230
3	0.7	0.015	0.032	4.847	23.659
	0.8	0.021	0.040	3.822	14.805
	0.9	0.018	0.066	2.372	5.842
4	0.7	0.001	0.024	6.439	41.729
	0.8	0.006	0.031	5.083	26.193
	0.9	−0.005	0.058	3.178	10.511

(2) Biases and MSEs of $\hat{t}_{0.05}$

D	π_1	Mixed distribution Biases	MSEs	Masked data model Biases	MSEs
2	0.7	0.011	0.026	2.060	4.322
	0.8	0.014	0.033	1.474	2.283
	0.9	0.015	0.053	0.815	0.869
3	0.7	0.003	0.018	2.400	5.907
	0.8	0.006	0.024	1.761	3.266
	0.9	0.002	0.047	0.911	0.983
4	0.7	0.000	0.017	3.030	9.411
	0.8	0.003	0.024	2.043	4.437
	0.9	0.000	0.050	0.816	0.987

f(y)

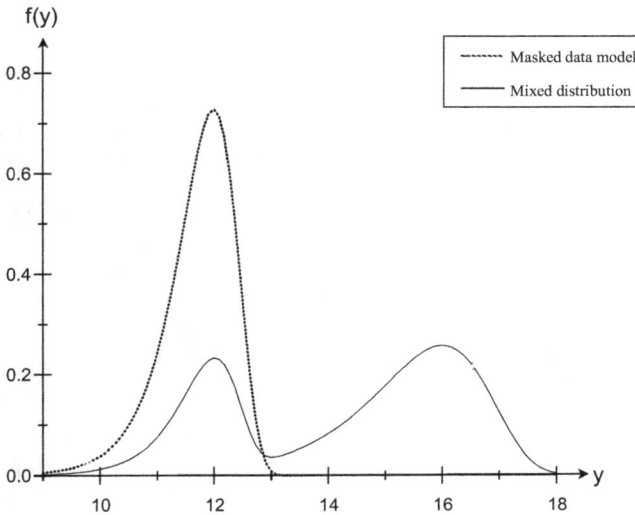

Fig. 1.   The pdfs of mixed distribution and the masked data model ($D = 4, p = 0.7$).

(2) MSEs of $\hat{t}_q$ obtained from the masked data model decreases as D decreases or π_1 increases. Figure 1 shows the pdfs of the mixed distribution and the masked data model. Since the pdf of the masked data model is unimodal, under the masked data model, the estimators from the mixed data with the weak degree of the bimodality are more accurate than those with the strong one.

5. Conclusions

We have proposed a method of estimating the lifetime distribution from the data obtained at higher-than-use condition when an extrinsic failure mode as well as intrinsic one exists. EM algorithm is used to estimate the parameters of the lifetime distributions and the mixing proportion simultaneously. Although we used Weibull distributions to demonstrate the method, it can be used for any location-scale distributions. Monte Carlo simulations show that the mixing proportion and the degree of separation between intrinsic and extrinsic failure modes are important in estimating the parameters, and the stress levels also have an important effect upon the estimators of qth quantile of the mixed lifetime distribution at use condition.

The accuracy of estimates greatly depends on stress levels and proportion of units tested at each stress level. Therefore, the problem of optimally designing ALTs under the two failure modes can be considered. A general limited failure population model (Chan and Meeker[3]) can be also considered for representing the situation where two failure modes coexist.

References

1. M. Aitkin and G. T. Wilson, "Mixture models, outliers, and the EM algorithm," *Technometrics* **22** (1980), pp. 325–331.

2. D. S. Bai and M. S. Kim, "Optimum simple step-stress accelerated life tests for Weibull distribution and type I censoring," *Naval Research Logistics* **40** (1993), pp. 193–210.
3. V. Chan and W. Q. Meeker, "A failure-time model for infant-mortaility and wearout failure modes," *IEEE Trans. Reliability* **48** (1999), pp. 377–387.
4. K. Croes, W. De Ceuninck, L. De Schepper and L. Tielemans, "Bimodal failure behavior of metal film resistors," *Quality and Reliability Engineering International* **14** (1998), pp. 87–90.
5. A. P. Dempster, N. M. Laird and D. R. Rubin, "Maximum likelihood from incomplete data," *J. Royal Statistical Society, Series B* **39** (1977), pp. 1–38.
6. S. Jiang and D. Kececioglu, "Graphical representation of two mixed-Weibull distributions," *IEEE Trans. Reliability* **41** (1992), pp. 241–247.
7. S. Jiang and D. Kececioglu, "Maximum likelihood estimates, from censored data, for mixed-Weibull distributions," *IEEE Trans. Reliability* **41** (1992), pp. 248–255.
8. C. M. Kim and D. S. Bai, "Analyses of accelerated life test data with EM algorithm," Technical report #2001-20 (Korea Advanced Institute of Science and Technology, Taejon, Korea).
9. J. F. Lawless, *Statistical Models and Methods for Lifetime Data* (John and Wiley & Sons, New York, 1982).
10. N. R. Mann, R. E. Schafer and N. D. Singpurwalla, *Methods for Statistical Analysis of Reliability and Life Data* (John and Wiley & Sons, New York, 1974).
11. A. Martin, P. O'sullivan and A. Mathewson, "Study of unipolar pulsed ramp and combined ramped/constant voltage stress on mos gate oxides," *Microelectron. Reliab.* **37** (1997), pp. 1045–1051.
12. G. J. Mclachlan and T. Krishnan, *The EM algorithm and Extensions* (John and Wiley & Sons, New York, 1997).
13. W. Q. Meeker and W. Nelson, "Optimum accelerated life-tests for the Weibull and extreme value distributions," *IEEE Trans. Reliability* **24** (1975), pp. 321–332.
14. M. S. Moosa, K. F. Poole and M. L. Grams, "EFSIM: An integrated circuit early failure simulator," *Quality and Reliability Engineering International* **12** (1996), pp. 229–234.
15. S. Mori, N. Arai, Y. Kaneko and K. Yoshikawa, "Polyoxide thinning limitation and superior ONO interpoly dielectric of nonvolatile memory devices," *IEEE Trans. Electron Devices* **38** (1991), pp. 270–276.
16. C. Mukhopadhyay and A. P. Basu, "Bayesian analysis of incomplete time and cause of failure data," *Journal of Statistical Planning and Inference* **59** (1997), pp. 79–100.
17. D. J. Murdoch, "Random walk approximation of confidence intervals," *Quality Improvement Through Statistical Methods* (1998), pp. 393–404.
18. W. Nelson, *Accelerated Testing: Statistical Models, Test Plans, and Data Analyses* (John and Wiley & Sons, New York, 1990).
19. W. Nelson and T. J. Kielpinski, "Theory for optimum censored accelerated life tests for normal and lognormal distributions," *Technometrics* **18** (1976), pp. 105–114.
20. W. Nelson and W. Q. Meeker, "Theory for optimum accelerated censored life tests for Weibull and extreme value distributions," *Technometrics* **20** (1978), pp. 171–177.
21. K. V. Sichart and R. P. Vollertsen, "Bimodal lifetime distributions of dielectrics for integrated circuits," *Quality and Reliability Engineering International* **7** (1991), pp. 299–305.
22. J. S. Usher, "Weibull component reliability-prediction in the presence of masked data," *IEEE Trans. Reliability* **45** (1996), pp. 229–232.
23. C. Wu, "On the convergence property of the EM algorithm," *Annals of Statistics* **11** (1983), pp. 95–103.

About the Authors

C. M. Kim received a Ph.D. in Industrial Engineering from Korea Advanced Institute of Science and Technology (KAIST) in 2001. He is currently a Senior Consultant in LG-CNS Inc. His research interests include accelerated life testing and lifetime data analysis.

D. S. Bai received a Ph.D. in statistics from Ohio State University in 1971. He taught statistics at the Ohio State University in 1971–1972 and 1981–1982 and at the State University of New York at Oneonta in 1972–1974. Since 1974 he has been a professor at KAIST. His research interests include statistical inference in stochastic processes, life testing and survival analysis, reliability theory, and quality control. He is a senior member of ASQ.

© 2025 World Scientific Publishing Company
https://doi.org/10.1142/9789819812547_0011

Chapter 11

THE EXACT RUN LENGTH DISTRIBUTION AND DESIGN OF THE S^2 CHART WHEN THE IN-CONTROL VARIANCE IS ESTIMATED[#]

PHILIPPE CASTAGLIOLA*

*Université de Nantes & IRCCyN UMR CNRS 6597
44475 Carquefou, France
philippe.castagliola@univ-nantes.fr*

GIOVANNI CELANO

*Università di Catania, Catania, Italy
gcelano@diim.unict.it*

GEMAI CHEN

*University of Calgary, Calgary, Alberta, T2N 1N4, Canada
gchen@math.ucalgary.ca*

When monitoring the process variability, it is a common practice that a Phase I data set is used to estimate the unknown in-control process standard deviation σ_0 or variance σ_0^2 to set up the control limits, then monitoring proceeds. Once the process is considered to be in-control, the estimated control limits are assumed as fixed. This practice ignores the effect of estimating the unknown in-control process variance σ_0^2. In this paper, we derive the exact run length distribution of the S^2 control chart when the in-control process variance σ_0^2 is estimated and find that $m = 200$ or more Phase I samples are needed to neglect the effect of using estimated control limits. New control limits when m is small are also derived.

Keywords: Process variance; S^2 chart; conditional distribution.

1. Introduction

Statistical Process Control (SPC) is a collection of statistical techniques that can provide a systematic monitoring of a manufacturing process in order to produce high quality final products. Among these techniques, the S^2 control chart is well known and widely used in practice. When an S^2 control chart is used to monitor the variance of a process variable of interest, a Phase I (i.e. past or trial) data set is used to compute an estimate $\hat{\sigma}_0^2$ of the unknown in-control process variance σ_0^2

---

*Corresponding author.

[#] This chapter appeared previously on the International Journal of Reliability, Quality and Safety Engineering. To cite this chapter, please cite the original article as the following: P. Castagliola, G. Celano and G. Chen, *Int. J. Reliab. Qual Saf. Eng*, **16**, 23–38 (2009), doi:10.1142/S0218539309003277.

and it is a common practice to replace the true variance σ_0^2 by $\hat{\sigma}_0^2$ when setting up the control limits for the S^2 control chart. By doing this, we clearly forget the natural variability induced by the estimation of σ_0^2, and the difference in chart performance between the case of using σ_0^2 and the case of using $\hat{\sigma}_0^2$ is either not noticed or is assumed as practically unimportant when the Phase I data set consists of 20 to 30 samples, with each sample having 4 to 5 measurements. Recently, some authors have studied the impact of the estimation of the in-control process parameters on the properties of the corresponding control charts. Most of these authors have focused on $\bar{X}$ type control charts: Del Castillo,[1] Chen,[2] Ghosh *et al.*,[3] Jones *et al.*,[4] Jones,[5] Quesenberry[6] and Yang *et al.*[7] Some other authors have focused on dispersion type control charts (S, S^2 and R control charts): Hillier,[8] Hawkins,[9] Quesenberry,[10] Chen[11] and Maravelakis *et al.*[12] For a literature review on control charts with estimated parameters, see Jensen *et al.*[13]

This paper is an extension of Chen[11] where the author derived, using appropriate approximations or exact results, the run length distributions of the S, S^2 and R control charts with estimated control limits, and graphically compared the run length distribution of the S control chart between the σ_0^2 known case and the σ_0^2 estimated case. Specifically, in this paper, we focus on the S^2 control chart and we further investigate the effect of estimating the control limits and go on to study the design of S^2 charts – an important issue not considered in Chen.[11] With respect to the available literature the breakthrough value of this manuscript thus consists of:

- demonstrating with great details that given the commonly used values of the sample size n, the minimum number m of Phase I samples needs to be at least 200 to treat the estimated control limits as if they were known. As a consequence, it becomes important to design charts with estimated parameters for those manufacturing processes which do not allow intensive sampling,
- extending the available results about the run length distributions to provide practitioners with some operative guidelines to implement control charts with estimated parameters in their specific manufacturing environments.

In Sec. 2, we define the S^2 control chart with estimated control limits and present an example illustrating the impact of the estimation of the in-control process variance σ_0^2 on the control limits. In Sec. 3, we derive the exact run length distribution, the average run length (ARL) and the standard deviation of the run length ($SDRL$) for the S^2 control chart with estimated control limits under the normality assumption. We also make a graphical comparison of the run length distribution of the S^2 control chart between the σ_0^2 known case and the σ_0^2 estimated case, using the maximum absolute differences between the two cases. In Sec. 4, we provide new tables of constants to achieve a desired in-control ARL when a small number m of Phase I samples are used to set up the control limits for the S^2 control chart.

2. The S^2 Control Chart with Estimated Control Limits

When an S^2 control chart is used to monitor the variance of a process variable of interest, a Phase I data set $X_{i,j}$, $i = 1, \ldots, m$, $j = 1, \ldots, n$ is used to estimate the unknown in-control process variance σ_0^2. If we assume that $X_{i,j} \sim N(\mu_0, \sigma_0)$, where μ_0 is the in-control process mean, σ_0^2 is the in-control process variance, and the samples are simple random samples themselves and are independent of each other, then an unbiased estimate $\hat{\sigma}_0^2$ of σ_0^2 is

$$\hat{\sigma}_0^2 = \frac{1}{m}(S_1^2 + S_2^2 + \cdots + S_m^2),$$

where S_i^2 is the ith sample variance

$$S_i^2 = \frac{1}{n-1}\sum_{j=1}^{n}(X_{i,j} - \bar{X}_i)^2 \quad \text{with } \bar{X}_i = \frac{1}{n}\sum_{j=1}^{n}X_{i,j}.$$

The estimated control limits for the S^2 control chart are

$$\begin{aligned} LCL &= K_L \times \hat{\sigma}_0^2, \\ UCL &= K_U \times \hat{\sigma}_0^2, \end{aligned} \tag{1}$$

where the constants K_U and K_L are such that the chosen Type I error probability α for the control chart is equally split, that is,

$$\begin{aligned} K_L &= \frac{F_{\chi^2}^{-1}\left(\frac{\alpha}{2}|n-1\right)}{n-1}, \\ K_U &= \frac{F_{\chi^2}^{-1}\left(1-\frac{\alpha}{2}|n-1\right)}{n-1}, \end{aligned} \tag{2}$$

where $F_{\chi^2}^{-1}(p|\nu)$ is the inverse cumulative distribution function (the pth percentile) of the χ^2 distribution with ν degrees of freedom. Due to the link between the χ^2 distribution and the gamma distribution, we also have

$$\begin{aligned} K_L &= F_\gamma^{-1}\left(\frac{\alpha}{2}\left|\frac{n-1}{2}, \frac{2}{n-1}\right.\right), \\ K_U &= F_\gamma^{-1}\left(1-\frac{\alpha}{2}\left|\frac{n-1}{2}, \frac{2}{n-1}\right.\right), \end{aligned} \tag{3}$$

where $F_\gamma^{-1}(p|a,b)$ is the inverse cumulative distribution function of the gamma(a,b) distribution. When the current or future random samples $Y_{i,j}$, $i = 1, 2, 3, \ldots$, $j = 1, \ldots, n$ are collected, we monitor the process variance by plotting the statistic $\hat{\sigma}_i^2$ versus i for $i = 1, 2, \ldots$, on a control chart with control limits LCL and UCL respectively, where $\hat{\sigma}_i^2$ is equal to:

$$\hat{\sigma}_i^2 = \frac{1}{n-1}\sum_{j=1}^{n}(Y_{i,j} - \bar{Y}_i)^2 \quad \text{with } \bar{Y}_i = \frac{1}{n}\sum_{j=1}^{n}Y_{i,j}.$$

In order to study both the out-of-control and the in-control process performance, we assume that $Y_{i,j} \sim N(\mu_0, \tau\sigma_0)$, where $\tau > 0$ is a constant reflecting the shift in dispersion. When $\tau = 1$, the process variability is in-control; otherwise the process variability has changed. It is worth noting that when $\tau < 1$ the out-of-control process condition corresponds to a reduction in process variability, which is usually the outcome of a correction action on the process itself. On the other hand, when $\tau > 1$ the out-of-control process condition may be the result of a deterioration in the process performance, which increases the process variability.

In order to illustrate the impact of estimation on the control limits of the S^2 chart, let us consider a procedure to be designed with the aim of monitoring the variability within a yogurt cup filling process: a Critical to Quality parameter of the filling process is the deviation of each cup filling from a specified target. We want to monitor the process dispersion by means of an S^2 control chart. Assume that the target is equal to 125 g. Table 1 contains some simulated Phase I and Phase II data

Table 1.   Simulated Phase I and Phase II data sets.

| Sample | Phase I Data set | | | | | | Phase II Data set | | | | | |
	$X_{i,1}$	$X_{i,2}$	$X_{i,3}$	$X_{i,4}$	$X_{i,5}$	$\hat{\sigma}_i^2$	$Y_{i,1}$	$Y_{i,2}$	$Y_{i,3}$	$Y_{i,4}$	$Y_{i,5}$	$\hat{\sigma}_i^2$
1	125.0	123.3	125.3	124.8	124.8	0.6030	123.9	124.4	124.7	124.7	125.7	0.4320
2	123.7	126.2	125.9	125.4	124.9	0.9670	124.2	125.4	126.9	125.4	123.8	1.4780
3	127.3	123.9	124.6	125.1	125.1	1.6200	124.8	126.9	126.2	127.1	125.4	0.9570
4	125.3	124.0	125.9	124.8	125.7	0.5830	124.6	123.9	125.4	126.1	124.7	0.7030
5	123.0	124.0	126.7	124.8	124.9	1.8570	125.3	127.2	124.5	124.4	125.0	1.2870
6	124.8	123.8	124.9	122.3	125.3	1.4570	125.3	124.1	127.1	124.4	124.2	1.5770
7	125.3	124.2	124.8	125.8	125.5	0.3970	125.6	125.1	124.3	125.1	124.5	0.2720
8	124.9	123.9	124.6	124.7	125.7	0.4180	124.3	126.7	127.0	126.5	125.3	1.2780
9	127.4	125.2	124.4	125.4	124.7	1.3820	126.4	124.2	125.0	123.7	126.4	1.5380
10	124.6	125.1	124.0	126.0	123.2	1.1320	126.9	124.8	127.7	126.5	124.5	1.9020
11	125.3	125.6	125.7	126.1	124.6	0.3130	123.7	124.9	122.3	125.1	125.0	1.4500
12	126.1	127.1	124.6	125.3	125.9	0.8700	124.4	125.5	125.7	125.5	124.1	0.5380
13	125.2	125.3	124.5	126.3	124.6	0.5170	124.2	125.7	124.1	122.3	123.2	1.6050
14	125.2	124.1	125.3	125.1	124.9	0.2320	124.6	124.7	124.1	127.0	125.9	1.3830
15	126.4	124.5	125.2	124.6	123.8	0.9500	124.8	124.6	125.7	126.8	124.5	0.9470
16	124.9	123.8	123.7	126.4	125.6	1.3470	125.7	125.8	124.9	126.4	125.5	0.2930
17	125.7	127.1	124.5	125.5	125.4	0.8780	123.8	127.6	125.1	125.3	125.1	1.8970
18	123.9	124.8	126.9	127.8	124.5	2.8170	126.4	124.7	126.1	125.3	124.6	0.6530
19	123.1	124.8	123.8	126.5	124.8	1.6450	127.1	124.8	125.8	125.1	125.5	0.7930
20	125.4	124.9	126.6	124.7	125.1	0.5630	124.5	126.1	127.1	124.4	124.2	1.6330
21	124.8	125.4	123.3	126.9	126.1	1.8650	126.4	126.1	125.1	124.8	124.9	0.5430
22	124.6	124.0	123.8	124.9	125.5	0.4730	126.0	121.3	123.7	125.2	126.5	4.4030
23	125.2	124.0	123.9	124.8	125.4	0.4680	125.1	126.9	126.8	125.0	125.2	0.9250
24	124.4	123.1	124.0	122.8	127.1	2.9070	123.9	124.7	125.2	123.2	125.9	1.1270
25	123.6	125.6	124.0	123.6	124.0	0.6880	125.6	126.6	126.2	126.0	124.8	0.4680
26	124.9	123.7	126.0	122.9	127.6	3.4670	124.9	127.1	125.1	121.1	124.4	4.7120
27	126.6	125.1	125.3	126.4	123.5	1.5370	125.1	124.4	124.0	124.8	125.4	0.3080
28	124.8	124.1	125.3	123.6	125.8	0.7870	125.4	124.6	125.5	126.3	122.4	2.2230
29	124.3	125.2	125.0	125.6	125.4	0.2500	126.1	125.0	124.2	126.0	125.8	0.6520
30	124.8	124.6	125.2	124.9	126.1	0.3470	125.6	125.6	126.5	125.3	127.1	0.5670

Using the first 15 samples to estimate variance

Using all 30 samples to estimate variance

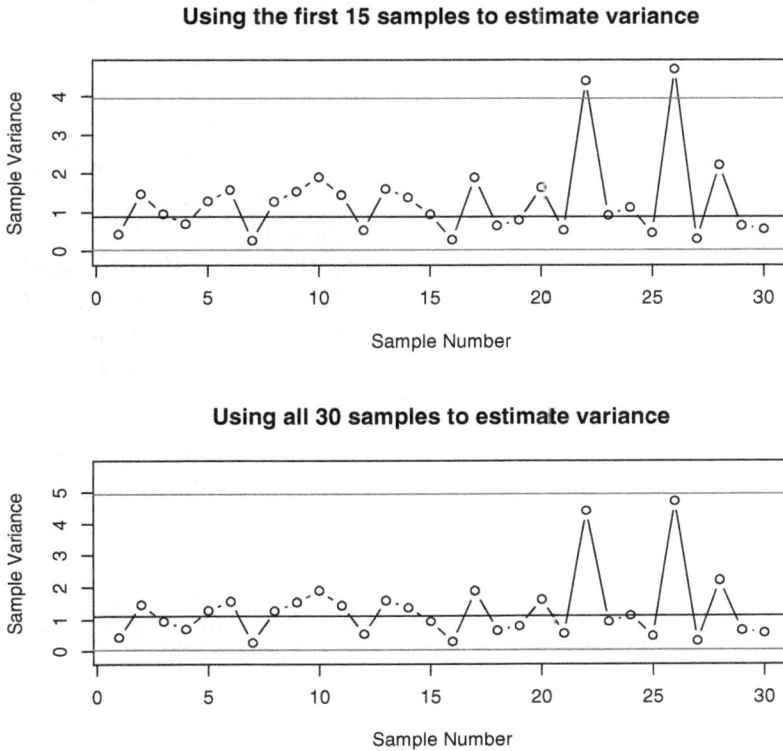

Fig. 1. The S^2 control charts based on two estimates of the process variance.

sets generated from the same normal ($\mu_0 = 125, \sigma_0 = 0.1$) distribution. To estimate the control limits, we consider two different cases: (i) using the first 15 samples out from the Phase I data set or (ii) using the entire Phase I data set, i.e. 30 samples. In the first case we find $\hat{\sigma}_0^2 = 0.8865$. Since the sample size is $n = 5$, for $\alpha = 0.0027$, we have $K_L = F_{\chi^2}^{-1}(0.00135|4)/4 = 0.0264$ and $K_U = F_{\chi^2}^{-1}(0.99865|4)/4 = 4.4501$, therefore the estimated control limits are $LCL = 0.0264 \times 0.8865 = 0.0234$ and $UCL = 4.4501 \times 0.8865 = 3.9451$. When the Phase II data, assumed in-control and simulated accordingly, are plotted using these estimated control limits (see the top plot in Fig. 1), samples #22 and #26 are out-of-control. In the second case, we use all 30 samples from the Phase I data set to estimate the in-control process variance to give $\hat{\sigma}_0^2 = 1.1112$, $LCL = 0.0294$ and $UCL = 4.9451$. When these estimated control limits are used to monitor the process (see the bottom plot in Fig. 1), there are no out-of-control samples. This example clearly highlights the impact of the estimation of σ_0^2 and particularly the impact of the number m of samples required in Phase I to have an accurate estimate of σ_0^2. In the $m = 15$ case, we probably underestimate the real value of σ_0^2, which generates the false alarms during the Phase II monitoring.

3. The Exact Distribution of the S^2 Chart Run Length when σ_0^2 is Estimated

Let $E_i = (\hat{\sigma}_i^2 < LCL)$ or $(\hat{\sigma}_i^2 > UCL)$ be the event that the plotting statistic $\hat{\sigma}_i^2$ lies outside the control limits of the S^2 control chart, and let W be the number of samples until the first event E_i occurs (i.e. the run length). It has been discussed in the literature (see Jensen *et al.*[13] for general references and Chen[11] for specific references) that because the control limits in equation (1) contain the estimate $\hat{\sigma}_0^2$, the events $\{E_i\}$ are no longer independent, therefore the run length W does not follow a geometric distribution. This results in the unknown performance properties for the S^2 chart when the in-control variance σ_0^2 is estimated. To study these properties, let $U = \hat{\sigma}_0^2/\sigma_0^2$ and let $h(u|m, n)$ denote the density function of U given by

$$h(u|m, n) = f_\gamma \left(u \left| \frac{m(n-1)}{2}, \frac{2}{m(n-1)} \right. \right),$$

where $f_\gamma(x|a, b)$ is the density function of the gamma(a, b) distribution, namely,

$$f_\gamma(x|a, b) = \frac{e^{-x/b}x^{a-1}}{b^a\Gamma(a)},$$

and let $G(x|n)$ denote the cumulative distribution function of $\hat{\sigma}_i^2/(\tau^2\sigma_0^2)$ given by

$$G(x|n) = F_\gamma \left(x \left| \frac{n-1}{2}, \frac{2}{n-1} \right. \right).$$

Then, for a given value of $\hat{\sigma}_0^2$, we have

$$P(E_i) = P(\hat{\sigma}_i^2 < K_L\hat{\sigma}_0^2) + P(\hat{\sigma}_i^2 > K_U\hat{\sigma}_0^2),$$

$$= P \left(\frac{\hat{\sigma}_i^2}{\tau^2\sigma_0^2} < K_L\frac{\hat{\sigma}_0^2}{\tau^2\sigma_0^2} \right) + P \left(\frac{\hat{\sigma}_i^2}{\tau^2\sigma_0^2} > K_U\frac{\hat{\sigma}_0^2}{\tau^2\sigma_0^2} \right),$$

$$= G \left(\frac{K_L u}{\tau^2} \middle| n \right) + 1 - G \left(\frac{K_U u}{\tau^2} \middle| n \right),$$

$$= p(u|n, \tau), \quad \text{say,}$$

where u denotes $\hat{\sigma}_0^2/\sigma_0^2$. Although the events $\{E_i\}$ are not independent, when conditioned on a specific value of $\hat{\sigma}_0^2$ they are independent. This implies that the conditional distribution of W is a geometric distribution and the conditional cumulative distribution function of W, given $\hat{\sigma}_0^2$, is

$$P(W \le w|\hat{\sigma}_0^2) = 1 - (1 - p(u|n, \tau))^w, \quad w = 1, 2, 3, \ldots.$$

The unconditional cumulative distribution function of W, which describes the performance of the S^2 control chart when σ_0^2 is estimated in an average sense, is found to be

$$F_W(w|m, n, \tau) = P(W \le w) = 1 - \int_0^\infty (1 - p(u|n, \tau))^w h(u|m, n)du.$$

In the same spirit, the first two unconditional moments of W can be found from

$$E(W) = \int_0^\infty \frac{1}{p(u|n,\tau)} h(u|m,n)du,$$

$$E(W^2) = \int_0^\infty \frac{2 - p(u|n,\tau)}{(p(u|n,\tau))^2} h(u|m,n)du.$$

The mean $E(W)$ is also called the average run length (ARL) and the standard deviation of run length W $(SDRL)$ is found from

$$SDRL = SD(W) = \sqrt{E(W^2) - (E(W))^2}.$$

Now we are ready to study the effect of estimating σ_0^2. Let V be the counterpart of W when the estimate $\hat{\sigma}_0^2$ in Eq. (1) is replaced with the true in-control process variance σ_0^2. Without loss of generality, we let the true process variance be $\sigma_0^2 = 1$.

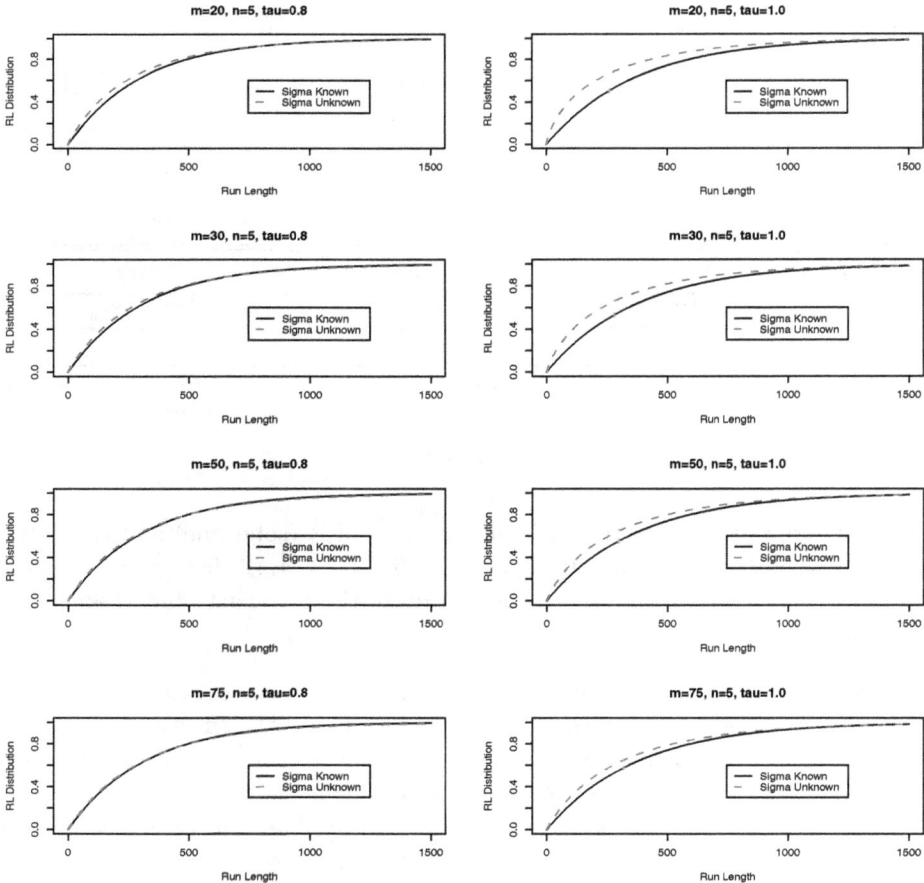

Fig. 2.   Run length distributions of the S^2 control chart for $n = 5$.

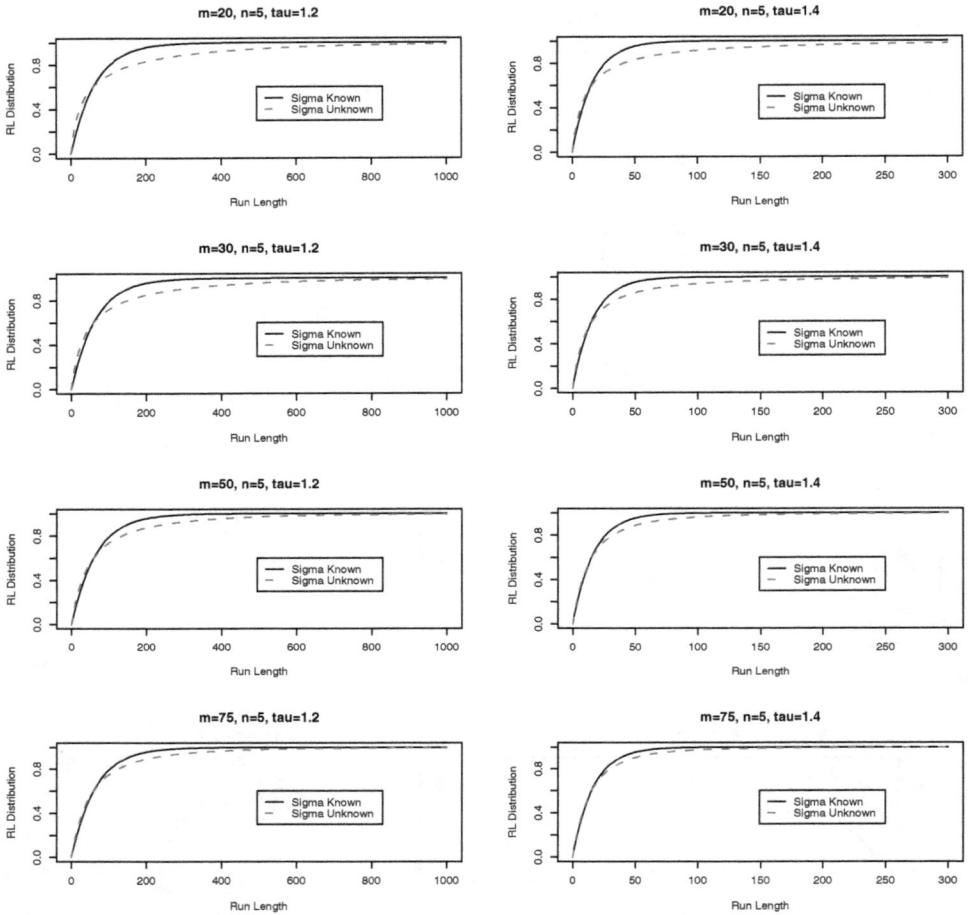

Fig. 2.   (*Continued*)

Figures 2 and 3 display the distribution functions of V (solid line) and W (dotted line) for $n = 5, 10$, $m = 20, 30, 50, 75$, and $\tau = 0.8, 1, 1.2, 1.4$, where $\tau = 1$ indicates that the process variance is in-control. We make the following observations from Figs. 2 and 3:

(1) The effect of estimating σ_0^2 depends on m and n. This effect is clearly noticeable for small m and n, but when m and/or n gets larger, the effect is reduced. For the typical range of n ($4 \leq n \leq 10$), this effect starts to be negligible when $m \geq 75$. But as we will show below, m has to be much larger than 75 in order to assume that both the σ_0^2 estimated case and the σ_0^2 known case are equivalent.

(2) When the process variance is in-control ($\tau = 1$), there is a higher probability to signal on the average a false alarm for the σ_0^2 estimated case than the σ_0^2 known case.

(3) If the process variance has decreased ($\tau < 1$), there is very little difference between the σ_0^2 estimated case and the σ_0^2 known case.

(4) If the process variance has increased ($\tau > 1$), there is a crossover, namely, initially there is a higher probability for the σ_0^2 estimated case to signal; then there is a higher probability for the σ_0^2 known case to signal.

The effect of estimating σ_0^2 can also be seen easily from Table 2 which displays the *ARL* and *SDRL* of the run length distribution of the S^2 control chart. The row $m = \infty$ corresponds to the σ_0^2 known case. In comparison with the σ_0^2 known case, estimating σ_0^2 will reduce *ARL* and *SDRL* when the process variance is in-control ($\tau = 1$), and will lead to larger *ARL* and *SDRL* when the process variance is out-of-control ($\tau \neq 1$).

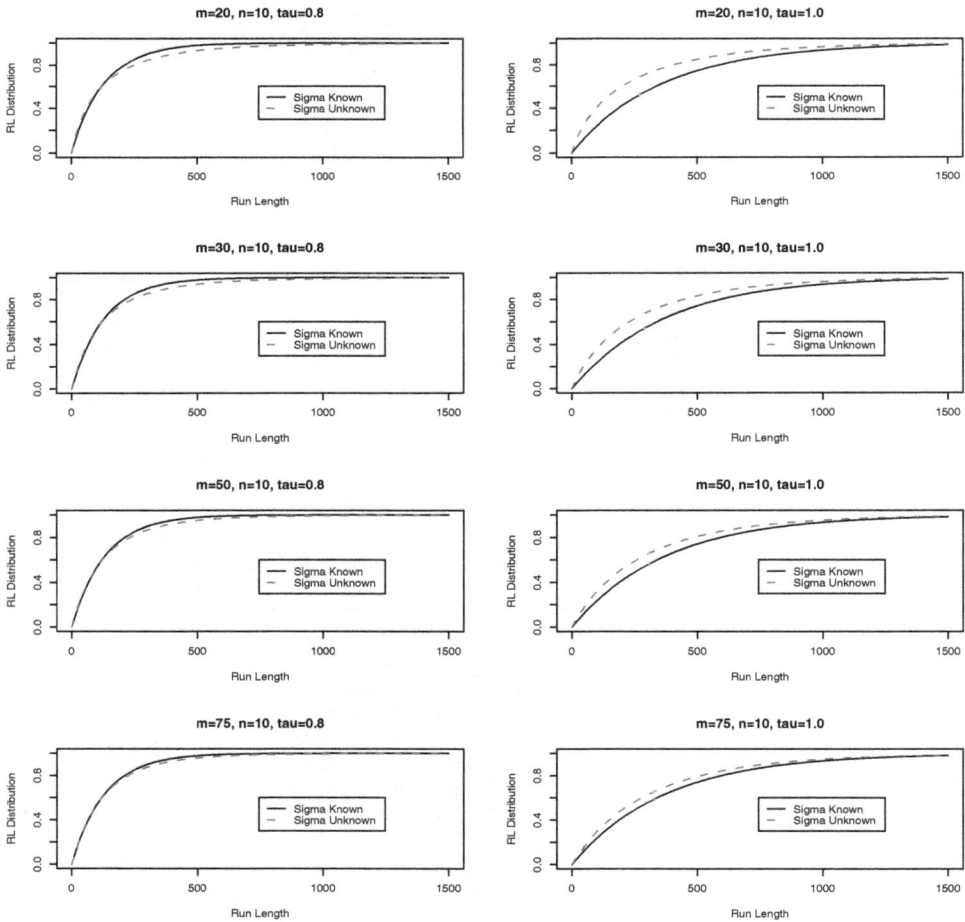

Fig. 3.   Run length distributions of the S^2 control chart for $n = 10$.

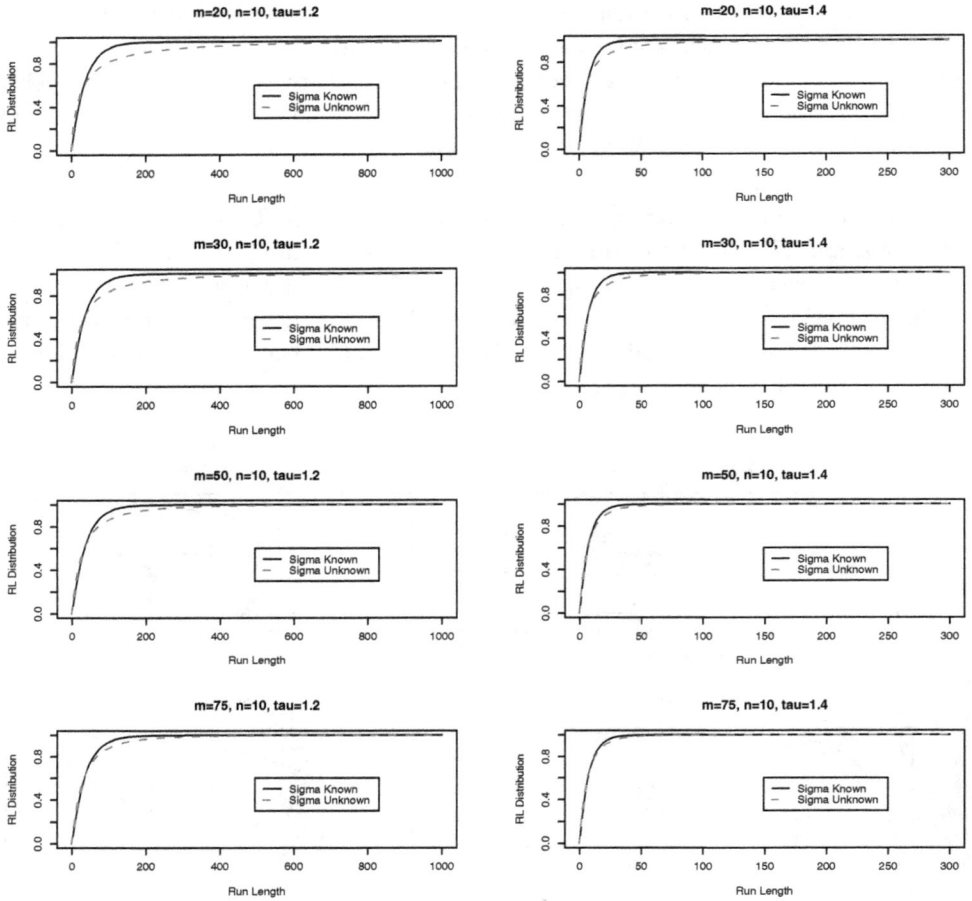

Fig. 3.   (*Continued*)

Table 2.   Average run length ARL and standard deviation $SDRL$ of the run length distribution of the S^2 control chart for $n = 5, 10$, and various values of m and τ. The σ_0^2 known case corresponds to $m = \infty$.

		ARL					SDRL				
n	m	$\tau = 0.6$	$\tau = 0.8$	$\tau = 1.0$	$\tau = 1.2$	$\tau = 1.4$	$\tau = 0.6$	$\tau = 0.8$	$\tau = 1.0$	$\tau = 1.2$	$\tau = 1.4$
5	20	109.8	317.0	325.0	83.2	20.3	120.0	335.6	366.4	127.0	28.4
5	30	107.1	316.2	336.9	77.3	19.0	113.4	330.4	369.2	107.1	23.6
5	50	105.1	314.0	348.3	72.2	18.1	108.5	323.4	371.0	89.7	20.3
5	75	104.1	312.4	354.9	69.6	17.6	106.2	318.8	371.4	80.8	18.9
5	∞	102.2	308.2	**370.4**	64.5	16.8	101.7	307.7	**369.9**	64.0	16.3
10	20	19.3	144.8	318.7	46.0	8.6	21.3	166.5	344.7	66.9	10.0
10	30	18.8	140.4	332.1	42.8	8.3	19.8	154.5	351.4	55.6	9.0
10	50	18.4	136.7	345.0	40.3	8.1	18.8	144.8	357.7	47.2	8.2
10	75	18.2	134.9	352.5	39.1	8.0	18.3	140.1	361.3	43.3	7.9
10	∞	17.8	131.4	**370.4**	36.9	7.8	17.3	130.9	**369.9**	36.4	7.2

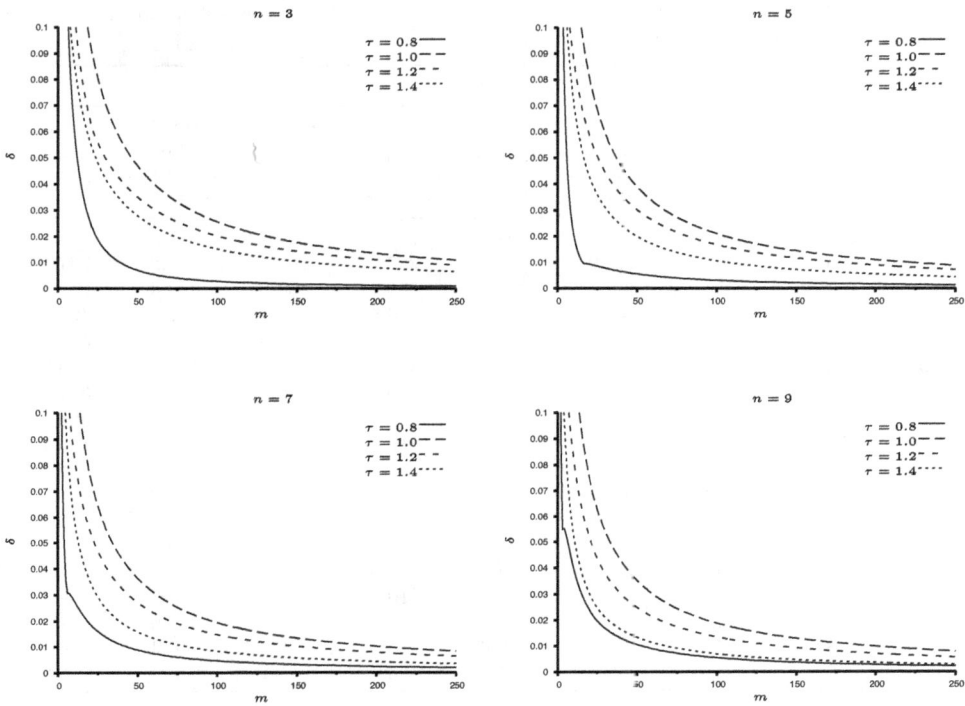

Fig. 4. Maximum absolute difference $\delta = \max_{w \geq 1} |F_V(w|n, \tau) - F_W(w|m, n, \tau)|$ for $n = 3, 5, 7, 9$, $\tau = 0.8, 1, 1.2, 1.4$ and $m = 1, 2, \ldots, 200$.

In order to better investigate how large the number m of Phase I samples should be in order to have a small difference between the cumulative distribution functions of V (σ_0^2 known) and W (σ_0^2 estimated), we have computed the maximum absolute difference $\delta = \max_{w \geq 1} |F_V(w|n, \tau) - F_W(w|m, n, \tau)|$ or equivalently $\delta = \max_{w \geq 1} |F_W(w|\infty, n, \tau) - F_W(w|m, n, \tau)|$, for $n = 3, 5, 7, 9$, $\tau = 0.8, 1, 1.2, 1.4$ and $m = 1, 2, \ldots, 200$. The results are computed in Fig. 4. We can see from Fig. 4 that for all the values of τ considered, if we want a small absolute difference (say no more than 2%) between the cumulative distribution functions of V and W, the minimum number m of Phase I samples should actually satisfy $m \geq 200$. From a practical point of view, this very large threshold value of m poses problems in terms of planning the sampling scheme—completing the phase I chart implementation can take too long a time. For instance, a 24 hours Phase I control chart implementation calling for $m = 200$ samples would require a sampling frequency of 7.2 minutes, whichever the selected sample size. This could be hard to handle in practice, very costly and these intensive inspection may generate unexpected autocorrelation structures between the collected samples. But we have to remember that the threshold $m = 200$ was obtained using the constants K_L and K_U (see Eq. (3)) which correspond to the $m = \infty$ case. For this reason and in order to relax

Table 3.   Constants (K'_L, K'_U) for $n = 3, 5$, $m = 10, 20, \ldots, 70$ and out-of-control ARL for $\tau = 0.5, 0.6, \ldots, 0.9, 0.95, 1, 1.05, 1.1, 1.2, \ldots, 2$.

	$n = 3$							
K'_L	0.0011	0.0012	0.0012	0.0013	0.0013	0.0013	0.0013	0.0014
K'_U	6.8030	6.7241	6.6914	6.6733	6.6616	6.6535	6.6475	6.6077
τ	$m = 10$	$m = 20$	$m = 30$	$m = 40$	$m = 50$	$m = 60$	$m = 70$	$m = \infty$
0.50	249.9	219.4	208.7	203.2	199.8	197.5	195.9	185.6
0.60	353.3	315.1	300.2	292.3	287.5	284.2	281.8	267.0
0.70	449.0	420.0	404.7	395.6	389.7	385.6	382.6	363.1
0.80	496.6	496.5	492.7	489.3	486.5	484.2	482.4	467.0
0.90	466.0	482.0	489.7	494.3	497.4	499.5	501.1	512.1
0.95	423.8	434.9	440.8	444.6	447.3	449.3	450.8	463.7
1.00	370.4	370.4	370.4	370.4	370.4	370.4	370.4	370.4
1.05	312.2	299.7	293.0	288.8	285.9	283.7	282.1	268.8
1.10	254.9	232.5	221.4	214.6	210.0	206.7	204.2	186.4
1.20	157.9	129.4	117.5	111.0	107.0	104.2	102.2	90.1
1.30	92.4	70.2	62.5	58.7	56.5	55.0	54.0	48.0
1.40	53.8	40.0	35.8	33.8	32.7	32.0	31.5	28.5
1.50	32.5	24.6	22.4	21.4	20.8	20.4	20.2	18.6
1.60	20.8	16.4	15.2	14.7	14.3	14.1	14.0	13.1
1.70	14.3	11.8	11.1	10.7	10.5	10.4	10.3	9.8
1.80	10.4	8.9	8.5	8.3	8.1	8.1	8.0	7.7
1.90	8.0	7.0	6.8	6.6	6.5	6.5	6.4	6.2
2.00	6.4	5.8	5.6	5.5	5.4	5.4	5.4	5.2

	$n = 5$							
K'_L	0.0236	0.0247	0.0252	0.0254	0.0256	0.0257	0.0258	0.0264
K'_U	4.5732	4.5239	4.5034	4.4919	4.4846	4.4795	4.4757	4.4501
τ	$m = 10$	$m = 20$	$m = 30$	$m = 40$	$m = 50$	$m = 60$	$m = 70$	$m = \infty$
0.50	73.3	62.6	59.0	57.1	56.0	55.3	54.8	51.4
0.60	146.6	124.9	117.6	113.9	111.6	110.1	109.0	102.2
0.70	260.1	225.5	212.5	205.8	201.7	198.9	196.9	184.5
0.80	388.8	362.0	347.9	339.5	334.0	330.1	327.3	308.2
0.90	445.3	450.6	451.6	451.6	451.3	451.0	450.7	445.8
0.95	422.7	432.0	436.6	439.4	441.2	442.5	443.6	451.0
1.00	370.4	370.4	370.4	370.4	370.4	370.4	370.4	370.4
1.05	301.4	286.7	279.2	274.5	271.3	268.9	267.1	253.5
1.10	230.1	204.7	192.8	185.9	181.3	178.1	175.7	159.6
1.20	117.4	92.3	83.0	78.3	75.5	73.7	72.3	64.5
1.30	56.0	42.1	37.9	35.9	34.8	34.0	33.5	30.5
1.40	28.1	21.7	19.9	19.1	18.6	18.3	18.1	16.8
1.50	15.7	12.8	12.0	11.6	11.4	11.2	11.1	10.5
1.60	9.8	8.4	8.0	7.8	7.7	7.6	7.5	7.2
1.70	6.8	6.0	5.8	5.7	5.6	5.5	5.5	5.3
1.80	5.0	4.6	4.4	4.4	4.3	4.3	4.3	4.2
1.90	4.0	3.7	3.6	3.5	3.5	3.5	3.5	3.4
2.00	3.2	3.1	3.0	3.0	2.9	2.9	2.9	2.9

Table 4.   Constants (K'_L, K'_U) for $n = 7, 9$, $m = 10, 20, \ldots, 70$ and out-of-control ARL for $\tau = 0.5, 0.6, \ldots, 0.9, 0.95, 1, 1.05, 1.1, 1.2, \ldots, 2$.

				$n = 7$				
K'_L	0.0649	0.0671	0.0680	0.0686	0.0689	0.0692	0.0693	0.0706
K'_U	3.7188	3.6806	3.6647	3.6558	3.6501	3.6461	3.6431	3.6232
τ	$m = 10$	$m = 20$	$m = 30$	$m = 40$	$m = 50$	$m = 60$	$m = 70$	$m = \infty$
0.50	26.4	22.4	21.1	20.4	20.0	19.8	19.6	18.4
0.60	67.2	56.4	52.9	51.1	50.0	49.3	48.8	45.6
0.70	152.9	128.1	119.8	115.6	113.1	111.4	110.1	102.5
0.80	295.9	261.2	246.4	238.4	233.4	230.0	227.5	211.8
0.90	416.2	411.1	406.6	403.2	400.7	398.7	397.1	384.1
0.95	417.0	423.7	426.5	428.1	429.1	429.8	430.3	433.1
1.00	370.4	370.4	370.4	370.4	370.4	370.4	370.4	370.4
1.05	294.0	277.6	269.3	264.3	260.8	258.3	256.4	242.1
1.10	212.2	185.0	172.8	165.8	161.3	158.2	155.9	140.8
1.20	92.2	70.8	63.6	60.1	58.0	56.7	55.7	50.0
1.30	38.2	29.2	26.6	25.3	24.6	24.2	23.8	22.0
1.40	17.8	14.4	13.4	13.0	12.7	12.5	12.4	11.7
1.50	9.8	8.4	8.0	7.7	7.6	7.5	7.5	7.2
1.60	6.2	5.5	5.3	5.2	5.1	5.1	5.1	4.9
1.70	4.3	4.0	3.9	3.8	3.8	3.7	3.7	3.6
1.80	3.3	3.1	3.0	3.0	3.0	2.9	2.9	2.9
1.90	2.6	2.5	2.5	2.4	2.4	2.4	2.4	2.4
2.00	2.2	2.1	2.1	2.1	2.1	2.1	2.1	2.0

				$n = 9$				
K'_L	0.1086	0.1116	0.1129	0.1136	0.1141	0.1144	0.1147	0.1163
K'_U	3.2500	3.2182	3.2048	3.1974	3.1926	3.1893	3.1868	3.1701
τ	$m = 10$	$m = 20$	$m = 30$	$m = 40$	$m = 50$	$m = 60$	$m = 70$	$m = \infty$
0.50	11.7	10.1	9.5	9.3	9.1	9.0	8.9	8.4
0.60	34.8	29.2	27.3	26.4	25.9	25.5	25.3	23.6
0.70	94.9	78.5	73.2	70.5	68.9	67.8	67.1	62.3
0.80	226.5	192.8	180.1	173.4	169.4	166.7	164.7	152.5
0.90	386.8	373.2	364.7	359.1	355.2	352.2	350.0	333.1
0.95	410.0	413.8	415.0	415.5	415.7	415.8	415.8	414.7
1.00	370.4	370.4	370.4	370.4	370.4	370.4	370.4	370.4
1.05	287.7	269.8	260.9	255.5	251.8	249.1	247.1	232.3
1.10	197.0	168.8	156.5	149.7	145.3	142.3	140.1	126.2
1.20	74.4	56.6	50.9	48.2	46.6	45.5	44.8	40.5
1.30	27.9	21.8	20.0	19.2	18.7	18.4	18.2	16.9
1.40	12.6	10.5	9.9	9.6	9.4	9.3	9.2	8.8
1.50	6.9	6.1	5.9	5.7	5.7	5.6	5.6	5.4
1.60	4.5	4.1	3.9	3.9	3.8	3.8	3.8	3.7
1.70	3.2	3.0	2.9	2.9	2.9	2.8	2.8	2.8
1.80	2.5	2.3	2.3	2.3	2.3	2.3	2.3	2.2
1.90	2.0	2.0	1.9	1.9	1.9	1.9	1.9	1.9
2.00	1.8	1.7	1.7	1.7	1.7	1.7	1.7	1.6

the $m = 200$ constraint, we will compute in the following section alternative constants K'_L and K'_U which take the value of m into account: this will let the quality practioner choose the value of m which is the more relevant for his particular case.

4. Modified Control Limits for the S^2 Control Chart with Estimated In-Control Variance

As we explained above, due to obvious economical considerations, the $m \geq 200$ requirement for the performance of the S^2 control chart with σ_0^2 estimated to be close to that of the S^2 control chart with σ_0^2 known can be hard to meet in practice. When a smaller number m of Phase I samples are available to estimate σ_0^2, the constants K_L and K_U used in Eq. (1) and given in Eq. (3) cannot provide the intended in-control $ARL = 1/\alpha$. In this section, we produce new constants K'_L and K'_U that take the number m of Phase I samples into account to guarantee a desired in-control ARL.

For a desired in-control $ARL_0 = 1/\alpha$, and for given m and n, we find (K'_L, K'_U) as the solution to $E(W) = ARL_0$ under the constraint $F_\gamma(K'_L | \frac{n-1}{2}, \frac{2}{n-1}) = 1 - F_\gamma(K'_U | \frac{n-1}{2}, \frac{2}{n-1})$. Table 3 and Table 4 list the K'_L and K'_U values for $ARL_0 = 370.4$, $n = 3, 5, 7, 9$ and $m = 10, 20, \ldots, 70$. For example, if $m = 20$ and $n = 5$, we have $K'_L = 0.0247$ and $K'_U = 4.5239$, while $K_L = 0.0264$ and $K_U = 4.4501$ for $m = \infty$ (i.e. σ_0^2 known). In Tables 3 and 4 we also give the out-of-control ARL's for shifts in dispersion represented by $\tau = 0.5, 0.6, \ldots, 0.9, 0.95, 1, 1.05, 1.1, 1.2, \ldots, 2$. For example, for $m = 20$, $n = 5$ and $\tau = 1.3$, the out-of-control ARL is 42.1, and for $m = \infty$, $n = 5$ and $\tau = 1.3$, the out-of-control ARL is 30.5. We see from Tables 3 and 4 that for fixed n and for most values of τ, the out-of-control ARL decreases as m increases and converges to the out-of-control ARL corresponding to the σ_0^2 known case. For $\tau = 0.9$ and $\tau = 0.95$ (very small decreases of the variability) the trend is reversed, that is, the out-of-control ARL increases as m increases.

5. Conclusions

In this paper we have investigated the impact of the estimation of the in-control variance σ_0^2, using a Phase I data set, on the properties of the S^2 control chart. This investigation was performed by evaluating the exact run length distribution, the ARL and the $SDRL$ of the S^2 control chart with estimated control limits and by comparing this run length distribution with the run length distribution of the S^2 control chart when σ_0^2 is assumed known. We have found that if a small difference (no more than 2%) between the distribution functions of the σ_0^2 known case and the σ_0^2 estimated case is desired, the minimum number m of Phase I samples should satisfy $m \geq 200$. In many or most applications, this is clearly impossible. To address this issue, we have produced new constants for the computation of the control limits for the S^2 control chart with estimated in-control variance and found the corresponding out-of-control ARL's. Using these new constants allows one to compute control limits for the S^2 control chart that achieve the desired in-control *ARL* when a small number m of Phase I samples are available.

Acknowledgements

The first author's rsearch is funded by the Projet International de Coopération Scientifique PICS-3753 entitled "Méthodes statistiques adaptatives pour la surveillance de la variabilité de procédés" of the CNRS (Centre National de la Recherche Scientifique). The third author's research is supported by a grant from the Natural Science and Engineering Research Council of Canada.

References

1. E. Del Castillo, Evaluation of the run length distribution of $\bar{X}$ charts with unknown variance, *Journal of Quality Technology* **28**(1) (1996) 116–122.
2. G. Chen, The mean and standard deviation of the run length distribution of $\bar{X}$ charts when control limits are estimated, *Statistica Sinica* **7** (1997) 789–798.
3. B. K. Ghosh, M. R. Reynolds Jr. and Y. Van Hui, Shewhart $\bar{X}$ charts with estimated variance, *Communications in Statistics — Theory and Methods* **18** (1981) 1797–1822.
4. L. A. Jones, C. W. Champ and S. E. Rigdon, The performance of exponentially weighted moving average charts with estimated parameters, *Technometrics* **43** (2001) 156–167.
5. L. A. Jones, The statistical design of EWMA control charts with estimated parameters, *Journal of Quality Technology* **34** (2002) 277–288.
6. C. P. Quesenberry, The effect of sample size on estimated limits for $\bar{X}$ and X control charts, *Journal of Quality Technology* **25**(4) (1993) 237–247.
7. Z. Yang, M. Xie, V. Kuralmani and K. Tsui, On the performance of geometric charts with estimated control limits, *Journal of Quality Technology* **34** (2002) 448–458.
8. F. S. Hillier, $\bar{X}$- and R- chart control limits based on a small number of subgroups, *Journal of Quality Technology* **1**(1) (1969) 17–26.
9. D. M. Hawkins, Self-starting cusum charts for location and scale, *The Statistician* **36**(4) (1987) 299–316.
10. C. P. Quesenberry, Spc Q charts for start-up processes and short or long runs, *Journal of Quality Technology* **23**(3) (1991) 213–224.
11. G. Chen, The run length distributions of the R, s and s^2 control charts when σ is estimated, *The Canadian Journal of Statistics* **26**(2) (1998) 311–322.
12. P. E. Maravelakis, J. Panaretos and S. Psarakis, Effect of estimation of the process parameters on the control limits of the univariate control charts for process dispersion, *Communications in Statistics — Simulation and Computation* **31** (2002) 443–461.
13. W. A. Jensen, L. A. Jones-Farmer, C. W. Champ and W. H. Woodall, Effects of parameter estimation on control chart properties: a literature review, *Journal of Quality Technology* **38**(4) (2006) 349–364.

About the Authors

Philippe Castagliola is graduated (PhD 1991) from the UTC (Université de Technologie de Compiègne, France). He is currently professor at the Université de Nantes, Institut Universitaire de Technologie de Nantes, France, and he is also a member of the IRCCyN (Institut de Recherche en Communications et Cybernétique de Nantes), UMR CNRS 6597. He is associate editor for the Journal of Quality Technology and Quantitative Management and for the International Journal of Reliability, Quality and Safety Engineering. His research activity includes developments of new SPC techniques (non normal control charts, optimized EWMA type control charts, multivariate capability indices, monitoring of batch processes, . . .).

Giovanni Celano received his PhD in 2003 from the University of Palermo discussing a thesis on sequencing of mixed model assembly lines. He is currently assistant professor in Technology and Manufacturing Systems at the University of Catania (Italy). His research is focused on statistical quality control, production scheduling and operations management applied to healthcare organizations. He is currently member of the Associazione Italiana di Tecnologia Meccanica (AITeM), the European Network of Business and Industry Statistics (ENBIS), the International Institute for Innovation, Industrial Engineering and Entrepreneurship (I4e2). He has co-authored about 70 papers in international journals and in proceedings of national and international conferences.

Gemai Chen is Professor, Department of Mathematics and Statistics, University of Calgary, Calgary, Alberta, Canada. He received his PhD in statistics from Simon Fraser University in 1991, and quality control is one of his many research areas including nonlinear time series, high dimensional data analysis and environmental statistics.

Chapter 12

OPPORTUNISTIC MAINTENANCE OPTIMIZATION FOR WIND TURBINE SYSTEMS CONSIDERING IMPERFECT MAINTENANCE ACTIONS[#]

FANGFANG DING[*] and ZHIGANG TIAN[†]

Concordia Institute for Information Systems Engineering
Concordia University Montreal, H3G 2W1, Canada
*fara.ding@gmail.com
†tian@ciise.concordia.ca

Currently corrective maintenance and time-based preventive maintenance strategies are widely used in wind power industry. However, few methods are applied to optimize these strategies. This paper aims to develop opportunistic maintenance approaches for an entire wind farm rather than individual components that most of the existing studies deal with. Furthermore, we consider imperfect actions in the preventive maintenance tasks, which address the issue that preventive maintenance do not always return components to the as-good-as-new status in practice. In this paper we propose three opportunistic maintenance optimization models, where the preventive maintenance is considered as perfect, imperfect and two-level action, respectively. Simulation methods are developed to evaluate the costs of the proposed opportunistic maintenance policies. Numerical examples are provided to demonstrate the advantage of the proposed opportunistic maintenance methods in reducing the maintenance cost. The two-level action method demonstrates to be the most cost-effective in different cost situations, while the imperfect maintenance policy, which is a simpler method, is a close second. The developed methods are expected to bring immediate benefits to wind power industry.

Keywords: Opportunistic maintenance; wind turbine; optimization; simulation; failure distribution.

1. Introduction

Wind, like solar and hydroelectric, is an important natural source that is essentially inexhaustible. Energy generated from wind is rapidly emerging as one of most important clean and renewable energy sources in the world. The huge potential and significant investment increase in generation capacity comes with the highly expected responsibility to manage wind farms to achieve the lowest operation and maintenance cost. To lower operation and maintenance cost as much as possible, more effective maintenance strategies need to be studied for successful future developments.

[#]This chapter appeared previously on the International Journal of Reliability, Quality and Safety Engineering. To cite this chapter, please cite the original article as the following: F. Ding and Z. Tian, *Int. J. Reliab. Qual. Saf. Eng*, **18**, 463–481 (2011), dci:10.1142/S0218539311004196.

A common goal of maintenance is to reduce the overall maintenance cost and improve the availability of the systems. The existing maintenance methods for wind power systems can be classified into failure-based, time-based, and condition-based maintenance (CBM). Failure-based (corrective) maintenance is carried out only after a failure occurs. In time-based maintenance, preventive maintenance is performed at predetermined time intervals. Condition-based maintenance applies the health condition prediction techniques to continuously monitor the components so that the components can be used the most effectively. However, with CBM, the availability of condition monitoring data is a big challenge for wind turbine systems today. Currently corrective maintenance and time-based preventive maintenance are widely used in wind power industry, which take advantage of ease of management, particularly in the case of extreme condition and high load associated with offshore farms. However, they have not been studied adequately and more effective methods need be developed.

In addition, the existing methods for wind turbine systems deal with individual components, but pay much less attention to the wind farm as a whole which generally consists of multiple wind turbine systems, and each wind turbine has multiple components with different failure distributions. Economic dependencies exist among wind turbine systems and their components in the wind farm. Therefore, opportunistic maintenance approach may be more cost-effective by taking advantage of already allocated resources and time. In opportunistic maintenance, whenever a failure occurs in the wind farm, the maintenance team is sent onsite to perform corrective maintenance, and take this opportunity to simultaneously perform preventive maintenance on multiple turbines and their components which show relatively high risks. The other disadvantage of existing research is that preventive maintenance is commonly considered as replacement, which is the perfect action to renew a component. In practice, preventive maintenance does not always return components to the as-good-as-new status. An imperfect preventive maintenance or minor repair action may reduce cost over the life cycle by taking advantage of fewer workload and balanced availability. Spinato *et al.* described that the repair actions for wind turbine components can be an addition to a new part, exchange of parts, removal of a damaged part, changes or adjustment to the settings, software update, and lubrication or cleaning.[1]

In the efforts to address the issues listed above, we propose a series of opportunistic maintenance optimization models, where preventive maintenance is considered as perfect, imperfect and two-level actions, respectively. These policies are defined by the age threshold value(s) at the component level. Based on failure distribution information of components, the age values of each component at each failure instant can be obtained, and the optimal policy corresponding to the minimum average cost can be decided. Simulation methods are developed to evaluate the costs of proposed opportunistic maintenance policies. Numerical examples will be provided to illustrate the proposed approaches. A comparative study with the widely used corrective maintenance policy will be used to demonstrate the advantage of the proposed opportunistic maintenance methods in reducing the maintenance cost significantly.

The remainder of the paper is presented as follows. Section 2 reviews the current literatures related to opportunistic maintenance in wind power industry. Section 3 introduces the proposed opportunistic maintenance optimization models, and develops the simulation methods to evaluate the total average costs. In Sec. 4, numerical examples are provided to illustrate the proposed approaches, and a comparative study is conducted to demonstrate the advantage of proposed opportunistic maintenance methods. Section 5 concludes the paper with some discussions on future work.

Abbreviations

CBM: Condition based maintenance
PM: Preventive maintenance
ANN: Artificial neutral network
MTTF: *mean time to failure*

Notations

C_E	The total maintenance cost per turbine per unit of time;
p	The component age threshold value;
q	The age reduction percentage threshold value;
C_{pv}	The variable preventive replacement cost for a component;
C_{pf}	The fixed preventive maintenance cost for a component;
C_p	The total preventive maintenance cost for a component;
C_f	The failure replacement cost for a component;
C_{fix}	The fixed cost for sending a maintenance team to the wind farm;
C_{Access}	The cost for accessing a wind turbine in the wind farm;
C_T	The total maintenance cost;
M	The number of wind turbines in a wind farm;
K	The number of components in a wind turbine;
α_k	Weibull distribution scale parameter for component k;
β_k	Weibull distribution shape parameter for component k;
$TA_{k,m}$	The absolute failure time of component k in turbine m;
$TL_{k,m}$	The generated lifetime of component k in turbine m by sampling failure distribution;
$FA_{k,m}$	The real lifetime of component k in turbine m;
$IP_{k,m}$	A value indicating if a preventive maintenance is to be performed on component k in turbine m;
IA_m	A value indicating if a preventive maintenance is to be performed in turbine m.

2. Review of Opportunistic Maintenance

For a multi-component system, a strategy called opportunistic maintenance can be used in many industries, and it is one of four maintenance strategies mentioned in Europe Wind Energy Report (2001) for European offshore wind farms. In opportunistic maintenance, the corrective activities performed on failed components can be combined with preventive activities carried out on other components. Opportunistic maintenance should be a considerable strategy for a wind farm because there are multiple wind turbines and a wind turbine has multiple components. Obviously economic dependencies exist among various components and systems in the farm. When a down time opportunity is created by the failed component, maintenance team may perform preventive maintenance for other components satisfying pre-specified decision rule. As a result, substantial cost can be saved comparing with separate maintenance.

Applications of various maintenance policies have been reported in the literature. Laggoune proposed a policy for a hydrogen compressor, in which the components have different failure distributions.[2] The maintenance decision on a component is made based on the cost comparison of performing or not performing replacements, and the conditional probability was used to calculate the expected cost. Crocker used age-related policy to optimize the cost of one part of military aero-engine.[3] They performed Monte Carlo simulation, and concluded that a potential benefit can be obtained from opportunistic maintenance performing on relatively cheap components. The policy presented by Mohamed-Salah *et al.* for a ball bearing system deals with the time difference between expected preventive maintenance time and failure instant.[4] The time difference factor is considered to decide an opportunistic maintenance action, and the threshold value is determined by cost evaluation. Kabir *et al.* studied a multi-unit system by assuming the lifetime of components following Weibull distribution with the same parameters.[5] They used a genetic algorithm technique to make an optimal decision for opportunistic replacement to maximize net benefit.

However, the opportunistic maintenance application studies in wind power industry are very few. Tian *et al.* proposed a condition based maintenance approach where a preventive replacement decision is made based on the failure probability during maintenance lead time.[6] Artificial neutral network (ANN) technique is used to predict component's remaining life at inspection times. Knowing that some specific turbine will fail and also the power production per day over a short summer period, Besnard specified the opportunity depending on both failure chance and real wind data.[7] They presented a cost objective function with a series of constraints, and accordingly an optimal maintenance schedule for a 5 turbines wind farm is suggested. Eunshin *et al.* presented optimal maintenance strategies for wind turbine systems under stochastic weather conditions, using simulation-based approaches.[8] In the field of engineering maintenance, imperfect maintenance has been studied for production systems[9] and systems connected in series configuration.[10]

3. The Proposed Opportunistic Maintenance Approaches

In this section, three opportunistic maintenance strategies for wind farms are proposed, where preventive maintenance are considered as perfect, imperfect and two-level action, respectively. Simulation methods for the cost evaluation of the proposed policies are presented. We first define the imperfect maintenance actions in the following subsection.

3.1. *Imperfect maintenance actions*

In this work, it is assumed that all the wind turbines under consideration are identical, and the degradation processes of the wind turbine components are mutually independent. The maintenance time is negligible. Detailed explanations of the imperfect action are given below.

After maintenance, the age of component is reduced by ratio $q(0 \leq q \leq 1)$, and the failure age is updated as-good-as-new with probability q and remains at the old lifetime with probability with $1 - q$, i.e.,

$$\text{Failure Age} = q \times T_{\text{Renew}} + (1 - q) \times T_{\text{Old}}. \tag{1}$$

For example, assume $q = 0.8$, a gearbox's lifetime is 20 years (T_{Old}) and its age before maintenance is 8 years. And randomly a new one has a lifetime of 25 years (T_{Renew}) if we replace it. Thus, this gearbox's new age after maintenance will be $8 - 8 \times 0.8 = 1.6$ years, and the imperfect maintenance action changes its failure age to $25 \times 0.8 + 2 \times 0.2 = 24$ years old.

Suppose the imperfect maintenance cost is a function of q, which is given as follows:

$$C_p = \begin{cases} q^3 C_{pv} + C_{pf} & 0 < q \leq 1 \\ 0 & q = 0 \end{cases} \tag{2}$$

where C_p is the imperfect maintenance cost, C_{pv} is the variable preventive replacement cost if a 100% imperfect maintenance (i.e., replacement) is performed, and C_{pf} is the fixed maintenance cost. The idea is that when there is no imperfect maintenance, the cost is of course 0. If there is an imperfect maintenance action, regardless of the age reduction percentage, the fixed cost C_{pf} will be incurred. The imperfect maintenance cost increases with the age reduction factor q in a nonlinear fashion. In the example shown in Eq. (2), we suppose it is in the form of $q^3 C_{pv}$, and it can take other forms in other applications. When the age reduction factor is equal to 1, the total preventive replacement cost is $C_{pv} + C_{pf}$, which is equivalent to 100% age reduction ($q = 1$).

3.2. *The proposed opportunistic maintenance optimization models*

As we mentioned earlier, the proposed opportunistic maintenance strategies are defined by the age threshold value(s), based on which a maintenance decision can be made on the components that reach this age value. Figure 1 generally illustrates

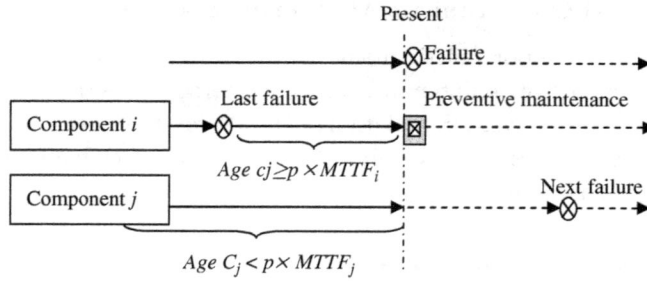

Fig. 1.   The proposed opportunistic maintenance concept.

the policy we propose. Suppose there is a failure occurring in the wind farm. The maintenance crew is sent to perform failure replacement, and they take this opportunity to perform preventive maintenance on other qualified components. Suppose component i will be performed a preventive maintenance action on because its age reaches the threshold, which is $p \times MTTF_i$ at this moment. The age of component j does not reach the threshold $p \times MTTF_j$ so that a maintenance task will not be performed and it will continue to operate till the next opportunity, or it may fail first in the farm. For the components decided to perform preventive maintenance on according to the two-level action method to be introduced, two age thresholds $p1$ and $p2$ are applied ($p2 > p1$). A preventive replacement is to be performed if its current age reaches the large age threshold $p2 \times MTTF_i$, otherwise an imperfect maintenance action is to be performed since its age reaches the small age threshold.

The proposed policies are based on the following assumptions or properties:

- The failures of all components in a turbine system follow Weibull distribution, and the failure rate increases over time. It is also assumed that the practitioners have no problem in obtaining the failure time distributions for the components.
- All wind turbines in the farm are identical, and the deterioration process of each component in a wind turbine system is independent.
- Any component failure leads to the turbine system failure.
- The corrective replacement time and preventive maintenance time are negligible. This is true in many applications, and it is also used to simplify the problem.
- Suppose there are M wind turbines in the wind farm, and we consider K critical components for each turbine.

In terms of preventive maintenance actions, we discuss them in three strategies presented in following subsections.

3.2.1. *Strategy* 1: *opportunistic maintenance with perfect action only*

- The maintenance policy is that corrective replacements are performed on demand, and we take this opportunity to perform preventive replacement (i.e., perfect action) on other components in the same and other turbine systems. The maintenance decision depends on the age of component at the failure instant. The detailed policy is explained as follows:

(i) Perform failure replacement if a component fails.
(ii) At the moment of failure, perform preventive replacement on component $k(k = 1, \ldots, K)$ in wind turbine $m(m = 1, \ldots, M)$ if $\text{age}_{k,m} \geq MTTF_k \times p$.
(iii) If the component does not undergo preventive maintenance, it will continue to be used until the next failure occurs in the wind farm.

• The brief objective function is

$$\min C_E(p), \tag{3}$$

where C_E is the total expected maintenance cost per turbine per day, and p is design variable. The objective is to determine the optimal age threshold p to minimize the total expected maintenance cost per turbine per day.

3.2.2. *Strategy 2: opportunistic maintenance considering imperfect actions*

• Under this maintenance policy, corrective replacements are performed on demand, and we take this opportunity to perform imperfect maintenance action on other components in the same and other turbine systems. The maintenance decision depends on the age of component at the failure instant. The detail policy is explained as follows:

(i) Perform failure replacement if a component fails.
(ii) At the moment of failure, perform imperfect preventive maintenance action, by reducing the component age by q, on component k $(k = 1, \ldots, K)$ in wind turbine $m(m = 1, \ldots, M)$ if $\text{age}_{k,m} \geq MTTF_k \times p$. The imperfect maintenance actions we defined in Sec. 3.1 are applied.
(iii) If the component does not undergo preventive maintenance, it will continue to be used until the next failure occurs in the wind farm.

• The brief objective function is

$$\min C_E(p, q), \tag{4}$$

where p, q are design variables. p represents the component's age threshold, and q is the percentage of age reduction. The objective is to determine the optimal age threshold p and age reduction ratio q to minimize the total expected maintenance cost per turbine per day.

3.2.3. *Strategy 3: opportunistic maintenance with two-level actions*

• Under this maintenance policy, corrective replacements are performed on demand, and we take this opportunity to perform two-level preventive maintenance actions on other components in the same and other turbine systems. In this policy, two-level actions are defined as imperfect and perfect level actions, which are determined by two age thresholds $MTTF \times p1$ and $MTTF \times p2$, where $p2 > p1$. The maintenance decision depends on the age of component at the failure instant. The detail policy is explained as follows:

(i) Perform failure replacement if a component fails.
(ii) At the moment of failure, perform imperfect preventive maintenance actions, by reducing the component age by q, on component k $(k = 1, \ldots, K)$ in wind turbine $m(m = 1, \ldots, M)$ if $MTTF_k \times p2 \geq \text{age}_{k,m} \geq MTTF_k \times p1$. And perform preventive replacement on this component if $\text{age}_{k,m} \geq MTTF_k \times p2$. It is implied that the older a component is, the more it tends to be replaced.
(iii) If the component does not undergo preventive maintenance, it will continue working until the next failure occurs in the wind farm.

• The brief objective function is

$$\text{s.t. min} \quad C_E(p1, p2),$$
$$0 < p1 < p2 < 1, \tag{5}$$

where $p1$, $p2$ are design variables corresponding to two age thresholds. The objective is to determine the optimal variable values to minimize the total expected maintenance cost per turbine per day.

3.3. *The solution method*

Due to the complexity of optimization problem in this study, we develop simulation methods to calculate the average cost C_E for each strategy. Suppose the failure distributions of components are known, and the age values of each component at each failure instant can be obtained. The simulation processes of the proposed strategies do not have large difference from one another, and those differences will be detailed in the corresponding simulation steps. Figure 2 shows the flow chart of the simulation procedures in general. Details are explained as follows.

Step 1. Simulation Initialization. Specify the maximum simulation iterations I. Specify the number of wind turbines M and components K in a system. Specify the maximum value of design variables, $p1$, $p2$, q, which correspond to different policies. For each component k, specify the cost values including the failure replacement cost C_{fk}, and preventive replacement cost C_{pv} and C_{pf}. The fixed cost C_{fix} and the access cost C_{Access} also need to be specified. The total cost is set to be $C_T = 0$, and will be updated during the simulation process. The Weibull distribution parameters α_k and β_k of each component are given, which are presented in Sec. 4.1. The absolute time, $TA_{k,m}$, is defined as the cumulative time of every failure for that component k in turbine m. At the beginning, generate the lifetimes $TL_{k,m}$ for each component in each turbine by sampling the Weibull distribution for component k with parameter α_k and β_k. Thus, the age values for all components are zero at the beginning, that is, $\text{Age}_{k,m} = 0$, and $TA_{k,m} = TL_{k,m}$ at the moment of the first failure. The other term $FA_{k,m}$ is applied in Strategy 2 and 3, which denotes the new failure age of the

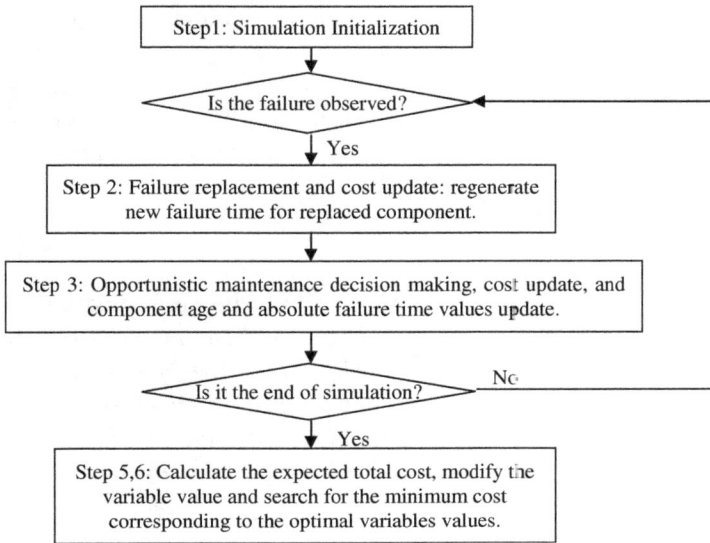

Fig. 2. Simulation process for cost evaluation for proposed opportunistic maintenance.

component k in turbine m after imperfect maintenance actions, and $FA_{k,m} = TL_{k,m}$ at the beginning.

Step 2. Failure replacement and cost update. Compare the values of all $TA_{k,m}$. The failure replacement of the ith iteration occurs at t_i, and $t_i = \min(TA_{k,m})$. The time to failure of the ith iteration is represented by Δt_i, and $\Delta t_i = t_i - t_{i-1} \cdot t_0 = 0$. The failure replacement cost C_{fk} is incurred due to the failed component k, and meanwhile the fixed cost of sending a maintenance team to the wind farm, C_{fix}, is incurred. The total cost due to failure replacement is updated as:

$$C_T = C_T + C_{fk} + C_{\text{fix}} \tag{6}$$

Regenerate a new lifetime $TL_{k,m}$ by sampling the Weibull distribution for this component with parameter α_k and β_k, and reset its age to 0. Its absolute time is moved to next failure, i.e.,

$$TA_{k,m} = t_i + TL_{k,m} \tag{7}$$

and its failure age (only applicable for Strategy 2 and Strategy 3) is updated as:

$$FA_{k,m} = TL_{k,m}. \tag{8}$$

The age of all the other components in M turbines at current point is summed up with Δt_i i.e.,

$$\text{Age}_{k,m} = \text{Age}_{k,m} + \Delta t_i. \tag{9}$$

Step 3. Opportunistic maintenance decision making, and cost, age and time value update. For the rest of the components in the systems M, opportunistic maintenance decisions can be made according to the proposed strategies described in Sec. 3.2. At the moment of failure replacement, three maintenance policies are applied, respectively. The total cost due to preventive maintenances is updated as:

$$C_T = C_T + \sum_{m=1}^{M} \left(\sum_{k=1}^{K} Cp_k \times IP_{k,m} + C_{\text{Access}} \times IA_m \right), \tag{10}$$

where $IP_{k,m} = 1$ if a preventive maintenance is to be performed on component k in turbine m, and otherwise it equals 0. $IA_m = 1$ if any preventive maintenance is to be performed on turbine m, and it equals 0 otherwise. Note that Cp_k represents the total preventive maintenance cost, and it varies with different q according to Eq. (2), where q can be considered to be the maintenance effort.

The changes of $TA_{k,m}$ and $\text{Age}_{k,m}$ for the three different policies are different, and they are explained as follows:

- Strategy 1, perfect maintenance only
  If $\text{Age}_{k,m} \geq MTTF_k \times p$, preventive replacement is performed on component k in turbine m. Generate a new lifetime $TL_{k,m}$ for this component with parameter α_k and β_k, and reset its age to 0. Its absolute time will be moved to next failure, i.e.,

$$TA_{k,m} = t_i + TL_{k,m}. \tag{11}$$

- Strategy 2, considering imperfect actions
  If $\text{Age}_{k,m} \geq MTTF_k \times p$, imperfect maintenance is performed on component k in turbine m. Generate a new lifetime $TL_{k,m}$, and its age, failure age and absolute time are updated as:

$$\text{Age}_{k,m} = \text{Age}_{km} \times (1 - q), \tag{12}$$

$$FA_{k,m} = q \times TL_{k,m} + (1 - q) \times FA_{k,m}, \tag{13}$$

$$TA_{k,m} = t_i + FA_{k,m} - \text{Age}_{k,m}. \tag{14}$$

- Strategy 3, two-level maintenance
  If $MTTF_k \times p2 \geq \text{Age}_{k,m} \geq MTTF_k \times p1$, imperfect maintenance is performed on the component. Generate a new failure time $TL_{k,m}$. Age, failure age and absolute time are updated similarly according to Eqs. (12)–(14).

  If $\text{Age}_{k,m} \geq MTTF_k \times p2$, preventive replacement is performed on the component. Generate a new life time $TL_{k,m}$, and reset its age to 0. Its failure age and absolute time are updated as:

$$FA_{k,m} = TL_{k,m},$$

$$TA_{k,m} = t_i + FA_{k,m}. \tag{15}$$

Note that in this strategy, q is a certain value determined as the optimal result in strategy 2, not a variable.

Step 4. After performing maintenance on all components, set $i = i + 1$. If i does not exceed the maximum simulation iteration I, repeat step 2 and step 3.

Step 5. Total expected cost calculation. When the maximum simulation iteration is reached, which is $i = I$, the simulation process for the cost evaluation with current variable value is completed. The total expected cost per wind turbine per day can be calculated as:

$$C_E = C_j. \tag{16}$$

If the maximum simulation iteration is not reached, repeat step 2, 3, 4 and 5. The expected total cost per turbine per day can be given by:

$$C_E = \frac{\sum_{i=1}^{I} \left(C_{\text{fix}} + Cf_m + \sum_{m=1}^{M} \left(\sum_{k=1}^{K} Cp_{k,m} \times IP_{k,m} + C_{\text{Access}} \times IA_m \right) \right)}{t_I \times M}, \tag{17}$$

where, $IP_{k,m} = 1$ if preventive maintenance is performed on the component k in turbine m, otherwise $IP_{k,m} = 0$. $IA_m = 1$ if any maintenance is performed on the wind turbine system m, otherwise $IA_m = 0$. t_i is the total length of I iterations.

Step 6. Search for the optimal value where the corresponding expected total cost per turbine per day C_E is minimized. As can be seen, once the optimal values of variables $p1$, $p2$ and q are found, the optimal maintenance strategies are determined.

4. Numerical Examples

4.1. *Opportunistic maintenance optimization using the proposed approaches*

In this section, examples are used to demonstrate the proposed opportunistic maintenance approaches for wind turbine systems. Consider 10 2-MW turbines in a wind farm at a remote site. To simplify the discussion, we study four key components in each wind turbine: the rotor, the main bearing, the gearbox and the generator.[9]

We assume that the failures of all the components follow Weibull distribution with increasing failure rate. All wind turbines in the farm are identical and the components deteriorate independently. The failure distribution parameters α(scale parameter) and β(shape parameter), and the cost data are given in Table 1.[12,13] The cost data includes the failure replacement costs (C_f), the variable preventive maintenance costs (C_{pv}), the fixed preventive maintenance cost (C_{pf}), the fixed cost of sending a maintenance team to the wind farm (C_{fix}), and the turbine access cost (C_{Access}).

The total maintenance cost can be evaluated using the proposed simulation method presented in Sec. 3.3. We present the optimization results for each proposed opportunistic maintenance strategy as follows.

Table 1. Failure distribution parameters and cost data for major components (k).

Component	α (day)	β(day)	C_f	C_{pv}	C_{pf}	C_{fix}	C_{Access}
Rotor	3000	3	112	28			
Bearing	3750	2	60	15	20	50	7
Gearbox	2400	3	152	38			
Generator	3300	2	100	25			

Strategy 1. Perfect maintenance only

As shown in Fig. 3, the obtained optimal threshold age value for a component is 70% of its mean lifetime, and the optimal average maintenance cost per unit time is $185.2/day.

Strategy 2. Considering imperfect maintenance

As can be seen in Fig. 4, the optimal maintenance plan is that performing the imperfect maintenance action on the component whose age reaches $p = 50\%$ of its average lifetime, and the best maintenance action is to reduce the age of component by $q = 60\%$. The optimal average maintenance cost per unit of time is $154.5/day. Figures 5 and 6 show the cost versus one variable plot while the other variable is kept at the optimal value, respectively.

Strategy 3. Two-level maintenance

Two-level maintenance defines that the low level maintenance is imperfect and the high level is perfect, and they are supposed to be performed at different age

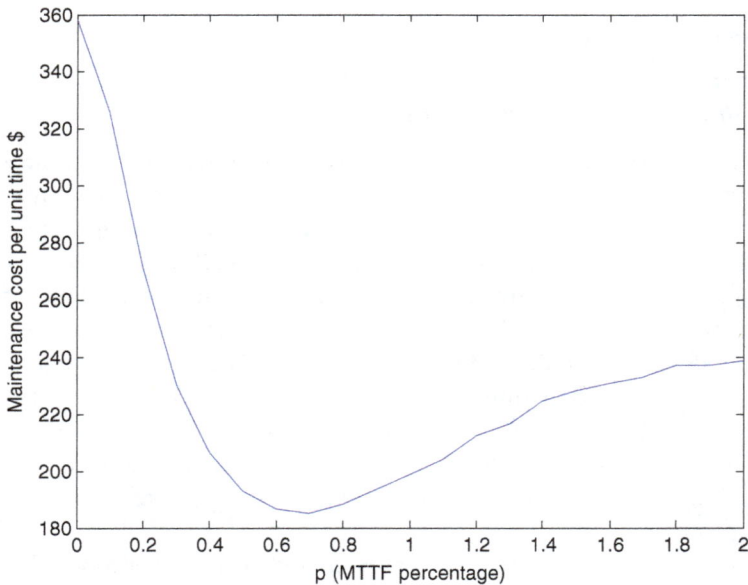

Fig. 3. Cost versus preventive replacement age threshold value (p).

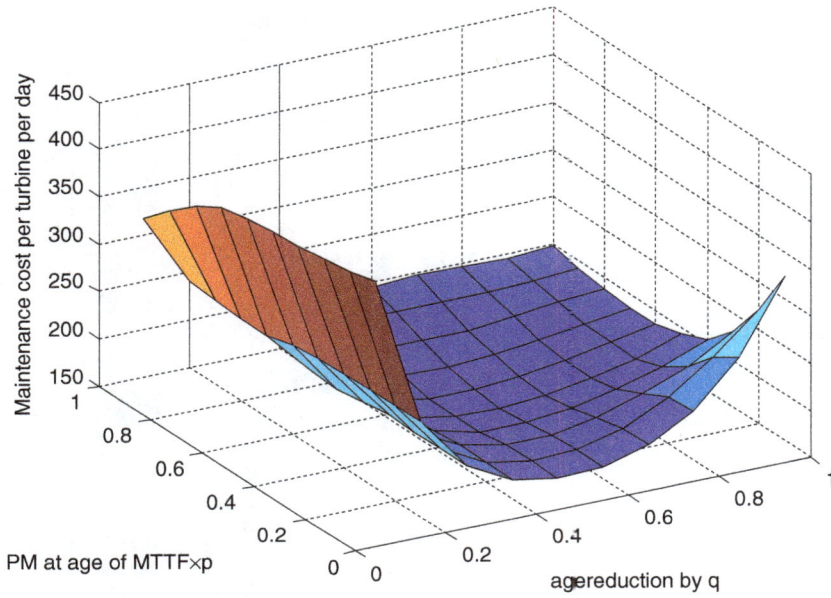

Fig. 4.   Cost versus maintenance age threshold value p and age reduction q.

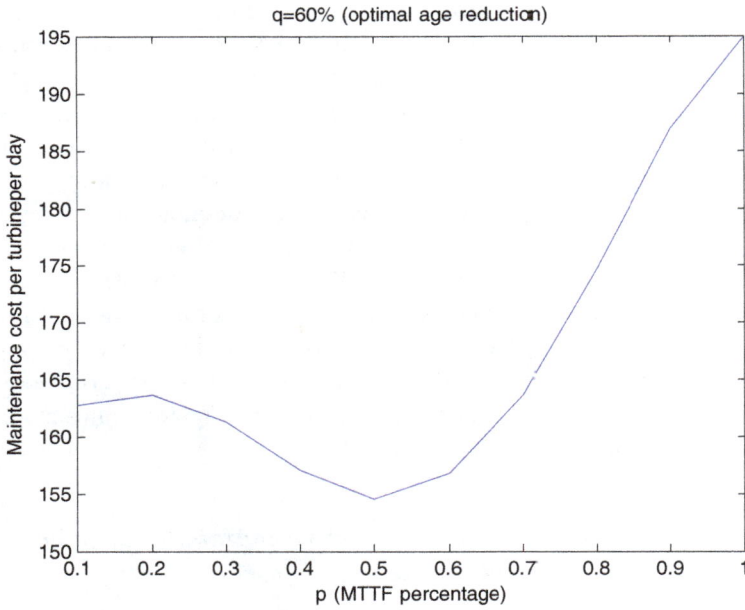

Fig. 5.   Cost versus $p(q = 60\%)$.

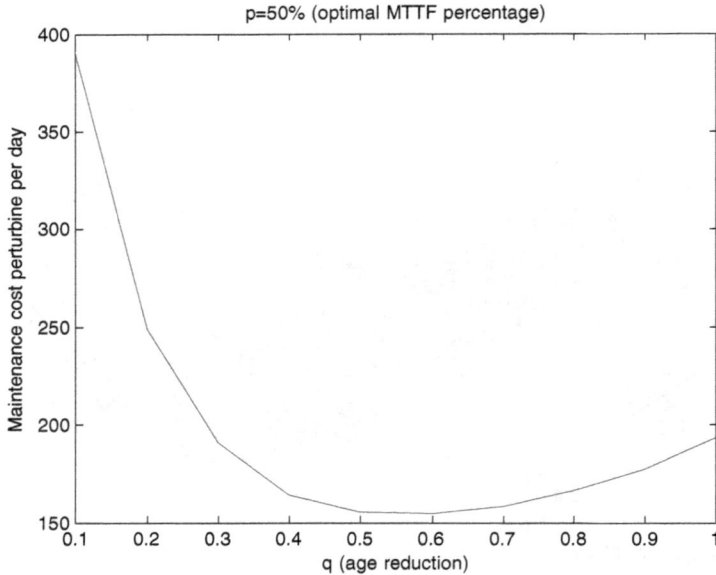

Fig. 6.   Cost versus $q(p = 50\%)$.

thresholds. A replacement will be considered when a component is older, while the imperfect action tends to be performed at the younger age. Based on the optimization result for strategy 2, the imperfect maintenance action of reducing the component's age by 60% is applied to this optimization problem.

As can be seen in Fig. 7, the two-level maintenance model leads to the minimum cost of \$153/day. Due to the constraint $p2 > p1$ in this proposed model, note that the costs of area of $p2 < p1$ in Fig. 6 are set to be null and does not need any concern in this study. The optimal policy is that at the moment of a failure in the farm, imperfect preventive maintenance action is taken on the component whose age is between 50% and 110% of its mean lifetime, and preventive replacement is performed on the component whose age exceeds 110% of its mean lifetime.

As can be seen from the optimization results for the three proposed opportunistic maintenance methods, the two-level action method is the most cost-effective comparing to the other two methods, and the imperfect maintenance method is the close second with a slightly worse performance.

4.2. Comparative study with the corrective maintenance policy under different cost situations

In this section, we investigate the advantage of the proposed opportunistic maintenance methods comparing to the corrective maintenance policy, where only failure replacement is performed when a component fails in the wind farm. In addition, we also investigate the methods under different cost situations.

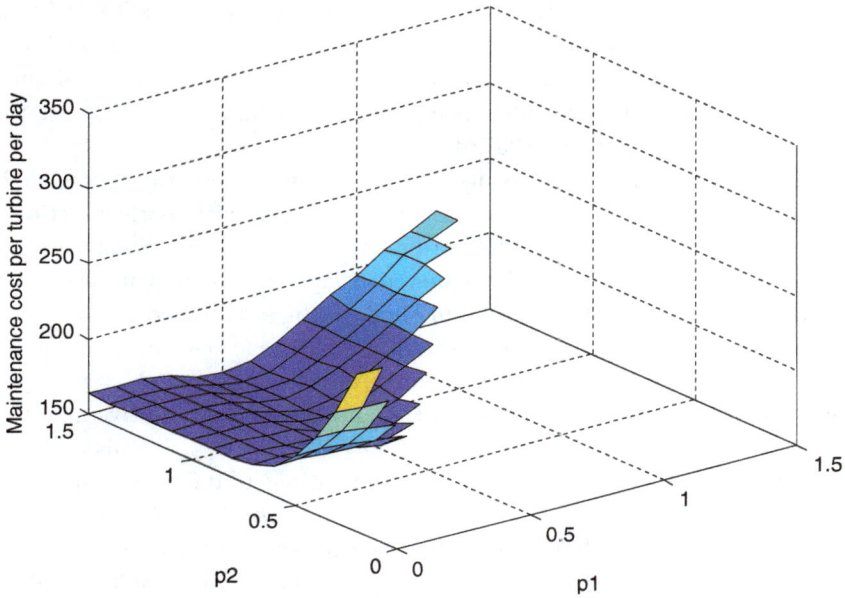

Fig. 7. Cost versus two age threshold values $p1$ and $p2$.

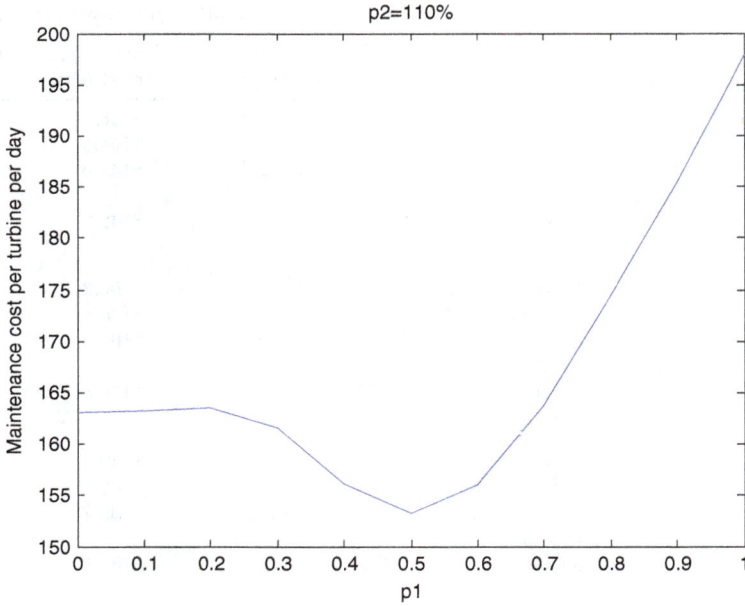

Fig. 8. Cost versus age threshold values $p1$ ($p2 = 110\%$).

We use the same failure replacement cost and fixed cost data in Table 1, and use a simulation method to evaluate the total average cost of corrective maintenance policy applied to the same wind farm, which consists of 10 2-MW wind turbines. By applying the corrective maintenance policy, the total average cost per turbine per unit of time is found to be $237/day.

In Sec. 4.1, the optimization results show the optimal average cost of $185.2/day, $154.5/day and $153/day for the proposed strategies with perfect action only, imperfect action and two-level action, respectively. Thus, significant cost savings of 21.9%, 34.8% and 35.4% can be achieved comparing to the corrective maintenance policy, and the two-level action method produces the lowest cost.

In addition, we also investigate the optimized average cost under different cost situations. Previously the variable preventive maintenance cost C_{pv} was specified as one-fourth of failure replacement cost C_f, and the total preventive maintenance cost for a component has the function of $Cp = q^3 C_{pv} + C_{pf}$. In this comparative study, we compare the total cost with all combinations of different cost situations, among which cost function $Cp = q^2 C_{pv} + C_{pf}$ is considered, C_{pv} is modified as $C_f/2$, $C_f/4$, $C_f/8$, respectively, and the fixed preventive maintenance cost C_{pf}, the fixed cost C_{fix} and the access cost C_{Access} are modified as well. According to

Table 2.   Optimal cost of proposed opportunistic maintenance strategies.

			Minimum cost (/day)		
Cost parameters		C_p	Perfect action only	Considering imperfect action	Two-level action
$C_{pv} = 1/4\,C_f$	$C_{fix} = \$50000$	$q^2 C_{pv} + C_{pf}$	$185.2	$168.6	$168
$C_{pf} = \$20000$	$C_{Access} = \$7000$	$q^2 C_{pv} + C_{pf}$	$185.2	$154.5	$153
	$C_{fix} = \$30000$	$q^2 C_{pv} + C_{pf}$	$172.4	$148.5	$147.6
	$C_{Access} = \$7000$				
	$C_{fix} = \$30000$	$q^2 C_{pv} + C_{pf}$	$168.6	$142.2	$141.8
	$C_{Access} = \$4000$				
$C_{pv} = 1/2\,C_f$	$C_{fix} = \$50000$	$q^2\,C_{pv} + C_{pf}$	$220.7	$196.3	$195.3
$C_{pf} = \$20000$	$C_{Access} = \$7000$	$q^2\,C_{pv} + C_{pf}$	$220.7	$170.2	$169.5
	$C_{fix} = \$30000$	$q^2\,C_{pv} + C_{pf}$	$200	$163.5	$162.8
	$C_{Access} = \$7000$				
	$C_{fix} = \$30000$	$q^2 C_{pv} + C_{pf}$	$198.1	$156.8	$155.9
	$C_{Access} = \$4000$				
$C_{pv} = 1/8\,C_f$	$C_{fix} = \$50000$	$q^2\,C_{pv} + C_{pf}$	$160.3	$149.4	$148.5
$C_{pf} = \$20000$	$C_{Access} = \$7000$	$q^2\,C_{pv} + C_{pf}$	$160.3	$142.9	$141.7
	$C_{fix} = \$30000$	$q^2\,C_{pv} + C_{pf}$	$149.9	$136.7	$136
	$C_{Access} = \$7000$				
	$C_{fix} = \$30000$	$q^2\,C_{pv} + C_{pf}$	$145.2	$131.1	$130
	$C_{Access} = \$4000$				
$C_{pv} = 1/4\,C_f$	$C_{pf} = \$5000$	$q^2\,C_{pv} + C_{pf}$	$157.8	$78.2	$77.6
$C_{fix} = \$50000$	$C_{pf} = \$10000$	$q^2\,C_{pv} + C_{pf}$	$167.9	$106.1	$105.5
$C_{Access} = \$7000$	$C_{pf} = \$15000$	$q^2\,C_{pv} + C_{pf}$	$177.5	$132.2	$131.4
	$C_{pf} = \$25000$	$q^2\,C_{pv} + C_{pf}$	$193.4	$173.2	$172.3

the optimization results, each combination shows that the two-level action method costs the lowest comparing to other proposed methods. However, the benefit of the two-level method is not significant comparing to the imperfect action method. Considering that the imperfect maintenance method is simpler comparing to the two-level maintenance method, it might be more attractive in some applications.

In addition, there are some findings when the cost varies in certain situations. The details are presented as follows:

(i) Keeping the cost C_{pf} constant, the smaller C_{pv} is, the less the advantage of Strategy 2 is comparing to Strategy 1.
(ii) Keeping the cost C_{pv} constant, the smaller C_{pf} is, the more the advantage of Strategy 2 is comparing to Strategy 1.

The optimization results for the proposed strategies and cost situations are given in Table 2.

5. Conclusions

Maintenance optimization generates great benefits for many industries. However, it is relatively new for wind power industry, which has been growing very fast in recent years due to the highly increasing demand on clean and renewable energy. This paper aims to develop opportunistic maintenance approaches for an entire wind farm rather than individual components that most of the existing studies deal with. Furthermore, we consider imperfect actions in the preventive maintenance tasks, which addresses the issue that preventive maintenance do not always return components to the as-good-as-new status in practice. In this paper we propose three opportunistic maintenance optimization models, where the preventive maintenance is considered as perfect, imperfect and two-level action, respectively. Simulation methods are developed to evaluate the costs of proposed opportunistic maintenance policies. Numerical examples are provided to demonstrate the advantage of the proposed opportunistic maintenance methods in reducing the maintenance cost. The two-level action method demonstrates to be the most cost-effective in different cost situations. The imperfect maintenance policy, which is a simpler method, is a close second and might be more attractive in some applications. The developed methods are expected to bring immediate benefits to wind power industry. Future work will be conducted to find analytical methods to evaluate cost more accurately in the opportunistic maintenance optimization problems, and investigate various imperfect actions with different effects on reliability.

References

1. F. Spinato, P. J. Tavner and G. J. W. van Bussel, Reliability of wind turbine sub-assemblies, *IET Renew. Power Gener.* **3**(4) (2008) 387–401.

2. R. Laggoune, A. Chateauneuf and D. Aissani, Opportunistic policy for optimal preventive maintenance of a multi-component system in continuous operating units, *Comput. Chem. Eng.* **33**(9) (2009) 1449–1510.
3. J. Crocker and U. D. Kumar, Age-related maintenance versus reliability centred maintenance: A case study on aero-engines, *Reliab. Eng. Syst. Saf.* **67**(2) (2000) 113–118.
4. A.-K. D. Mohamed-Salah and G. Ali, A simulation model for opportunistic maintenance strategies, in *Proc. Int. Conf. Emerging. Technologies and Factory Automation*, 18–21 October (1999), Barcelona, Spain, Vol. 1, pp. 703–708.
5. S. Kabir *et al.* Optimization model for opportunistic replacement policy using genetic algorithm with fuzzy logic controller, in *Proc. Int. Congr. Evolutionary Computation*, 8–12 December (2003), Canberra, Australia, Vol. 4, pp. 2834–2843.
6. Z. Tian, T. Jin, B. Wu and F. Ding, Condition based maintenance optimization for wind power generation systems under continuous monitoring, *Renew. Energ.* **36** (2011) 1502–1509.
7. M. P. Besnard *et al.* An optimization framework for opportunistic maintenance of offshore wind power system, in *Proc. IEEE PowerTech 2009 Conf.*, 28 June–2 July (2009), Bucharest, Romania.
8. B. Eunshin, N. Lewis and Ding Yu, Optimal maintenance strategies for wind turbine systems under stochastic weather conditions, *IEEE Trans. Reliab.* **59**(2) (2010) 393–404.
9. H. Z. Wang and H. Pham, Some maintenance models and availability with imperfect maintenance in production systems, *Ann. Oper. Res.* **91** (1999) 305–318.
10. H. Z. Wang and H. Pham, Availability and maintenance of series systems subject to imperfect repair and correlated failure and repair. *Eur. J. Oper. Res.* **174**(3) (2006) 1706–1722.
11. National instruments products for wind turbine condition monitoring, http://zone.ni.com/devzone/cda/tut/p/id/7676.
12. WindStats Newsletter. 2008–2009, 21(4)–22(3), Germany.
13. Hau E., *Wind Turbines: Fundamentals, Technologies, Application, Economics* (Springer, 2006).

About the Authors

Fangfang Ding is currently a Ph.D. student at the Department of Mechanical and Industry Engineering, Concordia University, Canada. She received the Bachelor of Science degree in Computer and Its Applications in 1999 from Wuhan University, China, and the Master of Applied Science degree in Quality System Engineering in 2010 from Concordia University, Montreal, Canada. Before pursuing the Master program in 2008 in Canada, she had been a Product Design Engineer and Quality Associate Manager in one of the biggest optical telecommunication company in China. Her research interests include reliability analysis, signal processing, condition monitoring, and maintenance planning and optimization.

Zhigang Tian is currently an Assistant Professor at Concordia Institute for Information Systems Engineering at Concordia University, Montreal, Canada. He received his Ph.D. degree in 2007 in Mechanical Engineering at the University of Alberta, Canada; and his M.S. degree in 2003, and B.S. degree in 2000 both in Mechanical Engineering at Dalian University of Technology, China. His research interests focus on reliability analysis and optimization, prognostics, condition monitoring, maintenance optimization and renewable energy systems. He is a member of IIE and INFORMS. He is an Associate Editor of the International Journal of Performability Engineerng. He is the recipient of the 2011 Petro-Canada Young Innovator Award (Technology, Industry, and the Environment).

Chapter 13

SELECTION OF MAINTENANCE STRATEGY FOR AIRCRAFT SYSTEMS USING MULTI-CRITERIA DECISION MAKING METHODOLOGIES[#]

ALIREZA AHMADI[*,‡], SUPRAKASH GUPTA[†,§],
RAMIN KARIM[*,¶] and UDAY KUMAR[*,‖]

*Division of Operation and Maintenance Engineering
Lulea University of Technology, Lulea, SE-97187, Sweden

†Department of Mining Engineering, Institute of Technology
Banaras Hindu University, Varanasi – 221005, India

‡alireza.ahmadi@ltu.se
§suprakash_gupta@yahoo.co.in
¶ramin.karim@ltu.se
‖uday.kumar@ltu.se

This paper, proposes the Multi-Criteria Decision Making (MCDM) methodology for selection of a maintenance strategy to assure the consistency and effectiveness of maintenance decisions. The methodology is based on an AHP-enhanced TOPSIS, VIKOR and benefit-cost ratio, in which the importance of the effectiveness appraisal criteria of a maintenance strategy is determined by the use of AHP. Furthermore, in the proposed methodology the different maintenance policies are ranked using the benefit-cost ratio, TOPSIS and VIKOR. The method provides a basis for consideration of different priority factors governing decisions, which may include the rate of return, total profit, or lowest investment. When the preference is the rate of return, the benefit-cost ratio is used, and for the total profit TOPSIS is applied. In cases where the decision maker has specific preferences, such as the lowest investment, VIKOR is adopted. The proposed method has been tested through a case study within the aviation context for an aircraft system. It has been found that using the methodology presented in the paper, the relative advantage and disadvantage of each maintenance strategy can be identified in consideration of different aspects, which contributes to the consistent and rationalized justification of the maintenance task selection. The study shows that application of the combined AHP, TOPSIS, and VIKOR methodologies is an applicable and effective way to implement a rigorous approach for identifying the most effective maintenance alternative.

Keywords: Aircraft maintenance; Multi-Criteria Decision Making (MCOM); maintenance strategy; AHP; TOPSIS; VIKOR; benefit-cost ratio; maintenance decision making; maintenance effectiveness.

[#]This chapter appeared previously on the International Journal of Reliability, Quality and Safety Engineering. To cite this chapter, please cite the original article as the following: A. Ahmadi, S. Gupta, R. Karim and U. Kumar, *Int. J. Reliab. Qual. Saf. Eng*, **17**, 223–243 (2010), doi:10.1142/S0218539310003779.

1. Introduction

Maintenance accounts for approximately 11 percent of an airline's employees and 10–15 percent of its operating expenses.[1] A large portion of the direct and indirect maintenance costs in the whole life cycle stems from the consequences of decisions made during the initial maintenance programme development. Since the decision made for developing the initial scheduled maintenance programme strongly affects the aircraft safety, availability performance, and lifecycle cost, it is essential to select the most effective maintenance options that assure system effectiveness. To this end, this paper suggests that, rather than use a decision diagram approach, one should use a rigorous approach which not only considers the maintenance strategies offered by ATA MSG-3,[2] but also allows consideration of other available technologies such as Prognostic Health Management (PHM). However, to make rational and justifiable decisions concerning maintenance, one needs to have a clear idea of what the advantages and disadvantages of each maintenance strategy are.[3] Moreover, in maintenance strategy formulation, Reliability, Availability, Maintainability and Safety (RAMS) characteristics and related consequences in system effectiveness should be taken into account.[4–6] Every maintenance strategy has its inherent merits and demerits. To evaluate the appropriateness of a maintenance strategy, one must formulate a set of evaluating criteria that will adequately assess the effectiveness and efficiency of the maintenance strategy and the cost of implementing it. Moreover, these assessments require knowledge of various factors which indicate the strengths and preferability of maintenance strategies, according to the associated evaluating criteria.

Due to a long list of contributory factors and attributes, inadequacy and uncertainty in the required information, and lack of modelling support for tangible and intangible cost and benefit factors, justification of a maintenance alternative is a critical and complex task.[7] However, the experiences of field experts provide an effective database supporting this estimation. In this process of decision making, the decision makers have to face making numerous and conflicting evaluations. In fact, the management of the large number of tangible and intangible attributes that must be taken into account represents the main complexity of the problem.[8]

Since maintenance decision making is often characterized by the need to satisfy multiple objectives, the formulation of multi-criteria decision models is a worthwhile topic of future research work in inspection (maintenance) problems.[9] To this end, the Multi-Criteria Decision Making (MCDM) approach has been proposed in the literature, and has gained impetus in the field of maintenance strategy selection to provide support in the decision making process.[7,8,10–12] MCDM aims at highlighting conflicting evaluations and deriving a way to come to a compromise in a transparent process. Multi-criteria optimization is the process of determining the best feasible solution according to the established criteria (representing different effects). Al-Najjar *et al.* (2003)[13] have proposed a fuzzy logic-based maintenance approach, setting failure causes as the criteria and ranking different maintenance approaches on the basis of their capability of detecting changes in the criteria, while Labib (2004)[14] has used failure frequency and downtime as the criteria. Almeida and Bohoris (1995)[10] discuss the application of decision making theory to maintenance with particular attention paid to multi-attribute utility theory. Triantaphyllou *et al.* (1997)[11] suggest the use of the Analytical Hierarchy Process (AHP) for the selection of a maintenance strategy considering four maintenance criteria: cost, reparability, reliability, and availability. Kumar *et al.*, (2010),[15] introduced an AHP based method to assess the risk of rail defects. Bevilacqua and Braglia (2000)[7] also used AHP for selecting the maintenance strategy for an Italian oil refinery based on four important criteria, namely cost, damages, applicability, and added value. Martorell *et al.* (2005)[16] introduced an Integrated Multi-Criteria Decision Making (IMCDM) approach based on RAMS and cost criteria, to assess changes to the technical specification and maintenance-related parameters, with respect to the constraint conditions. They reviewed and emphasized the benefits of an IMCDM approach when tackling RAMS as a multi-objective optimization problem in a nuclear power plant. Bertolini and Bevilacqua (2006)[8] presented an integrated AHP and goal programming (GP) approach to selecting the best maintenance policies for the maintenance of centrifugal pumps in an oil refinery. They took into account the failure occurrence, its severity, and its detectability as evaluating criteria and the budget and maintenance time as constraint conditions. Arunraj and Maiti (2010)[17] use an AHP and GP approach in which the risk of equipment failure and cost of maintenance are considered as criteria for maintenance selection in a benzene extraction unit of a chemical plant. In addition, the Fussell–Vesely importance measure is utilized for measuring the risk contribution of different equipment.

AHP is a flexible approach and allows individuals or groups to shape ideas and define problems by making their own assumptions and deriving the desired solution from them. It takes into consideration the relative importance of factors in a system and enables people to rank the alternatives based on their goals. However, this method is often criticized for its inability to deal adequately with the uncertainty and imprecision associated with the mapping of the decision makers' perception to crisp numbers.[18] Moreover, most of the time, the choice of a maintenance alternative decision is governed by the preferences of the decision makers. In general, the preferences are guided mainly by three factors:

(a) Maximizing the total profit — aiming to increase business and investment, with no bar,
(b) Maximizing the benefit-cost ratio — looking for the maximum percentage of return, indicating rational investment restriction, and
(c) Minimizing the investment — aiming at the highest utilization of resources and an investment crunch.

In fact, practical problems are often characterized by several incommensurable and conflicting (competing) criteria, and there may be no solution satisfying all the criteria simultaneously. Therefore, the solution is a set of non-inferior solutions or a compromise solution according to the decision maker's preferences. The TOPSIS (Technique for Order Preference by Similarity to an Ideal Solution) and VIKOR (VlseKriterijumska Optimizacija I Kompromisno Resenje, a Serbian term meaning Multi-criteria Optimization and Compromise Solution, see Sec. 4.) are MCDM methods which have been applied for solving many diverse real-world multi-attribute decision making problems. They are based on an aggregating function representing "closeness to the ideal, which originated in the compromise programming method". Both TOPSIS and VIKOR are based on the calculation of distances from the Positive Ideal Solution (PIS) and the Negative Ideal Solution (NIS).[19] Chu et al. (2007)[20] are in favour of using VIKOR when there are a larger number of decision makers, and otherwise they recommend the use of TOPSIS. Comparatively, the AHP method has a lesser distinguishing capability, whereas both VIKOR and TOPSIS are good at clarifying the differences between alternatives.[20] Moreover, the core process of AHP is pairwise comparison, which is significantly restrained by human information processing capacity.[21] TOPSIS and VIKOR processes are not restricted by human capacity restrictions and therefore can be easily accommodated with a large number of attributes and alternatives. Shyjith et al. (2008)[22] have also given a combined AHP and TOPSIS- based maintenance strategy selection methodology for the process industry.

This paper proposes a methodology for the selection of a maintenance strategy for non-safety category of failure, based on an AHP-enhanced benefit-cost ratio, TOPSIS and VIKOR. The methodology enables determination of the importance of the effectiveness appraisal criteria of a maintenance strategy by the use of AHP. Furthermore, in the proposed methodology the different maintenance policies are ranked using the benefit-cost ratio, TOPSIS and VIKOR. The proposed method has been tested through a case study within the aviation context for an aircraft system. The rest of the article is organized as follows. In Secs. 2, 3, 4 and 5, AHP, TOPSIS, VIKOR, and benefit-cost methods are discussed. In Sec. 6, the proposed methodology is described. In Sec. 7, the case study in the aviation sector is discussed and the paper ends with Sec. 8, which gives a discussion and conclusion.

2. Analytical Hierarchy Process (AHP)

The Analytical Hierarchy Process (AHP)[23] helps the analyst to organize the critical aspects of a problem into a hierarchical structure similar to a family tree. By reducing complex decisions to a series of simple comparisons and rankings and then synthesizing the results, AHP helps the analysts to provide a clear rationale for the importance of evaluating criteria.

AHP employs pairwise comparison in which experts compare the importance of two factors on a relatively subjective scale. In this way a judgment matrix of importance is built according to the relative importance given by the experts. Table 1 represents a pairwise comparison scale for the value rating of judgments and for

Table 1. Judgment scores in AHP.

Judgment explanation	Score
The two attributes are equally important.	1
One attribute is moderately more important than the other.	3 or (1/3)
One attribute is strongly more important than the other.	5 or (1/5)
One attribute is very strongly more important than the other.	7 or (1/7)
The evident greater importance of one attribute compared with the other is of the highest possible order (extremely more important).	9 or (1/9)

Note: 2,4,6,8 can be used as intermediate judgment values between adjacent scale values.

deriving pairwise ratio scales. It includes reciprocals ($a_{ij} = 1/a_{ji}$), which are equally often adopted for relative measurements or comparisons of factors. The geometric mean is the only averaging process that maintains the reciprocal relationship in the aggregate matrix. So, the weighted mean value for a group response is:

$$a_{ij} = \left(\prod_{k=1}^{n} w_k \cdot a_{ij}^k \right)^{1/\sum_{k=1}^{n} w_k}$$

where a_{ij}^k is the kth expert's paired comparison value, n is the number of experts, and w_k is the weight of the kth expert. In this study, it was assumed that all the experts have equal expertise in their judgments and therefore $w_k = 1 \, \forall \, k$.

Some degree of inconsistency may be introduced concerning the judgments due to a lack of adequate information, improper conceptualization, and mental fatigue. The AHP technique also allows the analysts to revaluate their judgments when the pairwise comparison matrix lacks consistency, as reflected by means of an inconsistency ratio. The judgments can be considered acceptable if and only if the inconsistency ratio is less than 0.1.[23] If the obtained value of the inconsistency ratio is not within an acceptable range, the experts may be asked to modify their judgments in the hope of obtaining a modified and consistent matrix.

In this paper, the AHP analysis outcome is the global priority, i.e. the importance value (w_j) of the different evaluating criteria, which is elicited from the aggregated pairwise comparison matrix of the experts' judgments. It is a vector, normalized to the unity, which allows identification of the importance of evaluating criteria with respect to the goal.

3. TOPSIS

The solution of a multi-attribute decision making (MADM) problem through TOPSIS is based on the simple logic that the best solution is furthest from negative ideal solution and preferably closest to the positive ideal solution. The alternatives are ranked by their distances from two cardinal points: the positive ideal solution (PIS) and the negative ideal solution (NIS). The different steps involved in TOPSIS are as follows[19,21]:

Step 1: The core of TOPSIS is the appraisal (decision) matrix D^k framed from the responses of the kth expert, $D^k = [x_{ij}^k]_{mxn}$, where i is the set of m alternatives and j is the set of n appraising attributes used to evaluate the alternative set i. x_{ij}^k is the score (appraisal rating) of the alternative i for attribute j given by expert k.

Step 2: The list of attributes in a real-world problem often contains conflicting, incommensurable, incompatible, unconformable, and unquantifiable attributes, which increases the complexity of MCDM problems. Therefore, it is mandatory to make the elements of the decision matrix unit-free to eliminate the scaling effect through normalization. This operation is performed column-wise to transform the attribute scores to a common norm or standard between 0 and 1 to allow the comparison of different attributes. The element r_{ij}^k of the normalized appraisal (decision) matrix (R^k) may be calculated using the expression for linear normalization to eliminate the effect of an evaluation unit for criteria, i.e.

$$r_{ij}^k = \left\{ \left(\frac{x_{ij}^k}{x_j^{k+} - x_j^{k-}} \middle| j \varepsilon J \right), \left(\frac{x_{ij}^k}{x_j^{k-} - x_j^{k+}} \middle| j \varepsilon J' \right) \right\}$$

where,

$$x_j^{k+} = \left\{ \left(\max_i x_{ij}^k \middle| j \varepsilon J \right), \left(\min_i x_{ij}^k \middle| j \varepsilon J' \right) \right\}$$

$$x_j^{k-} = \left\{ \left(\min_i x_{ij}^k \middle| j \varepsilon J \right), \left(\max_i x_{ij}^k \middle| j \varepsilon J' \right) \right\}$$

Here J is the set of benefit criteria and J' is the set of cost criteria.

Step 3: Appraisal attributes have varying importance and they influence the decision as per their importance value. The importance of an attribute is evaluated though AHP. The importance of each attribute (w_j) is elicited from the pairwise comparison matrix of the experts' judgments following the AHP methodology. The weight-normalized decision matrix V^k is formulated by multiplying the elements of the normalized matrix (R^k) by the corresponding weight (w_j) of the attributes, i.e.:

$$V^k = [v_{ij}^k]_{m \times n} = [w_j \times r_{ij}^k]_{m \times n}$$

Step 4: The aggregated decision matrix (V) is formulated through the aggregation of all the experts' decision matrices. The aggregated decision matrix (V) is the group decision matrix. The geometric mean will combine the judgments of all the experts. The general formula for calculating the element of the aggregated decision matrix (V) is:

$$v_{ij} = \left(\prod_{k=1}^n v_{ij}^k \right)^{1/k}$$

Step 5: The two cardinal points in the solution space are the positive ideal solution (PIS), composed of all the best criteria, and the negative ideal solution (NIS), composed of all the worst criteria. Therefore, PIS (v^+) contains all the highest scores of the benefit criteria and all the lowest scores of the cost criteria. NIS $(v-)$ contains all the lowest scores of the benefit criteria and all the highest scores of the cost criteria. The PIS and NIS of a group of experts are:

$$v^+ = \left\{ v_j^+ \middle| j = 1, 2, \ldots, n \right\} = \left\{ v_1^+, v_2^+, \ldots, v_n^+ \right\}$$

$$= \left\{ \left(\max_i v_{ij} \middle| j \varepsilon J \right), \left(\min_i v_{ij} \middle| j \varepsilon J' \right) \right\}$$

$$v^- = \left\{ v_j^- \middle| j = 1, 2, \ldots, n \right\} = \left\{ v_1^-, v_2^-, \ldots, v_n^- \right\}$$

$$= \left\{ \left(\min_i v_{ij} \middle| j \varepsilon J \right), \left(\max_i v_{ij} \middle| j \varepsilon J' \right) \right\}$$

Step 6: The decision alternatives are non-inferior solutions and at a distance from the cardinal points, i.e. PIS and NIS. The separation measures from the cardinal points are calculated through Minkowski's L_P metric. The separation measure of the ith alternative from PIS is (D_i^+) and that from NIS is (D_i^-), where:

$$D_i^+ = \left\{ \sum_{j=1}^n (v_j^+ - v_{ij})^p \right\}^{1/p} \qquad D_i^- = \left\{ \sum_{j=1}^n (v_{ij} - v_j^-)^p \right\}^{1/p}$$

where p is an integer ≥ 1. For $p = 2$ the metric is a Euclidean distance.

Step 7: Ranking of the decision alternatives is performed on the basis of their relative closeness index (C_i^*) in respect of the ideal solution. The relative closeness of the ith alternative with respect to PIS is calculated from the following expression:

$$C_i^* = \frac{D_i^-}{D_i^+ + D_i^-}$$

where $0 \leq C_i^* \leq 1$. If the alternative solution i is the positive ideal solution, then $C_i^* = 1$. The alternatives are ranked according to the descending order of C_i^* values.

4. VIKOR

VIKOR is a compromise decision making method in multi-criteria environments. This technique ranks the alternatives based on two measures: the utility measure (the weighted distance from the ideal solution) and the regret measure (the weighted distance from the negative-ideal solution). The VIKOR index for each alternative is calculated from these measures. The alternative with the least VIKOR index is the best alternative, as it has the maximum group utility and the least regret.[24−26] This method includes the following steps, of which Step 1 and Step 2 are the same as the corresponding steps in TOPSIS.

Step 3: Two cardinal values of each criterion represent the best and the worst values. The association of the best criteria values gives the ideal solution (IS), while the set of all the worst values gives the anti-ideal solution (AIS).[26] Therefore, IS (I^{k+}) contains all the highest scores of the benefit criteria and all the lowest scores of the cost criteria. AIS (I^{k-}) contains all the lowest scores of the benefit criteria and all the highest scores of the cost criteria. The I^{k+} and I^{k-} of the kth expert are:

$$I^{k+} = \left\{ r_j^{k+} \middle| j = 1, 2, \ldots, n \right\} = \left\{ r_1^{k+}, r_2^{k+}, \ldots, r_n^{k+} \right\}$$

$$= \left\{ \left(\max_i r_{ij}^k \middle| j \varepsilon J \right), \left(\min_i r_{ij}^k \middle| j \varepsilon J' \right) \right\}$$

$$I^{k-} = \left\{ r_j^{k-} \middle| j = 1, 2, \ldots, n \right\} = \left\{ r_1^{k-}, r_2^{k-}, \ldots, r_n^{k-} \right\}$$

$$= \left\{ \left(\min_i r_{ij}^k \middle| j \varepsilon J \right), \left(\max_i r_{ij}^k \middle| j \varepsilon J' \right) \right\}$$

where J is the set of benefit criteria and J' is the set of cost criteria.

Step 4: The utility measure (S_i^k) of the ith alternative is calculated from all the criteria values and their relative weights (w_j). The regret measure (R_i^k) of the ith alternative gives the most influential criterion and its corresponding values. The values of S_i^k and R_i^k are calculated using the expressions given below:

$$S_i^k = \sum_{j=1}^n w_j \times \frac{r_j^{k+} - r_{ij}^k}{r_j^{k+} - r_j^{k-}}$$

where

$$R_i^k = \max_j \left[w_j \times \frac{r_j^{k+} - r_{ij}^k}{r_j^{k+} - r_j^{k-}} \right]$$

Step 5: The group utility measure (S_i) and group regret measure (R_i) of an alternative are computed by the aggregation of all experts' S_i^k and R_i^k values. The group measures for each alternative are calculated through the geometric mean of all the individual expert's measures, as given below:

$$S_i = \left(\prod_{k=1}^n S_i^k \right)^{1/k} \quad \text{and} \quad R_i = \left(\prod_{k=1}^n R_i^k \right)^{1/k}$$

Step 6: The VIKOR index (Q_i) for the ith alternative is computed by the relation

$$Q_i = v \times \frac{S_i - S^-}{S^+ - S^-} + (1 - v) \times \frac{R_i - R^-}{R^+ - R^-}$$

where $S^- = \min_i S_i$, $S^+ = \max_i S_i$, $R^- = \min_i R_i$ and $R^+ = \max_i R_i$ and v is a weighting factor of the preferences for decision-making strategy. If the decision making is carried out on the basis of "consensus", v is 0.5, and v is >0.5 when the "voting by majority" rule is followed. v is <0.5 when the decision is made with a "veto". Here the term $\frac{S_i - S^-}{S^+ - S^-}$ is the scaled distance from the ideal solution and measures the overall closeness of alternative i, and the second term $\frac{R_i - R^-}{R^+ - R^-}$ gives the scaled distance for the most influential criterion, indicating its closeness to the desired value.

Step 7: The ranking of the decision alternatives is performed by sorting the (S_i), (R_i), and (Q_i) values in increasing order, which results in three ranking lists denoted as $S_{[.]} R_{[.]}$ and $Q_{[.]}$.

Step 8: The proposed decision alternative I_1, having the lowest Q_i value, will be a compromise solution if the following two conditions are satisfied.

C1: Alternative I_1 has an "acceptable advantage" when

$$Q_{[2]} - Q_{[1]} \geq \frac{Q_{[m]} - Q_{[1]}}{m - 1}$$

C2: Alternative I_1 has an "acceptable stability in the decision-making process" when it possesses the best rank in terms of S_i and/or R_i values.

If either of these two conditions is not met, more than one alternative solution is proposed. Both alternatives I_1 and I_2 are proposed when only condition C_2 is not met. Alternatives $I_1, I_2, \ldots, I_l$ are the proposed compromise solution set when condition C_1 is not satisfied and l is the maximum value as long as the following relation is satisfied.

$$Q_{[l]} - Q_{[1]} < \frac{Q_{[m]} - Q_{[1]}}{m - 1}$$

Here the compromise solution sets are in closeness to each other.

5. Benefit-Cost Ratio

Any decision has several favourable and unfavourable concerns to consider. The favourable sure concerns are positive value and are called benefit, such as business enhancement, planning flexibility, and reduction in maintenance cost. The unfavourable ones are negative and are called costs, such as maintenance investment and its associated costs. Each of these concerns contributes to the merit of decision and must be evaluated (rated) individually on a set of evaluating criteria. These criteria are measured in different units and scales and have different importance. Here the importance of the evaluating criteria is assigned through AHP and a normalized performance appraisal matrix will ease out the effect of units and scaling. The methodology is as follows:

A performance appraisal matrix or aggregated decision matrix is framed following Step 1 to Step 4 described in Sec. 3.

Now the combined benefit index of the ith alternative is $\sum v_{ij} | j \in J$ and the combined cost index of the ith alternative is $\sum v_{ij} | j \in J'$. Therefore, the rate of return, i.e. benefit-cost ratio, can be calculated by dividing the synthesized benefit value by the associated synthesized cost value for each alternative:

$$\frac{\sum(v_{ij}|j \in J)}{\sum(v_{ij}|j \in J')}$$

The rate of return is an index to quantify the amount of gain or loss generated from a specific maintenance alternative, according to the specified evaluating criteria.

6. Proposed Methodology

For the selection of a maintenance strategy, this paper proposes a methodology where the importance of the effectiveness appraisal criteria of the maintenance strategy is assigned by AHP and different maintenance policies are ranked using the three methods, i.e. the benefit-cost ratio, TOPSIS and VIKOR. The proposed decision making method includes two levels of an organization, i.e. the managerial level and the engineering level. The managerial level defines the goals and the associated evaluating criteria, and also performs the pairwise comparison to assign the importance of the evaluating criteria. The assignment of the importance value for the evaluating criteria, from a managerial point of view, is carried out by applying AHP. These importance values will be used for the whole analysis. The engineering level selects a failure mode, defines applicable maintenance alternatives and assesses the effectiveness of each alternative after due consideration of the positive and negative consequences of choosing any one of them from the standpoint of various evaluating criteria. At this level the analyst performs a multi-criteria ranking of the alternatives.

Compromise solutions are proposed based on the preferences of the decision makers. These preferences may be (a) the rate of return, (b) the total profit or (c) the lowest investment. When the preference is the rate of return, the benefit-cost ratio may be used, and for the total profit TOPSIS is appropriate. In cases where the manager has specific preferences, such as "the alternative solution should preferably include the lowest maintenance investment in comparison with that of other maintenance alternatives", VIKOR can be adopted.

Therefore, the ranking of the alternatives is carried out using the three methods, i.e. the benefit-cost ratio, TOPSIS and VIKOR. Finally, the analyst provides a ranking of the alternatives based on different points of view and the manager can select one that suits his preferences well.

6.1. The steps of the proposed methodology

The following steps form the proposed methodology, as shown in Fig. 1:

Step 1: Building a hierarchical structure for both the benefit and cost criteria. By decomposing these criteria, an attempt is made to prioritize, simplify the problem, and come down from the goals to specific and easily controlled factors.

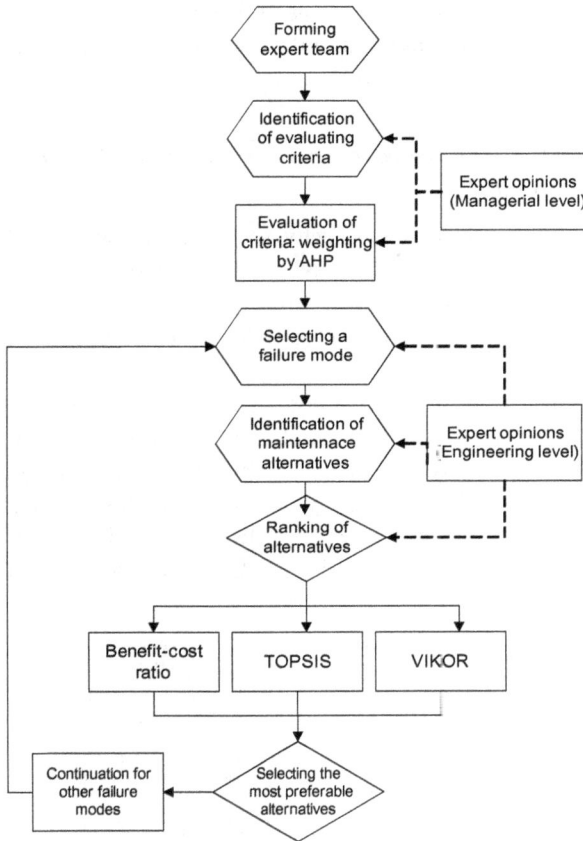

Fig. 1.   Proposed decision flow diagram.

Step 2: Calculating the weight of each evaluating criteria of each level from the pairwise comparison matrix, framed from the responses of managerial level experts, through the implementation of AHP. It is necessary to check the consistency of each matrix in this step and, if the consistency ratio exceeds 10%, the experts are asked to modify their judgments.[23] Considering that the opinions of different experts are used for pairwise comparison, it is essential to calculate the mean weight for each sub-criterion.

Step 3: At the engineering level, a failure mode should be selected, the applicable maintenance alternatives should be defined, and the effectiveness of each alternative should be assessed after due consideration of the positive and negative consequences of each maintenance alternative from the standpoint of various evaluating criteria.

Step 4: At this level the analyst should rank the different alternatives by using the three mentioned methods.

Step 5: Comparison of all the alternatives and their ranking by the three methods. The most consistent and justified task will be selected by the Maintenance Review Board (MRB). Thereafter, the analysis continues for the next failure mode.

6.2. *Development of the hierarchical structure of the maintenance selection criteria*

Structuring the problem into a hierarchy serves two purposes. First it provides an overall view of the complex relationship of variables inherent in the problem, and second, it helps the decision maker in making judgments concerning the comparison of elements that are homogenous and on the same level of the decision hierarchy.[17]

Adopting a more holistic view in selecting maintenance strategies, it has been decided that it would be more beneficial and business-oriented to consider the benefit-cost ratio as a measure for the overall effectiveness that an applicable maintenance strategy can gain. The hierarchy developed in this study is a five-level tree in which the top level represents the goal of the analysis, i.e. "selection of the most effective maintenance strategy" (see Fig. 2). Considering the role of maintenance

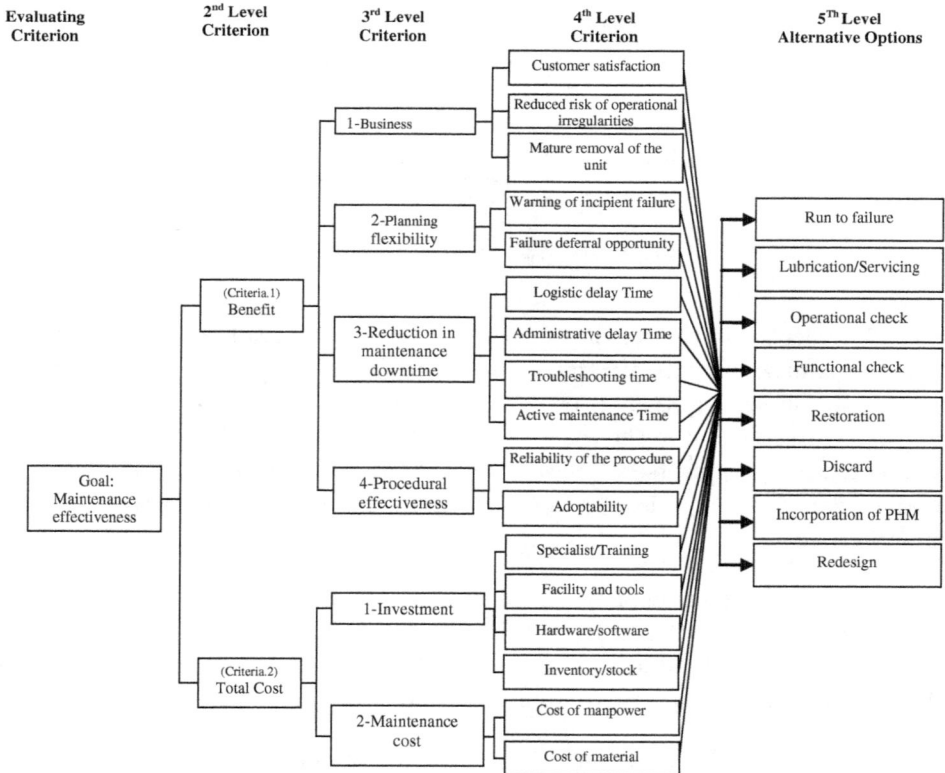

Fig. 2.   Hierarchy of evaluating criteria.

as added value and contributor to the business enhancement, the evaluation criteria that will influence the main goal have been defined as *benefit* and *total cost*, which are included at the second level of the hierarchy. These criteria are then broken down into the sub-criteria which form the third and fourth level. Finally, the lowest level, i.e. the fifth level, comprises the alternative maintenance policies. The relevant factors (sub-criteria) defining the criterion of benefit are identified as business enhancement, planning flexibility, reduction in maintenance downtime, and procedural effectiveness. It has been decided that the business enhancement factor itself should be evaluated based on the customer satisfaction, the reduced risk of operational irregularities, and the mature removal of the unit.

Moreover, planning flexibility should be evaluated based on the "warning of incipient failure" and the "failure deferral possibility". Furthermore, it has been decided that the maintenance downtime factor itself should be evaluated based on the logistic delay time, administrative delay time, troubleshooting time, and active maintenance time. Finally, the procedural effectiveness should be evaluated based on the "reliability of the procedure" and "adoptability". For the *total cost* incurred by the selected maintenance strategy, two factors (sub-criteria) have been selected, namely "investment" and "cost of maintenance". The investment should be evaluated according to the requirements for specialists/training, facility and tools, hardware/software, and inventory/stock. Likewise, the associated cost of maintenance will be evaluated based on the cost of manpower and the cost of material. The alternative maintenance strategies considered in this study include those offered by ATA MSG-3[2]. (i.e. Operational/Visual Check, Inspection/Functional Check, Restoration, Discard, Combination of Strategies, Run to Failure and Redesign) and the use of Prognostics and Health Management (PHM) provisions.

7. Case Study

Of confidential reasons, information related to company and the studied system has been masked. The company is European aircraft manufacturer and the component is part of the fuel system in an aircraft. Due to the high level of the redundancy, the failure of the studied component does not have any safety effect, and the major concern is the operational and economic consequences of failure. The performance of the five maintenance alternatives was evaluated using 16 evaluating attributes divided into two groups i.e. benefit and cost criteria (see Fig. 2). The list of evaluating attributes was decided in consensus with the subject matter experts. Through the discussion with the subject matter experts, it was decided that the criterion "customer satisfaction" could be evaluated through the criterion "reduced operational irregularities". Hence the former criterion was removed from the analysis. It was decided also that, among the available maintenance strategies mentioned in Section 6.2, lubrication/servicing, a combination of strategies and redesign were not applicable for this study, and the rest were considered. The responses of the experts were analyzed and the maintenance alternatives were ranked using three methodologies: (i) the benefit-cost ratio, (ii) TOPSIS and (iii) VIKOR.

7.1. Collection of judgments from experts

Two different sets of questionnaires were developed to collect the opinions of the participants. For pairwise comparison between the two criteria, benefit and total cost", on the second level of evaluating criteria, the following question was asked:

Question A: "To select the most effective maintenance strategy for the units/ systems whose failure affects the normal operation of aircraft, we have identified two main criteria: (1) the benefit that the selected maintenance strategy creates and (2) the total cost under the specific maintenance strategy. In your opinion, with respect to the overall goal "selection of the most effective maintenance strategy", which of these two criteria is of greater importance (or priority) in the selection of an appropriate maintenance strategy according to the following scale of importance?"

Likewise, a pairwise comparison was made between the criteria on the third and fourth levels of evaluating criteria. For example, for pairwise comparison among the four criteria of benefit on the third level, the following question was asked:

Question B: "The relevant factors defining the criteria of benefit are identified as the contribution to the business, increased planning flexibility, reduction in maintenance downtime, and enhancement of procedural effectiveness. In your opinion, with respect to sub-criterion 1, benefit, how important is sub-attribute 1 (business) when compared with sub-attribute 2 (planning flexibility)?" The question was repeated, after adaptation, for the other attributes.

The experts, on the basis of their knowledge and experience, gave their opinions in the multiple choice questionnaire that formed the basis of the pairwise comparison matrix for the given criteria. Then the verbal and qualitative responses were quantified and translated into a quantitative value/score by the use of a discrete 9-point scale (see Table 1). Using AHP the importance of all the 16 attributes was calculated from the judgment matrix of the individual experts and subsequently aggregated to obtain the group priority values as shown in Table 2. The experts chose to assign maximum importance to the "cost of manpower" (0.215), followed by "hardware/software" (0.180), among the cost criteria, and to "reduce risk of operational irregularities" (0.194), followed by "mature removal of unit" (0.113), among the benefit criteria. The managerial level experts assigned least importance to "adoptability" (0.04) among the benefit criteria and to "specialists/training" (0.18) among the cost criteria.

7.2. Ranking of the alternatives

In order to collect the engineering level data to rank the alternatives, a second set of questionnaires was developed. The purpose was to collect engineering judgments regarding the ability of each maintenance alternative to achieve the ideal level of each evaluating criterion. For example, to identify the ability of maintenance

Table 2. Results of AHP for calculation of weights of different evaluating criteria.

Sl no	Criteria	Aggregated weights of the criteria
1	Cost of manpower	0.215
2	Reduced risk of operational irregularities	0.194
3	Hardware/software	0.180
4	Mature removal of unit	0.113
5	Troubleshooting time	0.056
6	Warning of incipient failure	0.035
7	Failure deferral possibility	0.035
8	Cost of material	0.035
9	Inventory/stock	0.030
0	Reliability of the procedure	0.030
11	Facility and tools	0.022
12	Active maintenance time	0.018
13	Specialists/Training	0.018
14	Logistic delay time	0.009
15	Administrative delay time	0.007
16	Adoptability	0.004

alternatives to eliminate "the risk of operational irregularities", the following question was asked:

Question C: "What score $(0 - 100)$ do you assign to each maintenance strategy with respect to the criterion "reduced risk of operational irregularities"?" The question inquires about the extent to which the alternative is capable of reducing the risk of operational irregularities. If it is capable of eliminating them completely, a score of 100 will be assigned, and if it is unable to contribute any reduction, a score of "0" will be assigned. If the expert is not sure about a specific score, he can assign his judgments in a range, e.g. $65 - 80$.

The results of the TOPSIS calculation (see Table 3) show that the PHM alternative has the highest value for eight benefit criteria, whereas for the adoptability criterion it has the lowest value. At the same time it has the lowest value for four cost criteria. Therefore, for the majority of the criteria (12 out of 16), this alternative is close to PIS. In contrast, the alternative "run to failure" has the lowest value for all the ten benefit criteria and the highest value for one cost criterion, indicating its closeness to NIS.

Table 4 shows the weighted distance of each criterion from its ideal value by VIKOR, based on response of one of the experts. The PHM alternative has the least distance for nine benefit criteria and the highest distance for one cost criterion, showing its closeness to IS. However, for the alternative "run to failure", this distance is the highest for all the ten benefit criteria and the lowest for two cost criteria, indicating its nearness to AIS.

Table 5 shows the results of the benefit-cost analysis. The maintenance alternative "incorporation of PHM" has the maximum benefit index (0.486), followed by

Table 3.   Results of calculation by TOPSIS (aggregated decision matrix) for preferability of different maintenance alternatives.

	RRI	MRU	WIF	FDO	LDT	ADT	TST	AMT	ROP	ADO	SPE	FTO	HAW	INV	CMA	CMT
RTF	0.000	0.000	0.000	0.000	0.000	0.000	0.000	0.000	0.000	0.000	0.017	0.021	0.000	0.021	0.215	0.000
FUN	0.151	0.091	0.029	0.028	0.007	0.004	0.034	0.012	0.023	0.002	0.000	0.014	0.121	0.011	0.000	0.026
RES	0.108	0.069	0.015	0.018	0.003	0.002	0.030	0.011	0.025	0.001	0.000	0.000	0.180	0.022	0.049	0.000
DIS	0.116	0.052	0.006	0.000	0.005	0.003	0.051	0.015	0.028	0.003	0.016	0.022	0.180	0.028	0.094	0.000
PHM	0.193	0.113	0.035	0.035	0.007	0.007	0.055	0.017	0.020	0.000	0.000	0.014	0.000	0.000	0.000	0.021

Table 4.   Results of calculation by VIKOR for preferability of different maintenance alternatives (based on response of one of the experts).

	RRI	MRU	WIF	FDO	LDT	ADT	TST	AMT	ROP	ADO	SPE	FTO	HAW	INV	CMA	CMT
RTF	0.194	0.113	0.035	0.035	0.009	0.007	0.056	0.018	0.030	0.004	0.018	0.020	0.000	0.030	0.215	0.000
FUN	0.043	0.028	0.006	0.008	0.003	0.003	0.021	0.007	0.008	0.001	0.015	0.019	0.110	0.010	0.018	0.024
RES	0.086	0.057	0.020	0.020	0.006	0.004	0.021	0.007	0.005	0.002	0.000	0.000	0.180	0.021	0.036	0.016
DIS	0.065	0.071	0.024	0.032	0.005	0.003	0.007	0.002	0.002	0.000	0.015	0.022	0.180	0.026	0.108	0.000
PHM	0.000	0.000	0.000	0.000	0.000	0.000	0.000	0.000	0.000	0.001	0.000	0.011	0.100	0.000	0.000	0.035

Notes: **RTF**: Run-to-failure, **FUN**: Functional check, **RES**: Restoration, **DIS**: Discard, **PHM**: Incorporation of PHM, **RRI**: Reduced risk of operational irregularities, **MRU**: Mature removal of the unit, **WIF**: Warning of incipient failure, **FDO**: Failure deferral possibility, **LDT**: Logistic delay time, **ADT**: Administrative delay time, **TST**: Troubleshooting time, **AMT**: Active maintenance time, **ROP**: Reliability of the procedure, **ADO**: Adoptability, **SPE**: Specialists/Training, **FTO**: Facility and tools, **HAW**: Hardware/software, **INV**: Inventory/stock, **CMA**: Cost of manpower, **CMT**: Cost of material.

Table 5. Results of benefit-cost analysis and ranking of maintenance alternatives by benefit-cost ratio.

Alternatives	Benefit value	Cost value	Benefit-cost ratio	Rank
Run to failure	0.000	0.276	0.000	5
Functional check	0.385	0.174	2.210	2
Restoration	0.287	0.253	1.135	3
Discard	0.282	0.342	0.825	4
Incorporation of PHM	0.486	0.036	13.505	1

Table 6. Results of TOPSIS analysis and ranking of maintenance alternatives by C_i^* value.

Alternatives	Separation measures		Relative closeness Index(C_i^*)	Rank
	D_i^+	D_i^-		
Run to failure	0.324	0.182	0.360	5
Functional check	0.137	0.292	0.682	2
Restoration	0.215	0.219	0.505	3
Discard	0.234	0.188	0.446	4
Incorporation of PHM	0.027	0.370	0.932	1

the alternative "functional check" (0.385). The cost index value for the alternative "discard" is the maximum (0.342) and that for the alternative "incorporation of PHM" is the minimum (0.036). The benefit-cost ratio is highest (13.505) for the alternative "incorporation of PHM", indicating that this alternative is the most rational choice, followed by "functional check" and "restoration". The benefit-cost ratio for the remaining two alternatives is less than unity, indicating that they are not preferable alternatives.

The results of the TOPSIS analysis (see Table 6) indicate that the maintenance alternative "incorporation of PHM" is the overwhelming choice, as indicated by the highest relative closeness index (0.932). Therefore, this alternative is nearest to PIS and furthest from NIS. The alternative "functional check" is the second preferred choice, with a relative closeness index (0.682), followed by "restoration". The alternative "run-to-failure" is the least preferred choice due to its closeness to NIS.

The aggregated results of the VIKOR analysis are tabulated in Table 7. Both the separation measures and the VIKOR index rank the alternative "incorporation of PHM" first, followed by "functional check". All the three methods show that the alternative "incorporation of PHM" is the most favourable choice, followed by "functional check". The criterion "hardware/software" is the most influential criterion for both of these alternatives, as it gives the highest value of the regret measure.

Table 7.   Results of VIKOR analysis and ranking of maintenance alternatives.

Alternatives	Separation measures		VIKOR index (Q_i)	Ranking by:			Decision ranking
	Utility measure (S_i)	Regret measure (R_i)		(S_i)	(R_i)	(Q_i)	
Run to failure	0.875	0.215	1.000	5	4	5	5
Functional check	0.308	0.121	0.312	2	2	2	2
Restoration	0.480	0.180	0.623	3	3	3	3
Discard	0.558	0.180	0.674	4	3	4	4
Incorporation of PHM	0.111	0.067	0.000	1	1	1	1

8. Discussion and Conclusions

For selection of the maintenance tasks within the MRB process, this paper presents a rigorous approach in which all the applicable maintenance strategies plus the incorporation of PHM are considered for review. It suggests the use of the MCDM method to enhance the expert judgment process involved in the use of ATA MSG-3, to assure the consistency and effectiveness of maintenance decisions. The AHP method has been used to determine the importance of the maintenance effectiveness appraisal criteria of a maintenance strategy. This provides an overall view of the complex relationship between evaluating variables inherent in the decision making process, and helps the decision maker in making judgments in the comparison of attributes and criteria that are homogenous and are on the same level of the decision hierarchy.

The method provides a basis for consideration of different priority factors governing decisions, which may include (a) the benefit-cost ratio i.e. rate of return (b) the total profit or, (c) the lowest investment. When the preference is the rate of return, the benefit-cost ratio is used, and for the total profit TOPSIS is applied, as it governs by majority rule, neglecting the side-effects. In cases where the decision maker has specific preferences, such as "the alternative solution should preferably include the lowest investment needed in comparison with that of other maintenance alternatives", VIKOR can be adopted, as it considers the smallest damages. Finally, the analyst provides a ranking of the alternatives based on different points of view and the Maintenance Review Board can select one that suits the preferences well.

The proposed methodology has been verified through a case study for an aircraft system. The list of evaluating attributes was decided in consensus with field experts. The performance of the five maintenance alternatives was evaluated using 16 evaluating attributes. The results show that the alternative "incorporation of PHM" was found to be the most favourable choice, followed by "functional check", by all the three methods. The rest of the alternatives, i.e. "restoration", "discard", and "run to failure" have the same preferability as MSG-3 has suggested. However, it is noticeable that the ranking index for PHM strongly proffers that alternative in comparison with other alternatives, which shows the necessity of including this alternative among the decision alternatives.

It has been found that using the methodology presented in the paper, the relative advantage and disadvantage of each maintenance strategy could be identified in consideration of different aspects, and that justification of the maintenance task selection will be more consistent and rationalized. The study shows that using the combined AHP, TOPSIS, and VIKOR methodologies is an applicable and effective way to implement a rigorous approach for identifying the most effective maintenance alternative.

References

1. Airline Handbook, Air Transport Association of America, Washington, DC, USA www.air-transport.org. Air Transport Association of America, Inc. (ATA) (2000).
2. ATA MSG-3, *Operator/Manufacturer Scheduled Maintenance Development*. Pennsylvania: Air Transport Association of America (2007).
3. G. Waeyenbergh and L. Pintelon, Maintenance concept development: A case study, *Int. J. Production Economics* **89** (2004) 395–405.
4. U. Kumar, Maintenance strategies for mechanized and automated mining systems: A reliability and risk based approach, *Journal of Mine Metal and Fuels* **46** (1998) 343–347.
5. T. Markeset and U. Kumar, Design and development of product support and maintenance concepts for industrial systems, *Journal of Quality in Maintenance Engineering* **9** (2003) 376–392.
6. A. Ahmadi and U. Kumar, Cost based risk analysis to identify inspection and restoration intervals of hidden failures subject to aging, Accepted for publication in *IEEE Transaction on Reliability* (2010).
7. M. Bevilacqua and M. Braglia, The analytical hierarchy process applied to maintenance strategy selection, *Reliability Engineering and System Safety* **70** (2000) 71–83.
8. M. Bertolini and M. Bevilacqua, A combined goal programming — AHP approach to maintenance selection problem, *Reliability Engineering and System Safety* **91** (2006) 839–848.
9. A. H. C. Tsang, Condition based maintenance: tools and decision-making, *Journal of Quality in Maintenance Engineering* **1** (1995) 3–17.
10. A. T. de Almeida and G. A. Bohoris, Decision theory in maintenance decision making, *Journal of Quality in Maintenance Engineering* **1** (1995) 39–45.
11. E. Triantaphyllou, B. Kovalerchuk, L. Mann, and G. M. Knapp, Determining the most important criteria in maintenance decision making, *Journal of Quality in Maintenance Engineering* **3**(1) (1997) 16–24.
12. A. W. Labib, G. B. Williams and R. F. O'Conner, An intelligent maintenance model (system): an application of the analytic hierarchy process and fuzzy logic rule-based controller, *Journal of the Operation Research Society* **49** (1998) 745–757.
13. B. Al-Najjar and I. Alsyouf, Selecting the most efficient maintenance approach using fuzzy multiple criteria decision making, *International Journal of Production Economics* **84** (2003) 85–100.
14. A. W. Labib, A decision analysis model for maintenance policy selection using CMMS, *Journal of Quality in Maintenance Engineering* **10** (2004) 191–202.

15. S. Kumar, S. Gupta, B. Ghodrati and U. Kumar, An approach for risk assessment of rail defects. Accepted for publication in *International Journal of Reliability, Quality and Safety Engineering* (2010).
16. S. Martorell, J. F. Villanueva, S. Carlos, Y. Nebot, A. Sanchez, J. L. Pitarch and V. Serradell, RAMS+C informed decision-making with application to multi-objective optimization of technical specifications and maintenance using genetic algorithms, *Reliab Eng Syst Safe* **87** (2005) 65–75.
17. N. S. Arunraj and J. Maiti, Risk-based maintenance policy selection using AHP and goal programming, *Safety Science* **48** (2010) 238–247.
18. H. Deng, Multicriteria analysis with fuzzy pairwise comparison, *International Journal of Approximate Reasoning* **17** (2003) 109–125.
19. S. Opricovic and G. H. Tzeng, Compromise solution by MCDM methods: A comparative analysis of VIKOR and TOPSIS, *European Journal of Operational Research* **156** (2004) 445–455.
20. M. T. Chu, J. Shyu, G. H. Tzeng and R. Khosla, Comparison among three analytical methods for knowledge communities group-decision analysis, *Expert Systems with Applications* **33** (2007) 1011–1024.
21. H. S. Shih, H. J. Shyur and E. S. Lee, An extension of TOPSIS for group decision making, *Mathematical and Computer Modelling* **45** (2007) 801–813.
22. K. Shyjith, M. Ilangkumaran and S. Kumanan, Multi-criteria decision-making approach to evaluate optimum maintenance strategy in textile industry, *Journal of Quality in Maintenance Engineering* **14** (2008) 375–386.
23. T. L. Saaty, *The Analytic Hierarchy Process.* (McGraw-Hill, New York, 1980).
24. S. Opricovic and G. H. Tzeng, Extended VIKOR method in comparison with outranking method, *European Journal of Operational Research* **178** (2007) 514–529.
25. J. J. H. Liou and Y. T. Chuang, Developing a hybrid multi-criteria model for selection of outsourcing providers, *Expert Systems with Applications* **37** (2010) 3755–3761.
26. B. Vahdani, H. Hadipour, J. S. Sadaghiani and M. Amiri, Extension of VIKOR method based on interval-valued fuzzy sets, *Int J Adv Manuf Technol* **47** (2010) 1231–1239.

About the Authors

Alireza Ahmadi is PhD candidate at the Division of Operation and Maintenance Engineering, Luleå University of Technology, Sweden. He has received his Licentiate degree in Operation and Maintenance Eng. in 2007. His research topic is related to the application of RAMS in aircraft maintenance development.

Suprakash Gupta is Associate Professor in the Department of Mining Engineering, Institute of Technology, Banaras Hindu University, India. His research interests are reliability analysis and maintenance optimization. He has published and reviewed a numbers of related papers.

Ramin Karim is Assistant Professor at Luleå University of Technology (LTU). He has a B.Sc. in Computer Science, and both a Licentiate of Engineering and a Ph.D. in Operation & Maintenance Engineering. He has worked within the Information & Communication Technology (ICT) area for 18 years, as architect, project manager, software designer, product owner, and developer. Karim is responsible for the research area eMaintenance at LTU, and has published more than 15 papers related to eMaintenance.

Uday Kumar is Professor and Head of the Division of Operation and Maintenance Engineering, Luleå University of Technology, Luleå, Sweden. His research interests are reliability analysis and maintenance engineering. He has authored, reviewed and edited a number of papers related to reliability and maintenance engineering.

Chapter 14

REDEFINING FAILURE RATE FUNCTION FOR DISCRETE DISTRIBUTIONS*

M. XIE

Department of Industrial and Systems Engineering
National University of Singapore, Singapore
mxie@nus.edu.sg

O. GAUDOIN and C. BRACQUEMOND

Laboratoire IMAG-LMC
Institut National Polytechnique de Grenoble
Grenoble, France

For discrete distribution with reliability function $R(k)$, $k = 1, 2, \ldots, [R(k-1) - R(k)]/R(k-1)$ has been used as the definition of the failure rate function in the literature. However, this is different from that of the continuous case. This discrete version has the interpretation of a probability while it is known that a failure rate is not a probability in the continuous case. This discrete failure rate is bounded, and hence cannot be convex, e.g., it cannot grow linearly. It is not additive for series system while the additivity for series system is a common understanding in practice. In the paper, another definition of discrete failure rate function as $\ln[R(k-1)/R(k)]$ is introduced, and the above-mentioned problems are resolved. On the other hand, it is shown that the two failure rate definitions have the same monotonicity property. That is, if one is increasing/decreasing, the other is also increasing/decreasing. For other aging concepts, the new failure rate definition is more appropriate. The failure rate functions according to this definition are given for a number of useful discrete reliability functions.

Keywords: Discrete distribution; discrete failure rate; aging property; discrete reliability function.

1. Introduction

Discrete distributions are useful when measurements are taken at discrete time. For items like switches or devices that are observed or used for every fixed duration of operation, discrete distributions will provide better modeling, analysis and interpretation. For example, see Refs. 9 and 11.

Let the random variable K be discrete lifetime of a component and denote by $p(k)$ the probability that a failure will occur at time k. The corresponding reliability

---

*This chapter appeared previously on the International Journal of Reliability, Quality and Safety Engineering. To cite this chapter, please cite the original article as the following: M. Xie, O. Gaudoin and C. Bracquemond, *Int. J. Reliab. Qual. Saf. Eng*, **9**, 275–285 (2002), doi:10.1142/S0218539302000822.

function is given by

$$R(k) = \Pr(K > k) = \sum_{j=k+1}^{\infty} \Pr(K = j) = \sum_{j=k+1}^{\infty} p(j), \quad k = 1, 2, \ldots .$$

The quantity

$$\lambda(k) = \Pr(K = k \mid K \geq k) = \frac{\Pr(K = k)}{\Pr(K \geq k)} = \frac{R(k-1) - R(k)}{R(k-1)}, \quad k = 1, 2, \ldots \quad (1)$$

is commonly known as the discrete failure rate function, see e.g., Refs. 4, 6 and 13.

Although Eq. (1) is widely used in the literature, there are a few problems with this definition. For example, it is known that for continuous distribution, the failure rate is not a probability, but the failure rate defined as Eq. (1) is. As Eq. (1) is bounded, it cannot increase too fast. Furthermore, when we connect two or more discrete components in series, the failure rate function as in Eq. (1) will not be additive. These problems cause some confusion when discrete distributions are used in practice.

In this paper, another definition is discussed. This definition is in line with the continuous case and the above-mentioned problems are resolved. The paper is organized as follows. First we give a more detailed discussion of the problems associated with definition in Eq. (1). Then the definition is introduced with some studies on the properties of this definition. Particularly, we show that the discrete failure rate definition is additive for series system and it has the same monotonicity properties as that of Eq. (1). Hence if a distribution is increasing/decreasing failure rate (IFR/DFR) with respect to one definition, it is also IFR/DFR with respect to the other. However, for other aging concepts, they will be different and it will be shown that this definition also seems to be more appropriate. Finally some specific discrete distributions are studied.

2. Some Problems with the Definition of Eq. (1)

The definition of the failure rate as in Eq. (1) is different from that of continuous counterpart in many aspects. It can be noted that $\lambda(k) \leq 1$ and it has the interpretation of a probability. In fact, it is the conditional probability that an item will fail at time k given that it has survived to $k - 1$. Calling this the failure rate function might add to the confusion that is already common in industry that failure rate and failure probability are sometimes mixed up.

When a set of data is analyzed using a discrete counterpart of a continuous distribution, the estimated failure rate will not be close to that for the continuous case. For example, the discrete Weibull distribution[8] is given by

$$R(k) = q^{k^\beta}, \quad 0 < q < 1, \quad \beta > 0, \quad (2)$$

which is similar to the continuous Weibull distribution, and

$$\lambda(k) = \frac{R(k-1) - R(k)}{R(k-1)} = 1 - q^{k^\beta - (k-1)^\beta}. \quad (3)$$

Although this is increasing, decreasing or constant according to the value of β as for the continuous distribution, this value is always between 0 and 1. It can certainly not approximate that of continuous case when $\beta > 1$ for large k since the continuous failure rate function will approach infinity.

Another interesting problem is that the failure rates defined by Eq. (1) are not additive for series system. That is, if we have n discrete components in series,

$$\lambda(k) = \frac{R(k-1) - R(k)}{R(k)} = \frac{\prod_{i=1}^{n} R_i(k-1) - \prod_{i=1}^{n} R_i(k)}{\prod_{i=1}^{n} R_i(k-1)}$$

$$= 1 - \prod_{i=1}^{n} \frac{R_i(k)}{R_i(k-1)} = 1 - \prod_{i=1}^{n} [1 - \lambda_i(k)] \neq \sum_{i=1}^{n} \lambda_i(k).$$

Furthermore, $\lambda(k)$ cannot be a convex function. This is a serious problem when the interpretation of failure rate for actual data is concerned. The function $\lambda(k)$ can for example never grow linearly, not to say exponentially that is the case for components in the wear-out lifetime period. In practice, failure rate is related to the wear in engineering, and it is understood that it might increase exponentially in many cases.

There are a few other problems associated with the failure rate definition (1). The cumulative hazard function is usually defined as the negative logarithm of the survival function. On the other hand, as pointed by Lawless[6]:

$$H(k) = - \ln R(k) \neq \sum_{i=1}^{k} \lambda(i), \tag{4}$$

and hence $H(k)$ is not really a cumulative hazard function in discrete case (if Eq. (1) is used as the failure rate function).

Furthermore, a distribution is said to be increasing failure rate average (IFRA) if

$$[R(j)]^{1/j} \geq [R(k)]^{1/k}, \quad \text{for all } k \geq j \geq 0. \tag{5}$$

However, as pointed out by Shaked *et al.*,[14] unlike the continuous case, it is not equivalent to the cumulative average of failure rate comparison, i.e.,

$$\sum_{i=1}^{j} \lambda(i)/j \leq \sum_{i=1}^{k} \lambda(i)/k, \quad \text{for all } k \geq j \geq 0. \tag{6}$$

In fact, Eq. (6) will lead to another IFRA concept. Furthermore, two different concepts for new better than used were also introduced in Ref. 14 because of similar discrepancies.

3. Another Definition of Discrete Failure Rate Function

To distinguish between the discrete failure rate definition studied in this paper and the commonly used one, the function $r(\cdot)$ is used here.

Definition. For discrete distribution with reliability function $R(k)$, the failure rate function $r(k)$ is defined as

$$r(k) = \ln \frac{R(k-1)}{R(k)}, \quad k = 1, 2, \dots . \tag{7}$$

The background to the introduction of this definition is as follows. For continuous distribution, the failure rate function is defined as

$$r(t) = \frac{f(t)}{R(t)} = -\frac{d}{dt} \ln R(t).$$

Instead of using $R(k-1) - R(k)$ for $f(k)$ which leads to the expression (1) as the failure rate definition, we could use $\ln R(k) - \ln R(k-1)$ for $-d[\ln R(t)]/dt$ above and define the failure rate function as

$$r(k) = -[\ln R(k) - \ln R(k-1)] = -\ln \frac{R(k)}{R(k-1)} = \ln \frac{R(k-1)}{R(k)}.$$

This justifies the definition of failure rate function (7). Clearly, $r(k)$ is not bounded in this case.

It can be noted that Roy and Gupta[10] used this function and named it the second rate of failure. Other than that, this function has not attracted much attention.

3.1. *The relationship between $\lambda(k)$ and $r(k)$*

Ageing properties are very important in reliability studies, especially in decision making. The monotonicity of failure rate function is of particular interest. A distribution with increasing failure rate function is called IFR distribution and similarly, one with decreasing failure rate function is called DFR distribution. A lot of results have been derived for IFR/DFR distributions in discrete time based on the definition of $\lambda(k)$, see e.g., Refs. 3, 7 and 14. However, most of the results concerning IFR/DFR are still valid because of the following important property.

There is a simple relationship between the two functions, $\lambda(k)$ and $r(k)$:

$$r(k) = -\ln \frac{R(k)}{R(k-1)} = -\ln \left[1 - \frac{R(k-1) - R(k)}{R(k-1)} \right] = -\ln[1 - \lambda(k)]. \tag{8}$$

or

$$\lambda(k) = 1 - e^{-r(k)}. \tag{9}$$

Hence, the following interesting property is obvious. We state it separately because of its importance.

Property 1. The two concepts $r(k)$ and $\lambda(k)$ have the same monotonicity property. That is, $r(k)$ is increasing/decreasing if and only if $\lambda(k)$ is increasing/decreasing.

An important consequence of the above property is that to prove that a distribution is IFR/DFR it is sufficient to prove that either $\lambda(k)$ or $r(k)$ is increasing/decreasing as a function of k. These two functions uniquely determine each other, as well as for reliability function and probability mass function.

When $r(k)$ is small, we have that

$$\lambda(k) = 1 - e^{-r(k)} = \sum_{i=1}^{\infty} (-1)^{i+1} \frac{[r(k)]'}{i!} = r(k) + o(r(k))$$

where $o(r(k))$ is a term that approaches zero when $r(k)$ approaches 0. Hence $r(k)$ and $\lambda(k)$ are almost equal for small $r(k)$.

3.2. *Reliability and related functions*

The reliability function and other related functions can be easily determined when this failure rate function is given.

Property 2. Given the discrete failure rate function $r(k)$, the reliability function, cumulative hazard function $H(k)$, and mean residual function are determined as

$$R(k) = R(k-1)e^{-r(k)} = \cdots = \exp\left\{ -\sum_{i=1}^{k} r(i) \right\}, \tag{10}$$

$$H(k) = \sum_{i=1}^{k} r(i), \quad k = 1, 2, \ldots, \tag{11}$$

and

$$\mathrm{MRL}(k) = 1 + \sum_{j=k+1}^{\infty} \exp\left\{ -\sum_{i=1}^{j} r(i) \right\} \bigg/ \exp\left\{ -\sum_{i=1}^{k} r(i) \right\}$$

$$= 1 + \sum_{j=k+1}^{\infty} \left\{ \exp\left(-\sum_{i=k+1}^{j} r(i) \right) \right\}$$

assuming $R(0) = 1$ as usual.

Note that these are very similar to the case of continuous reliability function, for which we have that

$$R(t) = \exp\left\{ -\int_0^t r(s)\,ds \right\} = \exp\{-H(t)\}. \tag{12}$$

Note that the cumulative hazard function for given $r(k)$ is,

$$H(k) = \sum_{i=1}^{k} -\ln[1 - r(i)] = -\ln\left[\prod_{i=1}^{k} \{1 - r(i)\} \right] = -\ln R(k), \quad k = 1, 2, \ldots. \tag{13}$$

Because of Eq. (11), this is a proper definition of cumulative hazard function unlike that of Eq. (4).

3.3. *Additivity property for series system*

When we have n discrete components in series with failure rate function $r_j(k)$, we have that the system reliability is given by

$$R(k) = \prod_{j=1}^{n} R_j(k) = \prod_{j=1}^{n} \exp\left\{-\sum_{i=1}^{k} r_j(i)\right\} = \exp\left\{-\sum_{j=1}^{n}\sum_{i=1}^{k} r_j(i)\right\}$$

$$= \exp\left\{-\sum_{i=1}^{k}\sum_{j=1}^{n} r_j(i)\right\} = \exp\left\{-\sum_{i=1}^{k} r(i)\right\}$$

with $r(i)$ is the failure rate function for the system given by

$$r(i) = \sum_{j=1}^{n} r_j(i)$$

Hence, the failure rate Eq. (7) is additive for series systems, and this well-known and widely used property is now valid for discrete distributions.

3.4. *Increasing/decreasing failure rate average and new better/worse than used distribution*

Defining increasing/decreasing failure rate average (IFRA/DFRA) with $\lambda(k)$ causes a problem as it will lead to two different definitions.[14] A discrete distribution is said to be an IFRA distribution if Eq. (5) is satisfied.

Property 3. A distribution is an IFRA distribution if

$$\sum_{i=1}^{j} r(i)/j \leq \sum_{i=1}^{k} r(i)/k, \quad \text{for all } k \geq j \geq 0. \tag{14}$$

It can be seen that Eq. (14) is equivalent to

$$-\ln R(j)/j \leq -\ln R(k)/k, \quad \text{for all } k \geq j \geq 0$$

which is in turn equivalent to Eq. (5). Hence, unlike the case of using $\lambda(k)$, only one definition of IFRA is needed. When the equalities are reversed, we have the same conclusion for DFRA.

A distribution is called new better than used (NBU) if

$$R(j)R(k) \geq R(j+k), \quad \text{for all } j \geq 0,\ k \geq 0. \tag{15}$$

When the inequality if reversed, we have new worse than used (NWU).

It can be shown that Eq. (15) is equivalent to

$$\sum_{i=1}^{j} r(i) \geq \sum_{i=k+1}^{k+j} r(i), \quad \text{for all } j \geq 0,\ k \geq 0,$$

or

$$H(j) + H(k) \leq H(j+k), \quad \text{for all } j \geq 0, \ k \geq 0$$

which is similar to the continuous case. If $\lambda(k)$ is used in the above, then it will lead to a different definition as pointed out in Ref. 14. Hence the failure rate definition Eq. (7) is more appropriate for the study of this type of aging property as a single definition is sufficient.

4. Failure Rate Function for Some Specific Distributions

In this section, the failure rate functions $r(k)$ for some specific distributions are given. These distributions are useful in the modeling of discrete components.

4.1. *Discrete Weibull distribution*

The most well-known discrete Weibull model is the Nakagawa–Osaki model[8] and the reliability function is given in Eq. (2). The corresponding failure rate function is

$$r(k) = \ln \frac{R(k-1)}{R(k)} = \ln \frac{q^{(k-1)^{\beta}}}{q^{k^{\beta}}} = \ln q \cdot [(k-1)^{\beta} - k^{\beta}] \tag{16}$$

and this is increasing for $\beta > 1$ and decreasing for $\beta < 1$ as for the continuous Weibull distribution.

However, although Eq. (16) can grow linearly for $\beta = 2$ as the case for continuous distribution, it does not have a similar form as for the continuous distribution. In fact, Ref. 16 proposed a second type of discrete Weibull distribution with

$$\lambda(k) = ck^{\beta-1}, \quad k = 1, 2, \ldots, m,$$

where $m = \infty$ for $\beta = 1$ and for $\beta > 1$, m is the integer part of $c^{-1/(\beta-1)}$ to ensure that $\lambda \leq 1$. Because of the need for m, the distribution is more complicated. In fact, with the new definition of failure rate function, we could define a discrete Weibull distribution with

$$r(k) = ck^{\beta-1}, \quad k = 1, 2, \ldots$$

without constraints attached to this. This discrete Weibull distribution has the reliability function is given by

$$R(k) = \exp\left\{ -\sum_{i=1}^{k} r(i) \right\} = \exp\left\{ -\sum_{i=1}^{k} ci^{\beta-1} \right\}. \tag{17}$$

It is interesting to note that this is the third type of discrete Weibull introduced by Padgett and Spurrier.[9] Note that when $\beta = 1$, all three distributions reduces to the geometric distribution with constant failure rate.

4.2. *Discrete Pareto distribution*

The discrete Pareto distribution has the following reliability function

$$R(k) = \left(\frac{d}{k+d}\right)^c, \quad c, d > 0. \tag{18}$$

The corresponding failure rate function is then

$$r(k) = \ln\frac{R(k-1)}{R(k)} = \ln\frac{(d/k-1+d)^c}{(d/k+d)^c}$$

$$= c\ln\frac{k+d}{k-1+d} = c\ln\left[1 + \frac{1}{k-1+d}\right]$$

which is a decreasing failure rate function.

4.3. *Discrete logistic distribution*

The logistic model is useful when the failure rate is increasing but does not increase very fast at the beginning. The discrete version has a reliability function given by

$$R(k) = \frac{e^{-(k-c)/d} + e^{-k/d}}{1 + e^{-(k-c)/d}}, \quad d, c > 0. \tag{19}$$

The failure rate for this distribution is given by

$$r(k) = \ln\frac{R(k-1)}{R(k)} = \frac{e^{-(k-1-c)/d} + e^{-(k-1)/d}/1 + e^{-(k-1-c)/d}}{e^{-(k-c)/d} + e^{-k/d}/1 + e^{-(k-c)/d}}$$

$$= \ln\frac{e^{1/d}\left[1 + e^{-(k-c)/d}\right]}{1 + e^{-(k-1-c)/d}} = \ln\left(1 + \frac{e^{1/d} - 1}{1 + e^{-(k-1-c)/d}}\right)$$

and this is an increasing function of k.

4.4. *s-distribution*

The s-distribution was recently studied by Soler[15] and the model is derived in an interesting way through a stress-memory concept. The reliability function is given as

$$R(k) = \prod_{i=1}^{k}(1 - p + p\pi^i), \quad 0 < p, \ \pi < 1. \tag{20}$$

The corresponding failure rate function is given by

$$r(k) = \ln\frac{R(k-1)}{R(k)} = \ln\frac{\prod_{i=1}^{k-1}(1 - p - p\pi^i)}{\prod_{i=1}^{k}(1 - p - p\pi^i)} = -\ln(1 - p - p\pi^k).$$

The failure rate function is fairly simple and it is increasing. Note that when k approaches infinity, the failure rate approaches $-\ln(1 - p)$ and it is a model for asymptotically constant failure rate function.

4.5. *Generalized Salvia–Bollinger distributions*

Salvia and Bollinger[13] proposed two discrete distributions handling increasing and decreasing failure rate functions. They are further generalized in Padgett and Spurrier.[9] For the DFR case, the reliability function are given by

$$R(k) = \frac{\prod_{i=1}^{k-1}(i\alpha + 1 - c)}{\prod_{i=1}^{k-1}(i\alpha + 1)}, \quad \alpha > 0,\ 0 < c < 1. \tag{21}$$

The failure rate function is then

$$r(k) = \ln \frac{R(k-1)}{R(k)} = \ln \frac{\prod_{i=1}^{k-2}(i\alpha + 1 - c)/\prod_{i=1}^{k-2}(i\alpha + 1)}{\prod_{i=1}^{k-1}(i\alpha + 1 - c)/\prod_{i=1}^{k-1}(i\alpha + 1)}$$

$$= \ln \frac{(k-1)\alpha + 1}{(k-1)\alpha + 1 - c} = \ln \left(1 + \frac{c}{(k-1)\alpha + 1 - c} \right)$$

and this is clearly a decreasing function.

For the IFR case, the reliability function is

$$R(k) = \frac{c^k}{\prod_{i=1}^{k-1}(i\alpha + 1)}, \quad \alpha > 0,\ 0 < c < 1. \tag{22}$$

The failure rate function is then

$$r(k) = \ln \frac{R(k-1)}{R(k)} = \ln \frac{c^{k-2}/\prod_{i=1}^{k-2}(i\alpha + 1)}{c^{k-1}/\prod_{i=1}^{k-1}(i\alpha + 1)} = \ln \frac{(k-1)\alpha + 1}{c}$$

and this is clearly an increasing function.

5. Discussions

In fact, in the first paper by Nakagawa and Osaki,[8] the term failure "rate" (*with apostrophes*) was used together with Eq. (1). It seems that later on, this notation is adopted by others as the definition of failure rate for discrete distributions. This is partly due to its nice statistical interpretation of $\lambda(k)$ as the conditional probability that an item will fail at time k given that it has survived after $k - 1$. However, this leads to too many discrepancies between the discrete and continuous case.

In this paper, another definition on discrete failure rate function is given. Although it does not have the direct probabilistic interpretation, it is much more similar to that of continuous distributions, and hence easy to use in practice without causing any confusion. The application of the new failure rate function is similar to those for the continuous failure rate function. The estimation problem and properties will be investigated in further research and reported in future when there are interesting findings.

Acknowledgment

The authors would like that thank Dr. Marcel Chevalier of Schneider Electric, Grenoble, France, whose strong interest in using discrete Weibull distribution has lead the study presented in this paper. The research is partly supported by a research project "reliability analysis of engineering systems" (RP3992679) at National University of Singapore. This work is done while the first author was visiting INPG as Invited Professor.

References

1. N. Ebrahimi, "Classes of discrete decreasing and increasing mean-residual-life distributions," *IEEE Transactions on Reliability* **R-35** (1986), pp. 403–405.
2. F. M. Guess and D. H. Park, "Modeling discrete bathtub and upside-down bathtub mean residual life functions," *IEEE Transactions on Reliability* **R-37** (1988), pp. 545–549.
3. P. L. Gupta, R. C. Gupta, and R. C. Tripathi, "On the monotonic properties of the discrete failure rates," *Journal of Statistical Planning and Inference* **65** (1997), pp. 255–268.
4. J. D. Kalbfleisch and R. L. Prentice, *The Statistical Analysis of Failure Time Data* (Wiley, New York, 1980).
5. N. A. Langberg, R. V. Leon, J. Lynch, and F. Proschan, "Extreme points of the class of discrete decreasing failure rate life distributions," *Mathematics of Operations Research* **5** (1980), pp. 35–42.
6. J. F. Lawless, *Statistical Models and Methods for Lifetime Data* (Wiley, New York, 1982).
7. J. Mi, "Discrete bathtub failure rate and upside-down bathtub mean residual life," *Naval Research Logistics* **R-40** (1993), pp. 361–371.
8. T. Nakagawa and S. Osaki, "The discrete Weibull distribution," *IEEE Transactions on Reliability* **R-24** (1975), pp. 300–301.
9. W. J. Padgett and J. D. Spurrier, "Discrete failure models," *IEEE Transactions on Reliability* **R-34** (1985), pp. 253–256.
10. D. Roy and R. P. Gupta, "Classification of discrete lives," *Microelectronics and Reliability* **32** (1992), pp. 1459–1473.
11. D. Roy and R. P. Gupta, "Characterizations and model selections through reliability measures in the discrete case," *Statistics & Probability Letters* **43** (1999), pp. 197–206.
12. A. A. Salvia, "Some results on discrete mean residual life," *IEEE Transactions on Reliability* **R-35** (1996), pp. 359–361.
13. A. A. Salvia and R. C. Bollinger, "On discrete hazard functions," *IEEE Transactions on Reliability* **R-31** (1982), pp. 458–459.
14. M. Shaked, J. G. Shanthikumar, and J. B. Valdez-Torres, "Discrete hazard rate functions," *Computers and Operations Research* **R-22** (1995), pp. 391–402.
15. J. L. Soler, "Croissance de fiabilite des versions d'un logiciel," *Revue de Statistique Appliquee* **XLIV** (1996), pp. 5–20.
16. W. E. Stein and R. Dattero, "A new discrete Weibull distribution," *IEEE Transactions on Reliability* **R-33** (1984), pp. 196–197.

About the Authors

Min Xie received his Ph.D. in Quality Technology in 1987 from Linkoping University in Sweden. He is active in teaching and research in statistical analysis of quality and reliability problems. Dr Xie is a senior member of ASQ and IEEE. He was awarded the prestigious LKY research fellowship in 1991. He has published numerous papers and three books on quality and reliability engineering. He serves on the editorial board of several international journals and in the program committee for many international conferences.

Olivier Gaudoin is Associate Professor at the National Polytechnic Institute of Grenoble, France, where he teaches applied probability and statistics. His research interests are stochastic models and statistical methods for system reliability, especially software reliability, reliability of systems subjected to random stress, discrete reliability, goodness-of-fit testing and maintenance modelling.

Cyril Bracquemond graduated from the Joseph Fourier University, Grenoble, France, in 1998. He obtained his PhD in Applied Mathematics from the National Polytechnic Institute of Grenoble in 2001. He has now joined the Pechiney company as a Statistician Engineer.

https://doi.org/10.1142/9789819812547_0015

Chapter 15

ON OPTIMAL SETTING OF CONTROL LIMITS FOR GEOMETRIC CHART[#]

M. XIE,[*] T. N. GOH and V. KURALMANI

Department of Industrial and Systems Engineering
National University of Singapore, Kent Ridge, Singapore 119260
[*] *E-mail: xie_min@nus.edu.sg*

Control charts based on geometric distribution have shown to be very useful in the monitoring of high yield manufacturing processes and other applications. It is well known that the traditional 3-sigma limits will give too many false alarms and the probability limits should be used. This paper shows that the average time to alarm may even increase at the beginning when the process is deteriorated. A new procedure is established for the setting of control limits so that the average run length is maximized when the process is at the normal level. Hence the chart sensitivity can be improved. For the derivation of the control limits in this new procedure, a simple adjustment factor is suggested so that the probability limits can be used after the adjustment.

Keywords: Statistical Process Control; Average Run Length; Sensitivity Analysis; Optimal Control Limits; Geometric Distribution; High-quality Process Control.

1. Introduction

Recently, control charts based on geometric distribution have shown to be useful in a number of applications, see, e.g., Burke,[1] Nelson,[2] Kaminsky *et al.*,[5] Xie and Goh,[8] Glushkovsky[4] and Woodall.[7] For example, when the quality of the process is high, it is better to monitor the cumulative count of conforming (CCC) items and plot this number to monitor the process, see, e.g., Goh.[4] The use of cumulative count control chart was further studied by Xie and Goh.[8,9]

It is pointed out in Ref. 9 that the traditional 3-sigma limits such as that in Ref. 5 should not be used in this case because the geometric distribution is always skewed and normal approximation is not valid. In fact, in most of the publications, the exact probability limits have been adopted. However, although this will provide a fixed false alarm probability, it is shown in this paper that there is an undesirable property in the case of geometric chart. That is, the average time to an alarm when the process is deteriorated, in terms of average run length, may increase at the beginning. This means that it will take a longer time for an alarm to be raised when the process is deteriorated than when the process is normal.

[#] This chapter appeared previously on the International Journal of Reliability, Quality and Safety Engineering. To cite this chapter, please cite the original article as the following: M. Xie, T. N. Goh and V. Kuralmani, *Int. J. Reliab. Qual. Saf. Eng*, **7**, 17–25 (2000), doi:10.1142/S0218539300000031.

To eliminate this problem, a new procedure for determination of control limits for geometric distribution is developed. In this paper, the geometric chart properties are first studied, particularly with respect to average run length. The control limits providing maximum average run length when the process is in control is then derived. An adjustment factor is also proposed so that the simple formula for probability limits can be used after some adjustment.

2. Geometric Chart and Its Average Run Length

One important application of geometric chart is for the high-quality process. Instead of concentrating on the nonconforming items, the cumulative count of conforming items are counted. This quantity can be viewed as a random variable of a geometric distribution and the limits can be determined based on exact probabilities. This method is very suitable for automatic production environment where the product is inspected one after another and the count of conforming units are accumulated automatically.

Let X be a geometrically distributed quantity with parameter p, the distribution function is then $F(x) = 1 - (1-p)^x$. For a fixed false alarm probability α, the lower and upper control limits are (see, e.g., Ref. 8).

$$\text{LCL} = \frac{\ln(1 - \alpha/2)}{\ln(1 - p_0)} \tag{1}$$

and

$$\text{UCL} = \frac{\ln(\alpha/2)}{\ln(1 - p_0)}. \tag{2}$$

The operating characteristic (OC) function, defined as the probability that the count will fall within the control limits, is given by

$$\beta = F(\text{UCL}) - F(\text{LCL}) = (1-p)^{\text{LCL}-1} - (1-p)^{\text{UCL}}. \tag{3}$$

For different value of p, the OC curve provides a measure of the sensitivity of the control chart, that is, its ability to detect a process shift. As a measure of chart performance, the average run length (ARL) is usually defined as the average number of samples plotted before a point indicates an out of control condition. For the geometric chart, the ARL for a given fraction nonconforming p is

$$\text{ARL} = \frac{1}{1 - \beta} = \frac{1}{1 - (1-p)^{\text{LCL}-1} - (1-p)^{\text{UCL}}}. \tag{4}$$

However, for geometric chart, the ARL may initially increase when the process is deteriorated. This may lead to misinterpretation that the process is well in control, or even improved. In Table 1, some ARL values for $p = 50$ ppm and $\alpha = 0.0027, 0.005$ and 0.010 are shown. From the table, the ARLs are significantly lower

Table 1. Some ARL values at process average 50 ppm for various α.

Fraction Non-conforming	$\alpha = 0.0027$	$\alpha = 0.005$	$\alpha = 0.01$	Fraction Non-conforming	$\alpha = 0.0027$	$\alpha = 0.005$	$\alpha = 0.01$
1	1	1	1	110	337	182	91
10	4	3	3	120	309	167	84
20	14	11	8	130	285	154	77
30	51	35	22	140	265	143	72
40	163	97	54	150	247	134	67
50	**370**	**200**	**100**	**160**	**232**	**125**	**63**
60	505	266	129	170	218	118	59
70	504	268	132	180	206	111	56
80	458	246	122	190	195	106	53
90	411	221	110	200	186	100	50
100	370	200	100	210	177	96	48
110	337	182	91	220	169	91	46

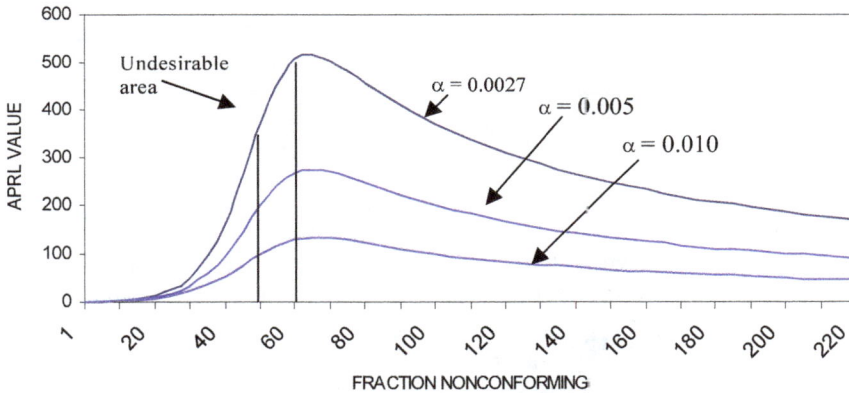

Fig. 1. ARL curves with $p = 50$ ppm for $\alpha = 0.0027$, 0.005 and 0.010 and it is clear that the maximum ARL is reached at a higher p value.

at $p = 50$ ppm irrespective of false alarm probabilities and they are increasing up to about one-third times of the desired p value, and then, it decreases.

This can also be seen in Fig. 1. The area under the ARL curve between the points $ARL(p_0)$ and $ARL(p_1)$ can be considered as undesirable area.

3. Maximizing ARL at a Fixed Process Level

As discussed in the previous section, the maximum of ARL is not at the normal process nonconforming level and this is not a desirable property. By properly adjusting the control limits, the undesirable area as indicated in Fig. 1 could be eliminated or minimized. In the following, an optimization method either to minimize this undesirable area or to maximize the ARL at the desired process average is proposed.

For geometric chart, the ARL is given by Eq. (4) and the maximum ARL at $p = p_0$ can be obtained by differentiating them and then solve for p. That is,

$$0 = (1 - p_0)^{L-1}\ln(1 - p_0)\frac{dL}{dp} - (1 - p_0)^{U-1}\ln(1 - p_0)\frac{dU}{dp}. \tag{5}$$

Solving the above equation, we have that the p value, where ARL reaches the maximum, is

$$p = 1 - \exp\left(\frac{\ln(1 - p_0)\ln\left[\frac{(\alpha/2)}{(1 - \alpha/2)}\right]}{\ln\left[\frac{\ln(1 - \alpha/2)}{\ln(\alpha/2)}\right]}\right). \tag{6}$$

Substituting this function into Eqs. (1) and (2), the new control limits are obtained as

$$\text{LCL} = \frac{\ln\left(1 - \frac{\alpha}{2}\right)\ln\left[\frac{\ln(1 - \alpha/2)}{\ln(\alpha/2)}\right]}{\ln(1 - p_0)\ln\left[\frac{(\alpha/2)}{(1 - \alpha/2)}\right]} \tag{7}$$

and

$$\text{UCL} = \frac{\ln(\alpha/2)\ln\left[\frac{\ln(1 - \alpha/2)}{\ln(\alpha/2)}\right]}{\ln(1 - p_0)\ln\left[\frac{(\alpha/2)}{(1 - \alpha/2)}\right]}. \tag{8}$$

A proof of maximum ARL at $p = p_0$ is given in the following proposition.

Proposition. For a given p_0 and specified α, supposed that the control limits are derived using Eqs. (7) and (8), the ARL is maximized at $p = p_0$.

Proof. We need to show that

$$\frac{d\text{ARL}}{dp} = 0 \quad \text{and} \quad \frac{d^2\text{ARL}}{dp^2} < 0. \tag{9}$$

From Eq. (4), we have that

$$\frac{d\text{ARL}}{dp} = \frac{(\text{LCL} - 1)(1 - p)^{\text{LCL}-2} - \text{UCL}(1 - p)^{\text{UCL}-1}}{(1 - \beta)^2}. \tag{10}$$

Substituting Eqs. (7) and (8) into Eq. (10) with $p = p_0$, the first condition can be shown.

Differentiating Eq. (10) again, we have that

$$\frac{d^2\text{ARL}}{dp^2} = \left[(\text{LCL} - 1)(\text{LCL} - 2)(1 - p)^{\text{LCL}-3} - \frac{\text{UCL}(\text{UCL} - 1)(1 - p)^{\text{UCL}-2}}{(1 - \beta)^2}\right]. \tag{11}$$

Since $1 \leq \text{LCL} < \text{UCL} < \infty$ and $0 < p$, $\beta \leq 1$, it is clear that this is negative. Hence, the proof is completed.

4. Implementation and Comparison

4.1. *An adjustment factor γ_α*

The Eqs. (7) and (8) does not have a direct probability interpretation. However, the new control limits can be derived by multiplying the probability limits with a constant. We call this constant an adjustment factor and it is denoted by γ_α. This adjustment factor is given by

$$\gamma_\alpha = \frac{\ln\left[\dfrac{\ln(1-\alpha/2)}{\ln(\alpha/2)}\right]}{\ln\left[\dfrac{(\alpha/2)}{(1-\alpha/2)}\right]}. \tag{12}$$

Since the adjustment factor γ_α is purely a function of false alarm probability alone, the original process average remains unchanged for new limits. In particular, this simple adjustment simply shifts the existing control limits by a factor of γ_α. A set of values is also presented in Table 2. A similar procedure has been proposed in Ref. 2 for p chart by doing simple adjustments on the limits and for which he used Cornish–Fisher expansion of quantiles to obtain a better normal approximation.

Table 2. Values of adjustment factor γ_α against false alarm probability α.

α	γ_α	α	γ_α	α	γ_α
0.0001	1.2315	0.0030	1.2881	0.0065	1.3051
0.0002	1.2411	0.0035	1.2914	0.0070	1.3069
0.0005	1.2551	0.0040	1.2942	0.0075	1.3085
0.0010	1.2669	0.0045	1.2968	0.0080	1.3100
0.0015	1.2743	0.0050	1.2991	0.0085	1.3115
0.0020	1.2799	0.0055	1.3013	0.0090	1.3129
0.0025	1.2844	0.0060	1.3033	0.0100	1.3155

4.2. *Comparison with existing limits*

The advantage of the new procedure is that the average run length will always decrease when the process is shifted from the normal value. The false alarm probability is further reduced. Some numerical results are shown here. For $p_0 = 50$ ppm (parts per million) and $\alpha = 0.0027$, the ARL curves are shown in Fig. 2. The numerical values for UCL, LCL and ARL are given in Table 3.

When comparing the ARL at $p = p_0$, we can find from Table 3 as 516 and 370 for the proposed and existing methods, respectively. Comparing these two values, we can point out that the proposed method will raise a false alarm signal at every

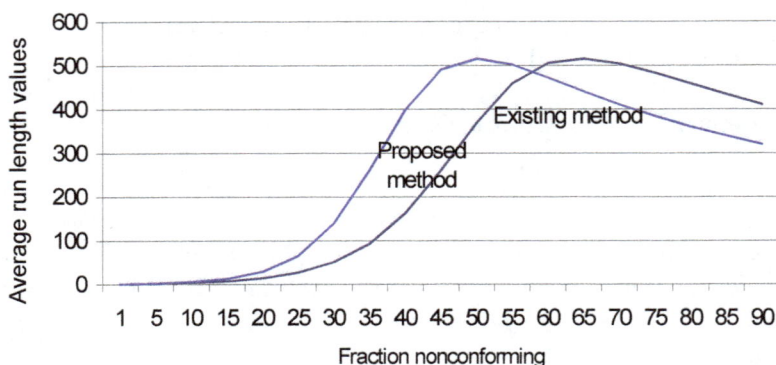

Fig. 2.   An example of ARL curves for existing and proposed methods for $p = 50$ ppm and $\alpha = 0.0027$.

Table 3. A comparison of ARL, LCL and UCL values for existing and proposed methods for given process average and $\alpha = 0.0027$.

Process Average (ppm)	LCL_e	UCL_e	ARL_e	LCL_P	UCL_P	ARL_P
10	135	660762	370	174	849690	516
20	68	330379	370	87	424843	516
30	45	220252	370	58	283227	516
40	34	165188	370	43	212419	516
50	27	132150	370	35	169935	516
60	23	110124	370	29	141611	516
70	19	94392	370	25	121381	516
80	17	82592	370	22	106208	516
90	15	73415	370	19	94406	516
100	14	66073	370	17	84965	516
200	7	33035	370	9	42480	516
300	5	22022	370	6	28319	516
400	3	16516	370	4	21238	516
500	3	13212	370	3	16990	516

516 points whereas the existing will do it at every 370 points plotted on the chart. Almost 40% is reduced in false alarm signal over the existing method.

When the process is shifted to 70 ppm, the average run length for the new procedure is 411 while the traditional procedure is 504. The proposed method simply needs 411 points to signal a deterioration whereas the existing method takes 23% more time to identify the shift.

4.3. An application example

In the example, the data set of Xie et al.[10] is used and also presented as Table 4 for $p = 500$ ppm and a false alarm probability $\alpha = 0.10$. For this example, the original

Table 4. A set of geometric quantities for given $p = 500$ ppm and $\alpha = 0.1$.

Nonconform-ing Items	ARL	Simulation Legend	Nonconform-ing Items	ARL	Simulation Legend
1	3706	$p = 500$ ppm	16	753	$p = 500$ ppm
2	9179	$p = 500$ ppm	17	3345	$p = 500$ ppm
3	78	$p = 500$ ppm	18	217	$p = 500$ ppm
4	1442	$p = 500$ ppm	19	3008	$p = 500$ ppm
5	409	$p = 500$ ppm	20	3270	$p = 500$ ppm
6	3812	$p = 500$ ppm	21	5074	Shift, $p = 50$ ppm
7	7302	$p = 500$ ppm	22	3910	$p = 50$ppm
8	726	$p = 500$ ppm	23	23310	$p = 50$ppm
9	2971	$p = 500$ ppm	24	11690	$p = 50$ppm
10	42	Shift, $p = 5000$ ppm	25	19807	$p = 50$ppm
11	3134	$p = 500$ ppm	26	14703	$p = 50$ppm
12	1583	$p = 500$ ppm	27	4084	$p = 50$ppm
13	3917	$p = 500$ ppm	28	826	$p = 50$ppm
14	3496	$p = 500$ ppm	29	9484	$p = 50$ppm
15	2424	$p = 500$ ppm	30	66782	$p = 50$ppm

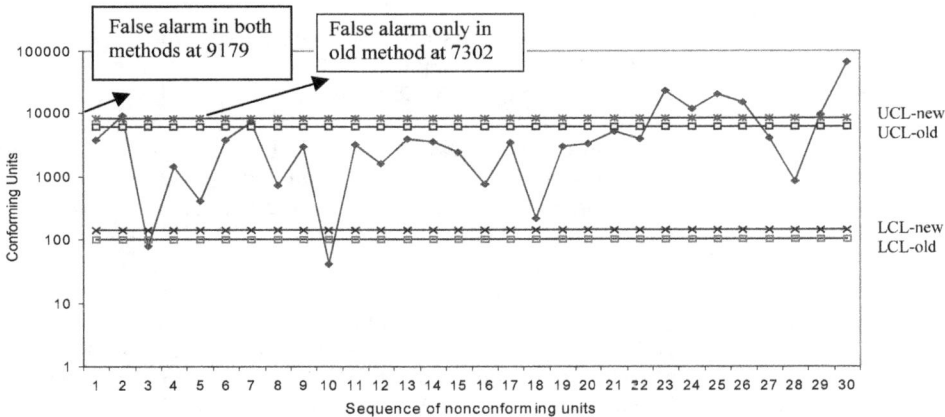

Fig. 3. The geometric chart showing both the existing and proposed control limits.

control limits are 102 and 5989 and proposed limits are 141 and 8274. Using these limits, the simulated values are plotted in Fig. 3.

From the chart, it is observed that there is only one false alarm signal encountered by the proposed limits whereas the existing method indicated two false alarm signals. Similarly, in identifying the process shift, both methods are showing same ARL when it is shifted to 50 ppm. This is a clear indication that the proposed method has less false alarm over the existing method and power of the limits.

5. Conclusions

Control chart for geometric distribution is shown to have an undesirable property of having a reduced average alarm rate when the process is shifted slightly. In this paper optimal control limits are derived to enhance the exiting procedure so that a minimum false alarm probability is achieved and the maximum average run length is reached at the process average. The proposed procedure is easily implemented and no extra calculation is needed compared with the existing method apart from an adjustment factor.

Acknowledgments

This research is supported by a research grant from the National University of Singapore for the project "Some practical aspects of SPC for automated manufacturing process" (RP3981625).

References

1. P. D. Bourke, "Detecting a shift in fraction non-conforming using run-length control charts with 100% inspection," *J. Qual. Technol.* **23** (1991), pp. 225–238.
2. G. Chen, "An improved *p* chart through simple adjustments," *J. Qual. Technol.* **30** (1998), pp. 142–151.
3. E. A. Glushkovsky, "On-line G-control chart for attribute data," *Quality and Reliability Engineering International* **10** (1994), pp. 217–227.
4. T. N. Goh, "A control chart for very high yield processes," *Quality Assurance* **13** (1987), pp. 18–22.
5. F. C. Kaminsky, J. C. Benneyan, R. D. Davis, and R. J. Burke, "Statistical control charts based on a geometric distribution," *J. Qual. Technol.* **24** (1992), pp. 63–69.
6. L. S. Nelson, "Control chart for parts-per-million conforming items," *J. Qual. Technol.* **26** (1991), pp. 239–240.
7. W. H. Woodall, "Control charts based on attribute data: Bibiliography and review," *J. Qual. Technol.* **29** (1997), pp. 172–183.
8. M. Xie and T. N. Goh, "Some procedures for decision making in controlling high yield processes," *Quality and Reliability Engineering International* **8** (1992), pp. 355–360.
9. M. Xie and T. N. Goh, "The use of probability limits for process control based geometric distribution," *International Journal of Quality and Reliability Management* **14** (1997), pp. 64–73.
10. W. Xie, M. Xie, and T. N. Goh, "A Shewhart-like charting technique for high yield processes," *Quality and Reliability Engineering International* **11** (1995), pp. 189–196.

About the Authors

M. Xie received his Ph.D. in Quality Technology from Linkoping University, Sweden in 1987. After four years of research and teaching at the same university, Dr Xie joined the National University of Singapore in 1991 as one of the first recipients of the Lee Kuan Yew Fellowship. He is active in research in quality, reliability and engineering statistics, and has published over 50 articles in refereed journals and two books in these areas. He is a senior member of both IEEE and ASQ.

T. N. Goh is a professor and head of Industrial and Systems Engineering Department at the National University of Singapore. He received his Ph.D. from the University of Wisconsin–Madison, USA. He specializes in the application of statistical methodologies for quality, reliability and productivity improvement in manufacturing, with more than two decades of experience in teaching, researching and industrial consulting. He is the Founder President of the Singapore Institute of Industrial Engineers and past President of the Singapore Institute of Statistics. Professor Goh is an elected member of the International Academy for Quality.

V. Kuralmani received his M. Phil and Ph.D. from Bharathiar University of India in 1988 and 1993, respectively. He was awarded senior research fellowship by the Council of Scientific & Industrial Research in India in 1993. Dr Kuralmani was a senior research fellow at Bharathiar University, and later research officer/manager in a private company in Singapore. He has published a number of research papers in refereed journals. Currently, he is doing research in the area of statistical process control and applied statistics.

https://doi.org/10.1142/9789819812547_0016

Chapter 16

COMMON-CAUSE FAILURES IN ENGINEERING SYSTEMS: A REVIEW*

B. S. DHILLON and O. C. ANUDE

Department of Mechanical Engineering
University of Ottawa
Ottawa, Ontario, K1N 6N5 Canada

This paper reviews selected published literature on common-cause failures. These publications have been collected from conference proceedings, engineering reliability and safety journals, textbooks etc. The collected literature is grouped into several categories and a comprehensive list of references is presented.

Keywords: Common-Cause Failures; Modeling Techniques; Reliability.

1. Introduction

General reliability theory and modeling usually support the concept that engineering equipment fail independently. This would be absolutely correct if one were to ignore the practical reality of the occurrence of common-cause failures and assume that all equipment failures are chance random events (often due to aging). If equipment failures were purely independent and random it would be possible to achieve near-perfect system reliability by merely increasing component redundancy.

Broadly speaking, a common-cause failure (CCF) event occurs when multiple items fail due to a single underlying cause.[154] In the worst-case scenario, the entire system may fail due to the same trigger event. Some of the reasons for common-cause failures include design errors and deficiencies, human errors in installation, operations and maintenance, functional deficiencies, extreme operating conditions such as high temperatures and humidity, external shocks created by earthquakes, fires, hurricanes, and floods etc. Typical examples of CCFs include multiple loss of aircraft jet engines due to bird ingestion, loss of power station emergency feedwater due to leaky values and multiple structural failures due to stress corrosion.[331]

At present, it is widely believed that the reliability and safety evaluation of engineering systems would be incomplete and would lead to overly optimistic results if common-cause failures possibilities and probabilities were discounted. This realization has led to intense research and investigation into the real and potential effects of common-cause failures on the reliability and safety of engineering systems

---

*This chapter appeared previously on the International Journal of Reliability, Quality and Safety Engineering. To cite this chapter, please cite the original article as the following: B. S. Dhillon and O. C. Anude, *Int. J. Reliab. Qual. Saf. Eng*, 1, 103–129 (1994), doi:10.1142/S0218539394000106.

ranging from the nuclear power reactor industry to the chemical and petro-chemical industrial sector. This trend is clearly demonstrated by the steady increase in the number of publications on CCFs within the last two decades or so as shown in Fig. 1.

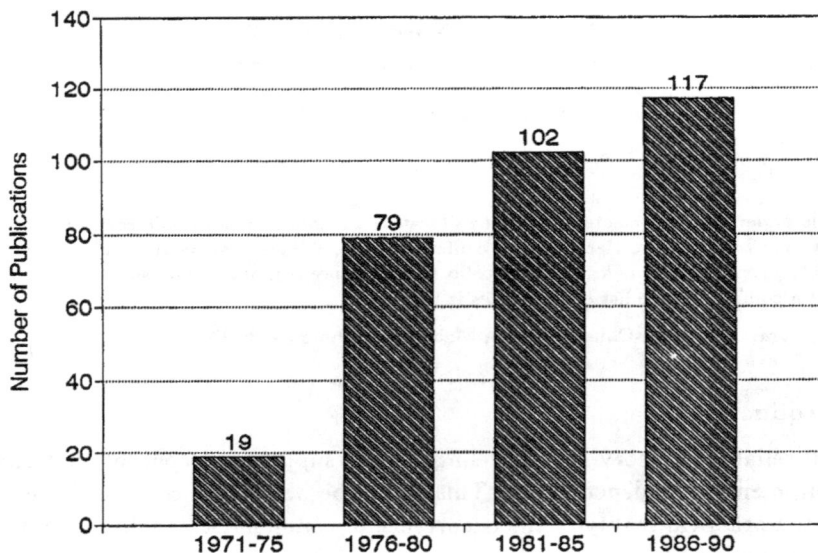

Fig. 1. Profile of publications on CCFs.

Prior to Epler's article[130] in 1969, classical reliability studies completely ignored the occurrence of common-cause failures in engineering systems and considered only random (often due to aging) failures. Some reasons for this practice, according to Edwards and Watson,[127] include: difficulties in identifying the many possible causes of common-cause failures by designers, operators, maintainers and reliability analysts; the lack of a widely acceptable interpretation and understanding of what constitutes a 'common-cause failure' leading to difficulties in the reporting and recording of data; and the relative rarity of common-cause failure events. This practice led to overly optimistic system reliability and safety predictions. However, as indicated by Fig. 1, the 1970s and 1980s witnessed a steady rise in number of studies, reports, and articles published on CCFs. As expected, most of these investigations, often probabilistic safety and risk studies (PSAs), were conducted in the power generation industry, particularly the nuclear power reactor industry, where the real and potential consequences of system failure/accidents due to a CCF are often catastrophic (e.g., the 1986 Chernobyl nuclear accident in the former Soviet Union).

This paper reviews selective publications listed in Ref. 96 (which contains only a bibliography) as well as publications that appeared after 1989 in addition to presenting a comprehensive list of references. The literature selected cover a wide range of issues relating to CCF analysis and research, raging from the ranging controversies surrounding CCF definition and quantification methods to the various CCFs defensive strategies that have been suggested to protect systems, particularly, the nuclear power reactor systems. Most of these publications appeared within the last twenty-two years (1969–1991). The sources of these publications include reliability engineering and safety journals, nuclear industry journals, conference proceedings, textbooks etc. We have attempted to provide an extensive reference list and sincerely apologise to the readers and authors if any relevant publications are omitted.

2. Classification of Literature

The literature has been classified, for convenience, into four broad categories: (i) common-cause failure modeling techniques, (ii) stochastic CCF analysis of systems, (iii) probabilistic safety and risk assessment (PSAs) in the power generation industry and, (iv) general. Table 1 contains the relevant references associated with each category and each of these categories is briefly discussed below.

2.1. *Common-cause failure (CCF) modeling techniques*

This category includes publications that dwelt on modeling techniques developed by various authors to quantify CCF rates and probabilities directly or indirectly using raw field data obtained from sources such as licensee event reports (LERs). Examples include the beta factor (β-factor), the binomial failure rate (BFR) and the multiple greek letter (MGL) models.

2.2. *Stochastic CCF analysis of systems*

This category covers published literature on the analysis of stochastic models of either existing or likely engineering systems susceptible to CCFs. Several new expressions and procedures for various reliability indices have been developed through the analysis of these stochastic models and significant contributions appeared in this area during the last decade.

2.3. *Probabilistic risk and safety assessment (PSAs)*

This category contains various publications associated with investigations conducted to assess the risk and safety implications of operating different types of power generating systems, particularly nuclear power reactors, with regard to CCFs.

2.4. *General*

This category includes publications that discussed other aspects of CCF analysis other than the ones mentioned in the above categories, such as fault-tree based CCF procedures and CCF computer codes.

Table 1. Classification of literature

<u>References</u>

(i) CCF Modeling   [13, 15, 17, 19, 62, 69, 134, 135, 137, 138, 141, 142, 164, 165, 169,
     Techniques   172, 175, 177, 186, 194, 199, 200, 205, 212, 232, 237, 238, 242, 243,
                     254–256, 258, 264, 268, 308, 323, 326]

(ii) Stochastic CCF [2, 30–32, 35, 40, 44–61, 74, 76, 77, 79–83, 85, 92–94, 95, 97–118,
      Analysis of   120, 157, 158, 160, 161, 170, 176, 184, 198, 199, 219, 222, 223, 230,
      Systems   259, 277, 285, 296, 300, 301, 302, 303, 304, 322]

(iii) PSAs   [1, 3, 5, 6, 9, 16, 18, 20–24, 27, 28, 35, 36, 40, 64, 65, 125, 131, 136,
                     145, 148, 166, 168, 173, 174, 183, 187, 189, 190, 191–193, 195–197,
                     201, 211, 213, 216, 218, 228, 235, 236, 239, 240, 241, 245, 248, 260,
                     270, 275, 276, 278–282, 292, 297, 310, 312–314, 316, 318, 321, 339,
                     340, 352]

(iv) General   [4, 7, 8, 10, 11, 12, 25, 26, 29, 33, 34, 37–39, 41–43, 63, 66–68, 70–73,
                     89–91, 119, 121–124, 126–130, 132, 133, 139, 140, 141, 143, 144, 146,
                     147, 149, 150–156, 159, 162, 163, 167, 171, 175, 178–182, 185, 187,
                     188, 193, 202–204, 206–210, 214, 215, 217, 220, 221, 224–227, 229,
                     231, 233, 234, 244, 246, 247, 249–253, 257, 261–263, 264–267, 269,
                     271–274, 283, 284, 286–291, 293–295, 298, 299, 305–307, 309, 311,
                     315, 317, 319, 320, 324, 325, 327, 328–338, 341–351, 353]

3. Review of Publications on CCFs

The following review is conducted according to the classification given in Table 1.

3.1. *Common-cause failure (CCF) modeling techniques*

Fleming and Mosleh[140] recognized two basic categories of CCF Models — the explicit models and the parametric models. In explicit modeling, the specific causes of multiple failures are modeled explicitly whereas only their impacts are modeled explicitly in parametric models. The ANS/IEEE PRA procedures guide[9] advocates the use of an appropriate combination of both modeling approaches in as many investigations as possible to achieve completeness. However, this position is not supported in the NREP procedures guide[262] which considers only the explicit models useful and recommends that parametric modeling should not be performed. This divergent of opinion is firmly rooted in the controversies and disagreements surrounding the search for a widely acceptable procedure(s) to quantify CCFs in engineering systems.

One of the earliest parametric models is the β-factor method.[134] Praised for its procedural simplicity, the β-factor was found to be excessively conservative when applied to redundant systems. There have been attempts to link the β-factor value directly to the degree of system protection and defense against CCFs[205,243] and some extensions of the β-factor have been proposed in an effort to increase its usefulness and coverage. Some of these include the Multiple Dependent Failure Fraction (MDFF),[308] developed for systems with three identical units and, modified by Hirschberg[186] to analyse systems with four identical units and the Multiple Greek Letter (MGL).[137] The β-factor's main limitation lies in its assumption of zero component survival probabilities in the event of a shock. In practice, it is used as a gauge to measure the impact of CCFs on redundant systems.

Vesely[323] proposed the Binomial Failure Rate (BFR) method. The BFR's strength lies in its ability to model higher component failure multiplicities but it has four CCF parameters (a possible demerit) and its assumption of independent component failures in the event of a shock could be difficult to defend in practice. The independent failures assumption of the BFR is somewhat softened by Atwood,[17] who incorporated the idea of having two kinds of shocks — the lethal shock which triggers off the failure of all system components and the non-lethal shock which does not. The Square Root Bounding Method, made popular by the U.S. Reactor Safety Study (WASH 1400),[318] is perhaps the most criticized approach in the literature.[124,126,170,175,226] Critics charge that it lacks real practical engineering foundation. Its adoption of the log-normal distribution to model failure probabilities is, at best, arbitrary.

Several new approaches, developed to address the shortcomings of the earlier techniques as well as to suit certain system-specific CCF features have emerged in the literature. Some of these include the Multinomial Failure Rate (MFR) model,[13] the Basic Parameter (BP)[252] and the Random Probability Shock (RPS) model.[199] A detailed tabulation of the key characteristics of these emerging approaches is presented in Ref. 252.

Several investigations have been conducted to compare the performances of these various approaches using actual reliability experience data. Fleming *et al.*[138] compared the performances of the BP, MGL and BFR models within the framework for system-level CCF analysis and their application to a three-train auxiliary feed-water system. Three methods — the β-factor, the square root bounding approach and a third approach based on the multi-variate exponential distribution developed by Marshall and Olkin[242] are compared by Fleming and Raabe.[142] Using the same field data from fifty groups of four diesel generators, Hirschberg[186] compared the performance of the β-factor method with three other approaches — the BFR, MDFF, and the MGL and concluded the MGL the best performer. References 177, 212 also report comparison studies.

Some stress-resistance models have been proposed in the literature. They include the Common Load Model (CML),[232] the Inverse Stress-Strength Interference Technique (ISSI),[165] and a time-dependent loading model suggested by Harris.[170] These stress-resistance models are quite useful and versatile and most, as observed by Heikkila,[175] were developed logically from first principles regarding stresses and resistances. Their main flaw could be that very precise engineering data regarding the physical and material properties of components must be available to enable an accurate determination of the distributions of loads and resistances. Thus, in practice, most of these techniques may not be easy and straight-forward to use.

Most of the techniques already mentioned, particularly, relate to situations when component states could be analysed as binary, i.e., either components are in a failed state or they are in an operational state. However, some attempts have been made to model explicitly, ambiguous component situations such as partial failures, incipient failures and potential failures. A notable article concerning this subject is by Han *et al.*[169] They suggest a new CCF model, the Trinomial Failure Rate model (TFR), capable of handling three component states: success; gray and; failure. The 'gray' component state is considered to be representative of all the ambiguous component situations.

3.2. *Stochastic CCF analysis of systems*

In an insightful article that dwelt on the assessment of the performance of repairable systems susceptible to chance failures and CCFs, Chu and Gaver[48] presented two Markov Models. Model *A* considered CCFs of a catastrophic nature while Model *B* assumed that CCFs merely increase the failure rates of components. They then derived formulas for some system reliability parameters. Formulas for relevant reliability indices for repairable systems susceptible to chance and common-cause failures are also presented in Refs. 54, 55, 105, 112, 118, 198. Dhillon and Anude[113-116] studied non-repairable systems with various redundant configurations composed of identical and non-identical components subject to CCFs and derived new reliability, availability and system mean time to failure expressions. The analyses of non-repairable systems subject to CCFs and chance failures are also reported in Refs. 55, 94, 95.

The stochastic analysis of multi-state device networks subject to both CCFs and chance failures are discussed in Refs. 51, 52, 76, 85, 111 and Markov models of redundant systems of *k-out-of-n:G* nature subject to CCFs are presented in Refs. 49, 50, 53, 60, 74, 80. Common-cause outages in power transmission lines are analysed using Markovian and related approaches in Refs. 30, 31, 32, 34, 117, 219, 300, 301, 302.

The determination of optimal maintenance schedules and optimal numbers of units in redundant networks on the basis of operating costs via stochastic approaches are studied in Refs. 84, 157, 161, 184, 222, 223, 352. Stochastically developed system replacement models are presented in Refs. 2, 36.

The strength of these stochastic approaches lies in their flexibility and versatility but they have serious drawbacks. Most are facilitated by incorporating some simplifying assumptions about system and sub-system transition phenomena such as failure and repair processes. Most of these assumptions are made, not out of any real concern to replicate reality, but purely for mathematical convenience. In addition, some of these approaches are too system-specific thereby limiting their coverage.

3.3. *PSAs in the power generation industry*

Epler,[130] in one of the earliest publications on CCFs, provided a list of simultaneous failure incidents in reactor protection instrumentation systems at the Oak Ridge National Laboratory (ORNL) and suggested the following as the main causes of CCFs:

changes in the characteristics of the system protected; unrecognized dependence on a common element; system unavailability by an accident and; human error. In a study of U.S. power reactor abnormal occurrence reports, Taylor,[313] determined that an overwhelming proportion of failures are CCFs including design, installation and operational errors.

Mosleh *et al.*[253] discussed the framework developed for the treatment of CCFs in risk and reliability analyses in a project jointly sponsored by the Electric Power Research Institute (EPRI) and the U.S. Nuclear Regulatory Commission (USNRC). In this study, a systematic guide for the identification, modeling and quantification of CCFs is developed. Issues such as logic model development, screening of CCF events, data analysis, system quantification and interpretation of results are addressed. This framework is reviewed by Rasmuson,[290] who touched upon related issues such as the importance of having plant-specific data, the quantification of CCFs and some approximations that can be used in the quantification process and data collection needs for CCF events.

Some PSAs conducted to estimate the common-cause failure probabilities of specific power reactor sub-systems are reported in the literature. For example, the probability safety assessment study of the High Temperature Gas Cooled Reactor conducted by the General Atomic Company in the United States which led to estimates of the failure probability of the cooling system is reported by Fleming and Hannaman.[136] Using a Marshall–Olkin based approach, and utilizing recorded field data from the nuclear power industry, Johnson and Vesely[216] estimated the failure probabilities of valve leakages. Mankamo and Kosonen,[241] in order to estimate the failure probability of the pressure relief function in the Teollisuuden Voima Oy (TVO) Plant Study in Finland, introduced an extended version of the Common-Load model (CLM) by incorporating a low probability extreme load distribution that enabled strong dependencies at high orders of failure multiplicities to be modeled. References 21, 27, 28, 183, 239, 240, 248, 270, 292 also dwelt upon estimating common-cause failure probabilities of specific power reactor sub-systems.

As expected, the literature is replete with various power reactor protection and defense strategies against CCFs. Epler[129] stressed the importance of applying diverse shut-down systems and periodic testing in reducing CCFs in the nuclear power reactor industry. Hayden[173] examined a broad category of failure mechanisms that can initiate CCFs in nuclear plant systems from reactor operating experiences and reports and outlined some ideas for the development of design and operating guidelines for reducing the probability of CCFs. System design principles adopted to ensure high system reliability and safety in German 1300MW Siemens/KWU nuclear power plants and reduce CCFs are presented by Schilling and Dörre.[297] Paula *et al.*[273] suggested the concept of a cause-defense matrix as a means of adequately representing the different impacts plant-specific defenses have on different types of failure causes and showed how qualitative cause-defense matrices may be used to perform comprehensive CCF analyses for any nuclear plant. Plant protection and defense issues are also addressed in Refs. 125, 196, 197, 271, 274, 307.

Hirschberg[195] presented a brief summary of the status of CCFs treatment in Nordic Countries and posited that activities concentrated upon: proper consideration of plant-specific protective measures against CCFs; the generation of CCF data which take into consideration the efficiency of such measures and; the quantification of CCF effects on the reliability and safety of systems with very high levels of redundancy. PSAs conducted on Scandinavian power plants are also reported in Refs. 27, 28, 188, 189, 190, 191, 193, 196, 197, 212, 213.

The major challenges faced by analysts when conducting PSAs lie in the areas of CCF identification and data generation and CCF classification and quantification. For example, questions such as: *which failures should be considered as CCFs and which should not? What CCF modeling approach would be most appropriate given a particular set of circumstances and data?* There are no easy answers. Some failure events are clearly CCFs, some others are not. Not all multiple failures are automatically CCFs and all single failures are not automatically independent. Most analysts conduct PSAs within the framework of their own perceptions of what constitutes a CCF. Thus, some of the techniques and approaches currently being used remain questionable and controversial.

3.4. *General*

There is no single and widely acceptable definition for CCFs. The literature contains several definitions.[24,41,143,294,318,336] Easterling[124] posited that perhaps difficulties in identifying CCFs stem from the fact that there are three types of statistical dependence involved: dependence among failure events themselves; dependence of failure events on the conditions under which an item is expected to perform; and dependence among these conditions. Most of these issues, disagreements and controversies associated with CCFs are amply addressed in Refs. 127, 151, 263, 305, 306.

Some fault-tree based procedures and computer codes have been developed to assess complex engineering systems susceptible to CCFs. Hudson and Gasca[201] presented a methodology for the prediction of accident sequence probabilities in a nuclear power plant as well as an analytical procedure for calculating fault-tree top event probabilities based on the correlated failures of primary events. A fault-tree based step-by-step procedure for assessing CCFs in complex systems such as nuclear power reactor plants is reported by Cate *et al.*[43] Worrell and Stack[344] outlined a procedure for identifying common-cause candidates (these are minimal cutsets whose primary events could be triggered by a special condition or are susceptible to the same secondary cause) using the set equation transformation system (SETS). Other fault-tree based procedures are presented in Refs. 4, 26, 67, 249, 283, 309.

Bickel and Caivano[29] described a simple and straightforward procedure for adapting the BFR point estimates (based on pooled United States Nuclear Industry Experience) for use in WAMCUT and SPASM (computer codes developed under EPRI sponsorship for Monte Carlo system evaluation). Fault-tree and qualitative CCF analyses of a power distribution box system using the MOCUS-BACFIRE

β-factor (MOBB) computer code is reported by Heising and Luciani.[178] Rasmuson *et al.*[294,295] discussed COMCAN IIA and COMCAN III computer codes developed to conduct automated CCF analysis in the Electric Power Industry. Computer codes for CCF analysis are also reported in Refs. 22, 38, 42, 43.

4. Conclusion

In 1979, Edwards and Watson[127] identified the following as problem areas in CCF analysis and research:

- the identification of the many possible causes of common-cause failures by professionals such as designers, operators, maintainers and reliability analysts.
- the appreciation of the various protective measures against CCFs that are available.
- the modeling techniques used in the reliability evaluation of systems susceptible to CCFs.
- the use of data from reported CCF events for the reliability evaluation of other systems.
- the lack of a widely acceptable definition of CCFs leading to difficulties and confusion in the reporting, recording and subsequent analysis of data.

Unfortunately, some fifteen years later, more heat than light has been shed on these areas and some controversy and confusion still remain in virtually all of these areas in spite of the fact that a lot of studies on CCFs have been conducted and published since then. In our view, the choice of what methods to adopt when quantifying CCFs will not be any easier in the future unless a widely acceptable CCF definition and relatively controversy-free procedure guides are presented. Everyone agrees that CCFs are a threat to the integrity of complex engineering systems but we do not seem to be united on *how best* to tackle this threat. In concluding, we echo the position of Smith and Watson[306] "unless the industry can recognize that the present state of affairs is chaotic and counterproductive, settle on a meaningful definition as a logical communication link, conduct appropriate research and then apply the indicated corrective actions, the CCF problem will remain confusing and controversial".

References

1. "A study of common-cause failures phase I: A classification system", Electrical Power Research Institute, report no. EPRI, NP 3383, research project 2169-1, Los Alamos Technical Associates Inc., New Mexico, U.S.A., Jan. 1984.
2. M. S. Abdel-Hameed, "Optimun replacement of a system subject to shocks", *Journal of Applied Probability* **23**, 107 (1986).
3. R. N. Allan *et al.*, "Modeling common-mode failures in the reliability evaluation of power system networks", IEEE Power Engineering Society Discussions and Closures from the Winter Meeting (New York, N.Y., Feb. 1979), pp. DISC. A 79 040-7.
4. R. N. Allan *et al.*, "An efficient computational technique for evaluating the cut/tie sets and common-cause failures", *IEEE Trans. Reliability* **30**, 101 (1981).

5. A. Amendola, ed., "Advanced seminar on common-cause failure analysis in probabilistic safety assessment", *Proc. of the ISPRA Course* (Kluwer, Dordrecht, The Netherlands, 1987).

6. A. Amendola, "Classification of multiple related failures", in *Advanced Seminar on Common-Cause Failure Analysis in Probabilistic Safety Assessment*, A. Amendola, ed. (Kluwer, Dordrecht, The Netherlands, 1989), pp. 31–46.

7. A. Amendola and A. Z. Keller, *Reliability Data Bases* (Kluwer, 1987).

8. J. A. Burns Jr., ed., American Society of Mechanical Engineers, "Advances in reliability and stress analysis", presented at the ASME Winter Annual Meeting (San Francisco, CA., December 1978).

9. American Nuclear Society and the Institute of Electrical and Electronics Engineers, *PRA Procedures Guide; A Guide to the Performance of Probabilistic Risk Assessments of Nuclear Power Plants*, sponsored by the U.S. Nuclear Regulatory Commission and Electric Power Research Institute, NUREG/CR-2300 (1983).

10. G. E. Apostolakis, "The effect of a certain class of potential common-cause mode analysis on reliability of redundant systems", *Nuclear Engineering and Design* **36**, 123 (1976).

11. G. E. Apostolakis, "On a certain class of potential common-mode failures", *Transactions of American Nuclear Society on Safety System Reliability Methods and Applications* **22**, 476 (1975).

12. G. E. Apostolakis, "On the reliability of redundant systems", *Transactions of the American Nuclear Society on Safety Systems Reliability Methods and Applications* **22**, 477 (1975).

13. G. Apostolakis and P. Moieni, "The foundation of models of dependence in probability safety assessment", *Reliability Engineering* **18**, 177 (1987).

14. G. E. Apostolakis and P. Moieni, "On the correlation of failure rates", reliability data collection and use in risk and availability assessment", *Proc. of the 5th EUREDATA Conference*, Heidelberg, Germany, (1986).

15. C. L. Atwood, "Data analysis using the binomial failure rate common-cause model", NUREG/CR-3737 (U.S. Nuclear Regulatory Commission, Washington D.C., 1983).

16. C. L. Atwood, "Common-cause and individual failure and fault rates for licensee event reports of pumps at U.S. commercial nuclear power plants", Report No. EGG EA 5289, EG and G Idaho Falls, Idaho (U.S. Nuclear Regulatory Commission, Washington D.C., 1980).

17. C. L. Atwood, "The binomial failure rate common-cause model", *Technometrics* **28**, 139 (1986).

18. C. L. Atwood, "Common-cause fault rates for pumps", Report No. NUREG/CR-2098, EGG-EA-5289 (U.S. Nuclear Regulatory Commission, Washington D.C., 1982).

19. C. L. Atwood, "Estimators for the binomial failure rate common-cause model", Report No. NUREG/CR-1401, EGG-EA-5112 (U.S. Nuclear Regulatory Commission, Washington D.C., 1981).

20. C. L. Atwood and J. A. Steverson, "Common-cause rates for diesel generators: Estimates based on licensee event reports at U.S. commercial nuclear plants, 1976–1978", Report No. NUREG/CR-2099, EGG-EA-5359, Rev. 1 (U.S. Nuclear Regulatory Commission, Washington D.C., 1982).

21. C. L. Atwood and J. A. Steverson, "Common-cause fault rates for valves: Estimates based on licensee event reports at U.S. commercial nuclear power plants, 1976–1980", EGG-EA-5485, Rev. 01 (1982).

22. C. L. Atwood and W. J. Suitt, "User's guide to BFR: A computer code based on the binomial failure rate common-cause model", Report No. 2729, EGG-EA-5502

(U.S. Nuclear Regulatory Commission Report, Washington D.C., 1983).

23. C. L. Atwood and R. M. Teresa, "Common-cause failure rates for instrumentation and control assemblies", Report No. NUREG/CR-3289, EGG-EA-2258 (U.S. Nuclear Regulatory Commission, Washington D.C., 1983).

24. G. M. Ballard, "Dependent failure analysis in PSA", in *IAEA International Conference on Nuclear Power Performance and Safety*, Vienna, Austria (1987).

25. J. V. Bank, "Catastrophic failure modes limit redundancy effectiveness", *IEEE Trans. Reliability* **32**, 409 (1983).

26. R. E. Barlow *et al.*, eds., "Reliability and fault-tree analysis: Theoretical and applied aspects of system reliability and safety assessment", in *Proceedings in Applied Mathematics Ser.: No. 6* (Society for Industrial and Applied Mathematics, 1975), 927pp.

27. M. Bengtz and S. Hirschberg, "Sensitivity studies of CCF contributions for motor operated valves in the Swedish PSAs", Report No. RAS-470 (88) 24 (ABB Atom Report RPC 88-91), (1988).

28. M. Bengtz *et al.* "NKA project risk analysis (RAS-470): Identification of common-cause failure events for motor operated valves in Swedish boiling water reactor plants", Final Report, RAS-470 (84) 4 (1986).

29. J. H. Bickel and J. D. Caivano, "Generation of common-cause failure rate distributions for Monte Carlo analysis using beta factor method", in *Proc. of the International Topical Meeting on Probabilistic Safety Methods and Applications*, EPRI NP-3912, Vol. 3, sessions 17–23, California (1985), pp. 179/1–179/6.

30. R. Billinton and Y. Kumar, "Transmission line reliability models including common-mode and adverse weather effects", *IEEE Trans. Power Apparatus and Systems* **100**, 3899 (1981).

31. R. Billinton *et al.*, "Application of common-cause outage models in composite system reliability evaluation", *IEEE Trans. Power Apparatus and Systems* **100**, 3648 (1981).

32. R. Billinton *et al.*, "Common-cause outages in multiple circuit transmission lines", *IEEE Trans. Reliability* **27**, 128 (1978).

33. R. Billinton and R. N. Allan, *Reliability Evaluation of Engineering Systems: Concepts and Techniques*, 2nd ed. (Plenum, 1992) 475.

34. R. Billinton and R. N. Allan, eds., *Reliability Evaluation of Power Systems* (Plenum, 1984) 436.

35. A. Birolini, "On the use of stochastic processes in modeling reliability problems", in *Lecture Notes in Economics and Mathematical Systems*, Vol. 252 (Springer-Verlag, New York, 1985), 105pp.

36. P. J. Boland and F. Proschan, "Optimun replacement of a system subject to shocks", *Operations Research* **31**, 697 (1983).

37. A. J. Bourne *et al.*, "Defences against common-mode failures in redundancy systems", Report No. SRD R 196, A Guide for Management, Designers and Operators, Safety and Reliability Directorate (United Kingdom Atomic Energy Authority, Warrington, U.K.).

38. G. R. Burdick, "COMCAN — A computer code for common-cause analysis", *IEEE Trans. Reliability* **26**, 100 (1977).

39. Canadian Society of Reliability Engineers (SRE), *Proc. of the 1980 Canadian SRE Reliability Symposium*, Ottawa, Ontario, Canada, 1980 (Pergamon), 170pp.

40. S. Canterella, "Treatment of multiple related failures by Markov method", *Advanced Seminar on Common-Cause Failure Analysis in Probabilistic Safety Assessment*, ed. A. Amendola (Kluwer, Dordrecht, The Netherlands, 1989), pp. 145–157.

41. A. Carnino, "View on the problems of 'Common-Modes' in CEA/Department of Nuclear Safety", *Proc. of the Second National Reliability Conference*, U.K. (1979), pp. 5A/3/1-5A/3/3.

42. C. L. Cate and J. B. Fussel, "BACFIRE — A computer code for common-cause failure analysis" (University of Tennessee, Knoxville, 1977).

43. C. L. Cate *et al.*, "A computer aided approach to qualitative and quantitative common-cause analysis for complex systems", *Proc. of the 8th Annual Pittsburgh Conference on Modeling and Simulation*, Vol. 1, Part 8 (1977), pp. 25–29.

44. K. C. Chae and G. M. Clark, "System reliability in the presence of common-cause failures", *IEEE Trans. Reliability* **35**, 32 (1986).

45. A. A. R. Chari, "A stochastic model for availability measure with common-cause failures", *IEEE Trans. Reliability* **35**, 570 (1986).

46. A. A. Chari, "A Markovian approach to system reliability and availability measures with common-cause failures", Ph.D thesis, Sri Venkateswara University, Tirupati, India, 1988.

47. A. A. Chari *et al.*, "Reliability analysis in the presence of chance common-cause shock failures", *Microelectronics and Reliability* **31**, 15 (1991).

48. B. B. Chu and D. P. Gaver, "Stochastic models for repairable redundant systems susceptible to common-mode failures", in *Proc. of the International Conference on Nuclear Systems Reliability Engineering and Risk Assessment*, Gatlinburg, Tennessee (1977), pp. 342–367.

49. W. K. Chung, "A K-out-of-N: G redundant system with cold standby units and common-cause failures", *Microelectronics and Reliability* **24**, 691 (1984).

50. W. K. Chung, "A N-unit redundant system with common-cause failures", *Microelectronics and Reliability* **19**, 377 (1979).

51. W. K. Chung, "A two non-identical three-state unit redundant system with common-cause failures and one standby unit", *Microelectronics and Reliability* **21**, 707 (1981).

52. W. K. Chung, "A K-out-of-N: G three-state redundant unit system with common-cause failures and replacements", *Microelectronics and Reliability* **21**, 589 (1981).

53. W. K. Chung, "A K-out-of-N redundant system with common-cause failures", *IEEE Trans. Reliability* **29**, 344 (1980).

54. W. K. Chung, "Reliability analysis of a repairable parallel system with standby involving human error and common-cause failures", *Microelectronics and Reliability* **27**, 269 (1987).

55. W. K. Chung, "Reliability analysis of repairable and non-repairable systems with common-cause failures", *Microelectronics and Reliability* **29**, 345 (1989).

56. W. K. Chung, "A reliability model for a K-out-of-N: G redundant system with multiple failure modes and common-cause failures", *Microelectronics and Reliability* **27**, 621 (1987).

57. W. K. Chung, "An availability calculation of K-out-of-N redundant systems with common-cause failures and replacement", *Microelectronics and Reliability* **20**, 517 (1980).

58. W. K. Chung, "A K-out-of-N: G redundant system with dependent failure rates and common-cause failures", *Microelectronics and Reliability* **28**, 201 (1988).

59. W. K. Chung, "An availability analysis of K-out-of-N: G redundant system with dependent failure rates and common-cause failures", *Microelectronics and Reliability* **28**, 391 (1988).

60. W. K. Chung, "A reliability analysis of a K-out-of-N: G redundant system with common-cause failures and critical human errors" *Microelectronics and Reliability* **30**, 237 (1990).

61. W. K. Chung, "Common-cause failures and critical human errors in repairable and non-repairable systems", *Microelectronics and reliability* **30**, 243 (1990).

62. A. C. Cohen and B. J. Whittens, "Parameter estimation in reliability and life span models", in *Statistics, Textbooks and Monographs*, No. 96 (Marcel Dekker, 1988), 312pp.

63. V. Colombari, *Reliability Data Collection and Use in Risk and Availability Assessment* (Springer-Verlag, New York, 1989), 898pp.

64. "Common-mode failure of incore instrumentation: Reactor operating experiences", Report No. ROE-73-7 (Oak Ridge National Laboratory, Oak Ridge, Tennessee, U.S.A., 1975).

65. "Common-mode failure of incore instrumentation: Reactor operating experiences no. ROE-73-7", USAEC Report No. ORNL/NSIC-64 (United States Atomic Energy Council, Oak Ridge National Laboratory, Oak Ridge, Tennessee, U.S.A., 1972), pp. 51–54.

66. S. Contini, "Algorithms for common-cause analysis — A preliminary report", TN 1.06.01.81.16 (CEC-JRC, ISPRA, Italy, 1981).

67. S. Contini, "Dependent failure modeling by fault-tree technique", in *Advanced Seminar on Common-Cause Failure Analysis in Probabilistic Safety Assessment*, ed. A. Amendola (Kluwer, Dordrecht, The Netherlands, 1989), pp. 145–157.

68. J. Cox and N. Tait, *Reliability Safety and Risk Management* (Butterworth-Heinemann, 1991), 256pp.

69. G. L. Crellin *et al.*, "Multiple component malfunction scenarios — A classification technique with emphasis on shared-cause events", International ANS/ENS Topical Meeting on Probablistic Safety Methods and Applications (1985).

70. G. L. Crellin *et al.* "A study of common-cause failures, part 2: a comprehensive classification system for component fault analysis", Interim Report EPRI-NP-3837, Research Project 2169-1 (1985).

71. G. L. Crellin *et al.*, "Defensive strategies for reducing susceptibility to common-cauase failures; Vol. 1", Defensive Strategies Report EPRI NP-5777 (1988).

72. S. Dai and M. Wang, *Reliability Analysis in Engineering Applications* (Van Nostrand Reinhold, 1992), 448pp.

73. S. J. David, *Reliability and Maintainability in Perspective: Technical, Management and Commerical Aspects* (Macmillian Press, London, 1991), pp. 158–163.

74. B. Dayal and J. Singh, "A 1-out-of-N: G system with common-cause failures and critical human errors", *Microelectronics and Reliability* **31**, 847 (1991).

75. B. S. Dhillon, "On common-cause failures — Bibliography", *Microelectronics and Reliability* **18**, 533 (1979).

76. B. S. Dhillon, "A K-out-of-N three-state device system with common-cause failures", *Microelectronics and Reliability* **18**, 447 (1978).

77. B. S. Dhillon, "Mechanical component reliability under environmental stress", *Microelectronics and Reliability* **20**, 153 (1980).

78. B. S. Dhillon, "Effects of weibull hazard rate on common-cause failure analysis of reliability networks", *Microelectronics and Reliability* **17**, 59 (1978).

79. B. S. Dhillon, "A common-cause failure availability model", *Microelectronics and Reliability* **17**, 583 (1978).

80. B. S. Dhillon, "A 4-unit redundant system with common-cause failures", *IEEE Trans. Reliability* **28**, 267 (1979)

81. B. S. Dhillon, "Unified availability model: A redundant system with mechanical, electrical, software, human and common-cause failures", *Microelectronics and Reliability* **21**, 653 (1981).

82. B. S. Dhillon, "A 1-out-of-N: G system with duplex elements", *IEEE Trans. Reliability* 169 (1979).

83. B. S. Dhillon, "Multi-state device redundant systems with common-cause failures and one standby unit", *Microelectronics and Reliability* **20**, 411 (1980).

84. B. S. Dhillon, "Optimal maintenance policy for systems with common-cause failures", in *Proc. of the 9th Annual Pittsburgh Confernce on Modeling and Simulation Pittsburgh*, Pennsylvania (1978), pp. 1121–1215.

85. B. S. Dhillon, "A system with two kinds of 3-state elements", *IEEE Trans. Reliability* **29**, 345 (1979).

86. B. S. Dhillon, *Reliability Engineering in Systems Design and Operation* (Van Nostrand Reinhold, New York, 1983).

87. B. S. Dhillon and C. Singh, *Engineering Reliability: New Techniques and Applications* (Wiley, New York, 1981), pp. 93–111, 167, 260.

88. B. S. Dhillon, *Mechanical Reliability: Theory, Models and Applications* (American Institute of Aeronautics and Austronautics, Washington D.C., 1988), pp. 127–129.

89. B. S. Dhillon, "Power system reliability safety and management", *Ann Arbor Science* (The Butterworth Group, Borough Green, Sevenoaks, England, 1983).

90. B. S. Dhillon, *Quality Control, Reliability and Enginnering Design* (Marcel Dekker, New York, 1985).

91. B. S. Dhillon, *Reliability in Computer System Design* (Ablex Publishing, Norwood, New Jersey, 1987), pp. 149–156.

92. B. S. Dhillon, "Stochastic analysis of a parellel system with common-cause failures and critical human errors", *Microelectronics and Reliability* **29**, 627 (1989).

93. B. S. Dhillon, "Mathematical modeling of common-cause failures and human errors in engineering systems", in *Proc. of the 7th International Conference of the Israel Society for Quality Assurance* (1988), pp. 2.1.1.1–2.1.1.5.

94. B. S. Dhillon and N. Yang, "Reliability and availability analysis of warm standby systems with common-cause failures and human errors", *Microelectronics and Reliability*, **32**, 561 (1992).

95. B. S. Dhillon and N. Yang, "Stochastic analysis of standby systems with common-cause failures and human errors", *Microelectronics and Reliability* **32**, 1699 (1992).

96. B. S. Dhillon and H. C. Viswanath, "On common-cause failures — Bibliography", *Microelectronics and Reliability* **30**, 1179 (1990).

97. B. S. Dhillon and H. C. Viswanath, "Stochastic analysis of common systems with common-cause failures", *Stochastic Analysis and Applications* (to appear).

98. B. S. Dhillon and H. C. Viswanath, "Reliability analysis of a non-identical unit parallel system with common-cause failures", *Microelectronics and Reliability* **31**, 429 (1991).

99. B. S. Dhillon and J. Natesan, "Probabilistic analysis of a pulverizer system with common-cause failures", *Microelectronics and Reliability* **22**, 1211 (1982).

100. B. S. Dhillon and C. L. Proctor, "Common-mode failure analysis of reliability networks", in *Proc. of the Annual Reliability and Maintainability Symposium* (1977), pp. 404–408.

101. B. S. Dhillon and S. N. Rayapati, "Reliability and availability Analysis of a parallel system with common-cause failures and human errors", in *Proc. of the International Conference on Nuclear Power Plant Aging, Availability Factor and Reliability Analysis*, San Diego, California, U.S.A. (1985), pp. 459–465.

102. B. S. Dhillon and S. N. Rayapati, "Human error and common-cause failure modeling of redundant systems", *Maintenance and Mangement International* **7**, 93 (1988).

103. B. S. Dhillon and S. N. Rayapati, "Common-cause failure and human error modeling of redundant systems with partially energized units", *Reliability Engineering* **19**, 1 (1987).

104. B. S. Dhillon and S. N. Rayapati, "Human error and common-cause modeling of redundant systems", *Microelectronics and Reliability* **26**, 1139 (1986).

105. B. S. Dhillon and S. N. Rayapati, "Common-cause failures in repairable systems", in *Proc. of the Annual Reliability and Maintainability Symposium* (1988), pp. 283–289.

106. B. S. Dhillon and S. N. Rayapati, "Probabilistic analysis of redundant systems with human errors and common-cause failures", *Stochastic Analysis and Applications* **4**, 367 (1986).

107. B. S. Dhillon and S. N. Rayapati, "Reliability modeling of redundant computer systems with common-cause failures", *Computers and Electrical Engineering* **14**, 125 (1988).

108. B. S. Dhillon and S. N. Rayapati, "Analysis of systems with human errors and common-cause failures", *International Journal of Modeling and Simulation* **9**, 124 (1989).

109. B. S. Dhillon and S. N. Rayapati, "Common-cause failure and human error analysis of redundant systems", in *Proc. of the Inter-Ram Conference for Electric Power Industry* (1986), pp. 398–405.

110. B. S. Dhillon and S. N. Rayapati, "Reliability and availability analysis of redundant systems with common-cause failures and human errors", *Proc. of the International Conference on Factory of the Future*, 1986, pp. 3–10.

111. B. S. Dhillon *et al.*, "Common-cause failure analysis of a three-state device system", *Microelectronics and Reliability* **19**, 345 (1979).

112. B. S. Dhillon and O. C. Anude, "Common-cause failure analysis of a non-identical unit parallel system with arbitrarily distributed repair times", *Microelectronics and Reliability* (to appear).

113. B. S. Dhillon and O. C. Anude, "Common-cause failure analysis of a parallel system with warm standby", *Microelectronics and reliability* (to appear).

114. B. S. Dhillon and O. C. Anude, "Common-cause failure analysis of a redundancy system with non-repairable units", *Microelectronics and Reliability* (to appear).

115. B. S. Dhillon and O. C. Anude, "Common-cause failure analysis of a redundant system wtih repairable units", *International Journal of Systems Science* (to appear).

116. B. S. Dhillon and O. C. Anude, "Common-cause failure analysis of a 1-*out-of-n* units system with non-repairable units", *Dynamic systems and applications* (to appear).

117. C. Dichirico and C. Singh, "Reliability analysis of transmission lines with common-mode failures when repair times are random", *IEEE Trans. Power Systems* **3**, 1012 (1988).

118. C. Dirchirico and C. Singh, "Availability analysis of two unit repairable parallel redundant system with common-cause failures", *Microelectronics and Reliability* **26**, 1183 (1986).

119. P. Dörre, *Possible Pitfalls in the Process of CCF Event Data Evaluation (PSA 1987)*, Vol. I (Verlag TUV Rheinland. GMBH, Koln, 1987), pp. 74–79.

120. P. Dörre, "Dependent failure — A multiple state stochastic description", *Reliability Engineering and System Safety* **35**, 225 (1992).

121. P. Dörre, "Pitfalls in common-cause failure data evaluation", in *Advanced Seminar of Common-Cause Failure Analysis in Probabilistic Safety Assessment*, ed. A. Amendola (Kluwer, Dordrecht, The Netherlands, 1989), pp. 205–219.

122. P. Dörre and R. Schilling, "Design defences against common-cause/multiple related failures", in *Advanced Seminar on Common-Cause Failure Analysis in Probabilistic Safety Assessment*, ed. A. Amendola (Kluwer, Dordrecht, The Netherlands, 1989), pp. 101–106.

123. L. Duckstein and E. J. Plate, eds., *Engineering Reliability and Risk in Water Resources* (Kluwer, 1987).

124. R. G. Easterling, "Probabilistic analysis of common-mode failures", *Proc. of the Topical Meeting on Probabilistic Analysis of Nuclear Reactor Safety*, Newport Beach, California, Vol. 3, May 1987, pp. X.7/1–X.7/12.

125. G. E. Edison, "Common-cause failure experience for reliability analysis and design guidelines in large nuclear breeder plants", in *Proc. of the Topical Meeting on Probabilistic Analysis of Nuclear Reactor Safety*, New Port, California (1987), pp. X.6/1–X.6/14.

126. G. T. Edwards, "Alleviation of CMF problems in protective systems", in *Proc. of the 2nd National Reliability Conference*, Birmingham, U.K. (1979), pp. 5A/2/1–5A/7/1.

127. G. T. Edwards and I. A. Watson, "A study of common-mode failures", Report No. SRD R 146 (Safety and Reliability Directorate, United Kingdom Atomic Energy Authority, Warrington, U.K. 1976).

128. J. Endrenyi, *Reliability Modeling in Electric Power Systems* (John Wiley, New York, 1978).

129. E. P. Epler, "Diversity and periodic testing in defense against common-mode failure", in *Proc. of the International Conference on Nuclear Systems Reliability Engineering and Risk Assessment*, Gatlinburg, Tennessee (1977), pp. 269–287.

130. E. P. Epler, "Common-mode failure considerations in the design of systems for protection and control", *Nuclear Safety* **10**, 38 (1969).

131. G. Ericsson and S. Hirschberg, "Treatment of common-cause failures in the barseback 1 safety study", in *Proc. of the 5th International Meeting on Thermal Reactor Safety*, Karlsruhe, Germany (Kerntechnische Gessellschaft eV, Bonn, 1984), pp. 1974–1982.

132. M. G. K. Evans *et al.*, "On the treatment of common-cause failures in system analysis", *Reliability Engineering* **9**, 107 (1984).

133. R. A. Evans, "Statistical independence and common-mode failures", *IEEE Trans. Reliability* **24**, 289 (1985).

134. K. N. Fleming, "A reliability model for common-mode failures in redundant safety systems", in *Proc. of the 6th Pittsburgh Conference on Modeling and Simulation*, Part 1, Vol. 6, Pittsburgh, U.S.A. (1975), pp. 579–581.

135. K. N. Fleming, "Parametric models for common-cause failure analysis", in *Advanced Seminar on Common-Cause Failure Analysis in Probabilistic Safety Assessment*, ed. A. Amendola (Kluwer, Dordrecht, The Netherlands, 1989), pp. 159–174.

136. K. N. Fleming and G. W. Hannaman, "Common-cause failures considerations in predicting HTGR cooling system reliability", *IEEE Trans. Reliability* **25**, 171 (1976).

137. K. N. Fleming and A. M. Kalinowski, "An Extension of the Beta Factor Method to Systems with Higher Levels of Redundancy" (Pickard, Lowe and Garrick, PLG-0289, 1983).

138. K. N. Fleming *et al.*, "A systematic procedure for the incorporation of common-cause events into risk and reliability models", *Nuclear Engineering and Design* **93**, 245 (1986).

139. K. N. Fleming *et al.*, "On the analysis of dependent failures in risk assessment and reliability evaluation", *Nuclear Safety* **24**, 637 (1983).

140. K. N. Fleming and A. Mosleh, "Common-cause data analysis and implications in system modeling", in *Proc. of the International Tropical Meeting on Probabilistic Safety Methods and Applications*, EPRI NP-3912-SR, Vol. 1, sessions 1–8, California (1975), pp. 3/1–3/12.

141. K. N. Fleming *et al.*, "Event classification and systems modeling of common-cause failures", in *Trans. American Nuclear Society Annual Meeting on Probabilistic Risk*

Analysis: Methods Development, New Orleans, Louisiana, U.S.A. (1984), pp. 519–521.

142. K. N. Fleming and P. H. Raabe, "A comparison of three methods for the quantitative analysis of common-cause failures", in *Proc. of the Tropical Meeting on Probabilistic Analysis of Nuclear Safety*, Vol. III, Newport Beach, California (1978), pp. X.3/1–X.3/12.

143. J. R. Fragola, "Common-mode/common-cause failure comments for discussion", in *Proc. of the 2nd National Reliability Conference*, Birmingham, U.K. (1978).

144. E. G. Frankel, *Systems Reliability and Risk Analysis* (Kluwer, 1983).

145. L. G. Frederick, "An analysis of functional common-mode failures in GE BWR protection and control instrumentation", Report No. NEDO-10189 (General Electric Company, U.S.A., 1970).

146. N. B. Fuqua, "Reliability engineering for electrical design", in *Electrical Engineering Ser.*, Vol. 34 (Marcel Dekker, 1986), 408pp.

147. J. B. Fussell and G. R. Burdick, ed., "Nuclear systems reliability engineering and risk assessment", in *Proceedings in Applied Mathematics Ser.*, No. 7 (Society for Industrial and Applied Mathematics, 1977), 849pp.

148. B. Gachot, "A probabilistic approach design for the ECCS of PWR", in *Proc. of the Annual Reliability and Maintainability Symposium* (1977), pp. 332–324.

149. A. M. Games, "Some aspects of common-cause failure analysis in engineering systems", Ph.D Thesis, University of Liverpool, U.K. (1986).

150. A. M. Games et al., "Multiple related component failure events", in *Proc. of Reliability 1985*, NCSR, Birmingham, U.K. (1985).

151. A. M. Games et al., "Exploitation of a component event data bank for common-cause failure analysis", in *Proc. of the International Topical Meeting on Probabilistic Safety Methods and Applications*, EPRI NP-3912-SR, Vol. 1, sessions 1–8, California (1975), pp. 1/1–1/11.

152. A. M. Games et al., "Common-cause failure investigation using the European reliability data system", *Reliability Engineering* **13**, 33 (1985).

153. W. C. Gangloff, "Common-mode failure analysis is 'in'", in *Electrical World* (1972), pp. 30–33.

154. W. C. Gangloff, "Common-mode failure analysis", *IEEE Trans. Power Apparatus and Systems* **94**, 27 (1975).

155. W. C. Gangloff and T. H. Franke, "An engineering approach to common-mode failure analysis", in *Proc. of the Development and Applications of Reliability Techniques to Nuclear Power Plants Symposium*, Paper SNI 3/9, Liverpool, England, 1974 (SRD R41).

156. D. L. Gano, "Root cause and how to find it", *Nuclear News* **30**, 39 (1987).

157. R. Garg and L. R. Goel, "Cost analysis of a system with common-cause failure and two types of repair facilities", *Microelectronics and Reliability* **25**, 281 (1985).

158. L. R. Goel et al., "A two (multi-component) unit parallel system with standby and common-cause failure", *Microelectronics and Reliability* **24**, 415 (1984).

159. H. Goldberg, *Extending the Limits of Reliability Theory* (John Wiley, New York, 1981), pp. 110–114.

160. A. K. Govil, "Availability of a complex system having shelf-life of the components and common-cause failures", *Microelectronics and Reliability* **22**, 685 (1982).

161. A. K. Govil, "Reliability of a standby system with common-cause failures and scheduled maintenance", *Microelectronics and Reliability* **21**, 269 (1981).

162. I. W. Grant and F. C. Clyde, eds., *Handbook of Reliability Engineering and Management* (McGraw-Hill, New York, 1988), pp. 13–23.

163. A. E. Green, *Safety Systems Reliability* (John Wiley, New York, 1983).

164. C. N. Guey, "A method for estimating common-cause failure probability and model parameters: The inverse stress-strength interference (ISSI) Technique", Energy Laboratory Report No. MIT-El 84-010 (1984).

165. C. N. Guey and C. D. Heising, "A method for estimating common-cause failure probability and model parameters: The inverse stress-strength interference (ISSI) technique", in *Proc. of the International Topical Meeting on Probabilistic Safety Standards and Applications*, EPRI NP-3912-SR, Palo Alto, California, Vol. 3, sessions 17–23 (1985), pp. 182/1–182/13.

166. H. Gyllenbaga *et al.*, "Treatment of dependencies in ringhals 1 and 2 safety studies", in *Workshop on Dependent Failure Analysis*, Vasteras, Sweden (1983).

167. E. W. Hagen, "Common-mode/common-cause failure: A review", *Nuclear Engineering and Design* **59**, 423 (1980).

168. E. W. Hagen, "Technical note: The Kahl relay common-mode failure", *Nuclear Safety* **20**, 579 (1979).

169. S. G. Han *et al.*, "The trinomial failure rate model for treating common-mode failures", *Reliability Engineering and System Safety* **25**, 131 (1989).

170. B. Harris, "Stochastic models for common-cause failures", in *Proc. of the International Conference on Reliability and Quality Control*, 185 (1986).

171. J. Hartung, "Common-cause failure theory", Report No. N00IT1000 145 (Atomics International Division, Rockwell International 1981).

172. J. Hartung, "A statistical correlation model and proposed general statment of theory for common-cause failures", in *Proc. of the International Meeting on Probabilistic Risk Assessment* (Port Chester, New York, 1981), pp. 239–246.

173. K. C. Hayden, "Common-mode failure mechanisms in redundant systems important to reactor safety", *Nuclear Safety* **17**, 686 (1976).

174. K. C. Hayden, "Common-mode failure mechanisms in nuclear plants protection systems", ERDA Report No. ORNL/TM-4984 (Oak Rridge National Laboratory, Oak Ridge, Tennessee, U.S.A., 1975).

175. M. Heikkila, "A model for common-mode failures", in *Proc. of the 2nd National Reliability Conference*, Birmingham, U.K. (1979), pp. 6C/6/1–6C/6/5.

176. C. D. Heising, "Development of unavailability expressions for one and two component systems with periodic testing and common-cause failures", *Reliabilty Engineering* **6**, 229 (1983).

177. C. D. Heising and C. N. Guey, "A comparison of methods for calculating system unavailabilty due to common-cause failures: The beta factor and multiple dependent failure fraction methods", *Reliability Engineering* **8**, 101 (1984).

178. C. D. Heising and D. M. Luciani, "Application of a computerized methodology for performing common-cause failure analysis (MOBB) code", *Reliability Engineering* **17**, 193 (1987).

179. C. D. Heising *et al.*, "Common-cause failure analysis", Energy Laboratory Report No. MIT-E182-038 (MIT, Cambridge, Massachusetts, 1982).

180. E. J. Henley and K. Hiromitsu, *Probabilistic Risk Assessment: Reliability Engineering, Design and Analysis* (Institute of Electrical and Electronics Engineers, 1992), 592pp.

181. E. J. Henley and K. Hiromitsu, *Reliability Engineering and Risk Mangement* (Prentice-Hall, New Jersey, 1985), pp. 120–127, 385–390.

182. E. J. Henley and K. Hiromitsu, *Designing for Reliability and Safety Control* (Prentice-Hall, New Jersey, 1985), pp. 18, 401–402.

183. N. F. Heylmun *et al.*, "Common-cause failure analyses for the primary heat transport system and reactor vessel", Report No. WARD-B69101-2 (Westinghouse Electric Corporation, Madison, Pennsylvania, U.S.A., 1976).

184. T. Hidaka, "Approximated Reliability of r-out-of-n (F) system with common-cause failure and maintenance", *Microelectronics and Reliability* **32**, 817 (1992).

185. R. Himanen *et al.*, "Defences against common-cause failures: Introduction to quantitative approach", in *Proc. of the 10th Annual Symposium of the Society of Reliability Engineers*, Scandinavian Chapter, Stavanger, Norway (Elsevier, London and New York, 1989).

186. S. Hirschberg, "Comparison of methods for quantitative analysis of common-cause failures — A case study", in *Proc. of the International Topical Meeting on Probabilistic Safety Methods and Applications*, EPRI NP-3912-SR, Vol. 3, sessions 17–23, California (1985), pp. 183/1–183/10.

187. S. Hirschberg and U. Pulkkinen, "Common-cause failure data: Experience from diesel generator studies", *Nuclear Safety* **26**, 305 (1985).

188. S. Hirschberg, "Workshop on dependent failure analysis, vasteras, Sweden (1983). Summary Report: Conclusions and Recommendations for Future Work", AB ASEA-ATOM Report KPA 83-212 (1983).

189. S. Hirschberg, "Experiences from dependent failure analysis in nordic countries", *Reliability Engineering and System Safety* **34**, 1991, pp. 355–388.

190. S. Hirschberg, ed., "NKA project 'risk analysis' (RAS-470): Summary report on common-cause failure data benchmark exercise", Final Report RAS-470 (86) 14 (ABB Atom Report RPA 86-241) (1987).

191. S. Hirschberg, "Retrospective analysis of dependencies in the Swedish probabilistic safety studies. Phase I: Qualitative overview", Report RAS-470 (87) 4 (ABB Atom Report RPC 87-36), July 1987.

192. S. Hirschberg *et al.*, "Retrospective quantitative analysis of common-cause failures and human interactions in Swedish PSA studies", in *Proc. of PSA '89 — International Topical Meeting on Probability, Reliability and Safety Assessment*, Pittsburgh, Pennsylvania (American Nuclear Society, La Grange Park, Illinois, 1989), pp. 258–269.

193. S. Hirschberg, "Project plan: Defences against common-cause failures (CCFs) and generation of CCF data. Pilot study for diesel generators (DGs)", ABB Atom Report RPC 89-60 (1989).

194. S. Hirschberg, "Comparison of methods for quantitative analysis of common-cause failures — A case study", in *Proc. of ANS/ENS International Topical Meeting on Probabilistic Safety Methods and Applications*, San Francisco, California (Electric Power Research Institute, 1985), pp.183/1–183/10.

195. S. Hirschberg, "Treatment of common-cause failures: The nordic perspective", in *Proc. of the ISPRA Advanced Seminar on Common-Cause Failure Analysis in Probabilistic Safety Assessment*, Ispra, Italy (Kluwer, Dordrecht/Boston/London, 1989), pp. 9–29.

196. S. Hirschberg and L. I. Tiren, "Design related defensive measures against dependent failures: ABB atom's approach", in *Proc. of the ISPRA Advanced Seminar on Common-Cause Failure Analysis in Probabilistic Safety Assessment*, Ispra, Italy (Kluwer, Dordrecht/Boston/London, 1989), pp. 71–100.

197. S. Hirschberg and L. Gunsell, "Defensive measures against external events and status of external events and status of external event analysis in Swedish probabilistic safety assessments", International Post — SMIRT 10 Seminar on Probabilistic Risk Assessment (PRA) of Nuclear Power Plants for External Events, Irvine, California (1989).

198. H. Hirshmann *et al.*, "Steady-state unavailability of a K-out-of-N system with total repair", *Reliability Engineering* **4**, 181 (1983).

199. P. Hoksad, "A shock model for common-cause failures", *Reliability Engineering and System Safety* **23**, 127 (1988).

200. J. M. Hudson and J. D. Collins, "The prediction of accident sequence probabilities in a nuclear power plant due to earthquake events", in *Proc. of the American Nuclear Society/European Nuclear Society Topical Meeting, Vol. II: Thermal Reactor Safety*, Knoxville, Tennessee (1980), pp. 762–772.

201. J. M. Hudson and J. Gasca, "Common-mode failure in nuclear power plants", in *Proc. of the Annual Reliability and Availability Symposium* (1981), pp. 149–155.

202. R. P. Hughes, "A new approach to common-cause failure", *Reliability Engineering* **17**, 211 (1987).

203. R. P. Hughes, "The relationship between common-cause failure and data uncertainties", in *International Atomic Energy Authority Seminar on Implications of Probabilistic Risk Analysis*, eds. M. V. Cullingford, S. M. Shah and J. H. Gittus, IAEA-SR-111/10 (1985), pp. 239–252.

204. R. P. Hughes, "A framework for dependent failure analysis", *Reliability Engineering and System Safety* **24**, 139 (1989).

205. R. A. Humphreys, "Assigning numerical value to the beta factor common-cause evaluation", in *Proc. of the 2nd National Reliability Conference*, Vol. 1, Birmingham, U.K. (1987), pp. 2C/5/1–2C/5/8.

206. P. Humphreys and A. M. Jenkins, "Dependent failures developments", in *Reliability Engineering and System Safety* **34**, 417 (1991).

207. P. Humphreys and B. D. Johnston, "SRD dependent failures procedures guide", Report SRD R418, UKAEA (1987).

208. P. Humphreys, "Design defences against multiple relayed failures", in *Advanced Seminar on Common-Cause Failure Analysis in Probabilistic Safety Assessment*, ed. A. Amendola (Kluwer, Dordrecht, The Netherlands, 1989), pp. 47–70.

209. P. Humphreys *et al.*, "Progress towards a better understanding of dependent failures by data collection, classification and improved modeling techniques", in *Proc. Reliability '87*, Vol. 1, Paper 2C/4, ed. A. M. Jenkins (Elsevier, London, 1987), pp. 2C/4/1–2C/4/14.

210. I. M. Jacobs, "The common-mode failure study discipline", *IEEE Trans. Nuclear Science* **17**, 594 (1970).

211. I. M. Jacobs, "The nature, analysis and impact of common-cause failures on the design and licensing of future LBRs", Report No. REM 80-22 (General Electric Co., 1980).

212. P. Jacobsson, "Comparison of two CCF models, alpha factor model and MGL model", Report RAS-470 (88) 29 (ABB Atom Report RPC 88-108), August 1988 (Revised October 1988).

213. P. Jacobsson, "Sensitivity study on diesel generator and pump CCF data in the Swedish PSAs", Report RAS-470 (88) 32 (ABB Atom Report RPC 88-160) (1988).

214. D. Jerwood and F. A. Georgiakodia, "Application of multivariate techniques to monitor system reliability and detect common-mode failure", in *Proc. of the 2nd National Reliability Conference*, Vol. 1, Birmingham, U.K. (1979), pp. 2B/5/1–2B/5/7.

215. D. John, ed., "The reliability of mechanical systems" (Mechanical Engineering Publications, London, 1988), pp. 81–85.

216. J. W. Johnson and W. E. Vesely, "Common-mode analysis of valve leakages", in *Procs. of the Topical Meeting on probabilistic analysis of nuclear safety*, Vol. 3, Newport Beach, California (1978), pp. X.5/1–X.5/8.

217. B. D. Johnston, "A structural procedure for dependent failure analysis (DFA)", *Reliability Engineering* **19**, 125 (1987).

218. M. E. Jolly and J. Wreathall, "Common-mode failures in reactor safety systems", *Nuclear Safety* **18**, 624 (1977).

219. G. E. Jorgensen *et al.*, "Common-mode forced outages of overhead transmission lines", *IEEE Trans. Power Apparatus and Systems* **95**, 859 (1976).

220. D. Kececioglu, *Reliability and Life Testing Handbook*, Vol. 1 (Prentice Hall, 1992), 600pp.

221. A. A. Lakner, *Reliability Engineering for Nuclear and Other High Technology Systems: A Practical Guide* (Elsevier, New York, 1985), pp. 129–130.

222. M. Lai and J. Yuan, "Periodic replacement model for a parallel system subject to independent and common-cause shock failures", *Reliability Engineering and System Safety* **31**, 355 (1991).

223. M. Lai and J. Yuan, "Cost-optimal replacement policy (N^*, n^*) for a parallel system with common-cause failure", *Reliability Engineering and System Safety* **32**, 339 (1991).

224. R. D. Leitch, *BASIC Reliability Engineering Analysis*, paperback text edition (Butterworth-Heinemann, 1988), 168pp.

225. E. E. Lewis, *Introduction to Reliability Engineering* (John Wiley, New York, 1987).

226. H. W. Lewis, (Chairman), Risk Assessment Review Group Report to the U.S. Nuclear Regulatory Commission, NUREG/CR-0400 (1978).

227. J. J. Lisboa, "Quantification of human error and common-mode failures in man-machine systems", *IEEE Trans. Nuclear Science* **35**, 907 (1988).

228. Los Alamos Technical Associates, Data Benchmark Test of a Classification Procedure for Common-Cause Failures, LATA-ER-02-02 (Rev. 1) (1983).

229. S. Osaki and J. Cao, eds., "Reliability theory and applications", in *Proc. of China-Japan Symposium* (World Scientific, 1987, 448).

230. M. Mahmoud *et al.*, "Availability analysis of a repairable system with common-cause failure and standby unit", *Microelectronics and Reliability* **27**, 741 (1987).

231. Y. Malmn and V. Rouhiainen, eds., "Reliability and safety of processes and manufacturing systems", in *Proc. of the 12th Annual Symposium of the Society of Reliability Engineers, Scandinavian Chapter*, Tampere, Finland (Elsevier, 1991), 380pp.

232. T. Mankamo, "Common-load model: A tool for common-cause failure analysis", Finnish Report No. 31 (Technical Research Center, Electrical Engineering Laboratory, Espoo, Finland, 1977).

233. T. Mankamo, "Dependent failure modeling", 6th Advances in Reliability Technology Symposium, University of Bradford, Vol. 2, U.K., NCSR R23 (1980).

234. T. Mankamo, "Common-mode failures", Finnish Report No. 18 (Technical Research Center, Electrical Engineering Laboratory, Espoo, Finland, 1976).

235. T. Mankamo, "Project plan: CCF analysis of high redundant systems; safety relief valve data analysis and reference application", Avaplan Oy Report (1989).

236. T. Mankamo and U. Pulkkinen, "Dependent failures of diesel generators", *Nuclear Safety* **23**, 32 (1982).

237. T. Mankamo, "SHACAM, shared cause model of dependencies — A review of the multiple greek method and a modified extension of the beta factor method", Avaplan Oy, Espoo, Finland (1985).

238. T. Mankamo, "Extended common load model, a tool for dependent failure modeling in highly redundant structures", Avaplan Oy, Espoo, Finland (1990).

239. T. Mankamo *et al.*, "CCF analyses of High Redundancy systems, SRV data analysis and reference BWR application", Technical Report SKI TR-91:6 (Swedish Nuclear Power Inspectorate, Stockholm, 1991).

240. T. Mankamo and U. Pulkkinen, eds., *Proc. of the CCF Workshop*, Lepolampi, Espoo, Finland (1984).

241. T. Mankamo and M. Kosonen, "Dependent failure modeling in highly redundant structures — Application to BWR safety valves", *Reliability Engineering and System Safety* **35**, 235 (1992).

242. A. W. Marshall and I. Olkin, "A multivariate exponential distribution", *Journal of the American Statistical Association* **62**, 30 (1967).

243. B. R. Martin and R. I. Wright, "A practical method of common-cause failure modeling", *Reliability Engineering* **19**, 185 (1987).

244. T. Matsuoka, "Component failure model dependent on time and causes", *Nuclear Engineering and Design* **75**, 109 (1982).

245. T. R. Meachum and C. L. Atwood, "Common-cause fault rates for instrumentation and control assemblies", Report No. NUREG/CR-2700, EGG-EA-5485 (U.S. Nuclear Regulatory Commission, Washington D.C., 1983).

246. T. Meslin, "Measures taken at design level to counter common-cause failures, a few comments conerning the approach to EDF", in *Advanced Seminar on Common-Cause Failure Analysis in Probabilistic Safety Assessment*, ed. A. Amendola (Kluwer, Dordrecht, The Netherlands, 1989), pp. 107–111.

247. T. Meslin, "Analysis of common-cause failures based on operating experience: Possible approaches and results", in *Advanced Seminar on Common-Cause Failure Analysis in Probabilistic Safety Assessment*, ed. A. Amendola (Kluwer, Dordrecht, The Netherlands, 1989), pp. 257–276.

248. J. H. Moody and S. M. Follen "Common-cause modeling of reactor trip breaker configuration", in *Proc. of the International Topical Meeting on Probabilistic Safety Methods and Applications*, EPRI NP-3912-SR, Vol. 3, sessions 17–23, California (1985), pp. 180/1–180/10.

249. B. M. E. Moret and M. G. Thomson, "Boolean difference techniques for time-sequence and common-cause analysis of fault trees", *IEEE Trans. Reliability* **33**, 399 (1984).

250. A. Mosleh, "Hidden sources of uncertainty: Judgement in collection and analysis of failure data", *Nuclear Engineering Design* **93**, 187 (1986).

251. A. Mosleh, "Dependent failure analysis", *Reliability Engineering and System Safety* **34**, 243 (1991).

252. A. Mosleh, "Common-cause failures: An analysis methodology and examples", *Reliability Engineering and System Safety* **34**, 249 (1991).

253. A. Mosleh et al., "Procedures for treating common-cause failures in safety and reliability studies, Vol. I: Procedural framework and examples", NUREG/CR-4780 (EPRI NP-5613) (Electric Power Research Institute, 1988).

254. A. Mosleh, "Estimation of parameters of common-cause failure models", in *Advanced Seminar of Common-Cause Failure Analysis in Probabilistic Safety Assessment*, ed. A. Amendola (Kluwer, Dordrecht, The Netherlands, 1989), pp. 175–203.

255. A. Mosleh et al., "Methodological advancements in procedures for common-cause failure analysis", in *Proc. of the ANS/ENS Topical Meeting on Probabilistic Safety Assessment*, Pittsburgh, U.S.A. (1989).

256. A. Mosleh and N. O. Siu, "A multiparameter, event based common-cause failure model", Paper M7/3, in *Proc. of the 9th International Conference on Structural Mechanics in Reactor Technology*, Lausanne, Switzerland (1987).

257. A. Mosleh and N. O. Siu, "On the use of uncertain data in common-cause failure analysis", in *Proc. PSA '87 — International Topical Conference on Probabilistic Safety Assessment Risk Management* (1987).

258. A. Mosleh *et al.*, "Procedures for treating common-cause failures in safety and reliability studies, Vol. 2: Analytical background and techniques", NUREG/CR-4780, EPRI NP-5613, PLG-0547 (1989).

259. J. Natesan and A. K. S. Jardine, "Stochastic behaviour of a single server n-unit pulverizer system with common-cause failures", *Microelectronics and Reliability* **24**, 1045 (1984).

260. R. J. Page *et al.*, "A common-cause analysis of the TREAT upgrade reactor protection system", in *Proc. of the International Topical Meeting on Probabilistic Safety Methods and Applications*, EPRI NP-3912-SR, Vol. 1, sessions 1–8, California (1985), pp. 64/1–64/10.

261. A. Pages and M. Gondran, Translated by E. Griffin, *System Reliability: Evaluation and Prediction in Engineering* (Springer-Verlag, New York, 1986), pp. 282–289.

262. S. A. Papzoglou *et al.*, "National reliability evaluation program (NREP) procedure guides", NUREG/CR-2815 (Brookhaven National Laboratory, 1983).

263. G. W. Parry, "Common-cause failure analysis: A critique and some suggestions", *Reliability Engineering and System Safety* **34**, 309 (1991).

264. G. W. Parry, "Incompleteness in data bases: Impact on parameter estimation uncertainty", presented at the Annual Meeting of the Society for Risk Analysis, Knoxville, Tennessee, U.S.A. (1984).

265. G. W. Parry *et al.*, "A cause-coupling defense approach to common-cause failures", presented at PSA '89, Pittsburgh, Pa., U.S.A. (1989).

266. G. W. Parry *et al.*, "Data needs for common-cause failure analysis", in *Proc. of the International Conference on Probabilistic Safety Assessment and Management*, Vol. 2, ed. G. Apostolakis (Elsevier, New York, 1991), pp. 847–858.

267. F. Paul, "Reliability assessment of technical systems by reference to examples from the field of process automation", *Siemens Power Engineering* **6**, 86 (1984).

268. H. M. Paula, *A Restructured Approach to the Partial Beta Factor Method* (JBF Associates, Knoxville, Tennessee, U.S.A., 1986).

269. H. M. Paula, "Comments on the analysis of dependent failures in risk assessment and reliability evaluation", *Nuclear Safety* **27**, 210 (1986).

270. H. M. Paula, "A probabilistic dependent failure analysis of a DC electric power system in a nuclear power plant", Report 37932-3341 (JBF Associates, Knoxville, Tennessee, 1987).

271. H. M. Paula and D. J. Campbell, "Task force draft report: A cause–defense methodology for common-cause failure analysis", Report JBFA-116-88 (JBF Associates, Knoxville, Tennessee, U.S.A., 1988).

272. H. M. Paula and D. J. Campbell, "Analysis of dependent failures events and failure events caused by harsh environmental conditions", Report JBFA-LR-111-85 (JBF Associates, 1985).

273. H. M. Paula *et al.*, "Qualitative cause defense matrices: Engineering tools to support the analysis and prevention of common-cause failures", *Reliability Engineering and System Safety* **34**, 389 (1991).

274. H. M. Paula *et al.*, "A cause–defense approach to the understanding and analysis of common-cause failures", NUREG/CR-5460 (SAND 89-2368) (Sandia National Laboratories, Albuquerque, New Mexico, U.S.A., 1990).

275. K. E. Peterson, "Analysis of common-cause failure data — Identification, the experience from the nordic benchmark", in *Advanced Seminar on Common-Cause Analysis in Probabilistic Safety Assessment*, ed. A. Amendola (Kluwer, Dordrecht, The Netherlands, 1989), pp. 235–241.

276. J. K. Plastiras, "Intersystem common-cause analysis of a diesel generator failure", *Risk Analysis* **6**, 463 (1986).

277. O. Platz, "A Markov model for common-cause failures", *Reliability Engineering* **9**, 25 (1984).

278. K. Porn, "Some comments on CCF quantification: The experience from the nordic benchmark", in *Proc. of the ISPRA Advanced Seminar on Common-Cause Failure Analysis in Probabilistic Safety Assessment*, Ispra, Italy (Kluwer, Dordrecht/Boston/London, 1989), pp. 243–256.

279. A. Poucet *et al.*, "CCF-BRE common-cause failure reliability benchmark exercise, EUR 11054 EN, Joint Research Centre, Ispra Establishment, I-21020, Ispra (VA) (Commission of the European Communities, Luxembourg, 1987).

280. A. Poucet, "Experience and results of the common-cause failure reliability benchmark exercise", in *Advanced Seminar on Common-Cause Failure Analysis in Probabilistic Safety Assessment*, ed. A. Amendola (Kluwer, 1989), pp. 221–234.

281. A. Poucet *et al.*, "Common-cause failure probability benchmark exercise", Final Report, CEC-JRC, Ispra, Italy, EUR 11054 EN.

282. A. Poucet *et al.*, "Summary of the common-cause failure reliability benchmark exercise", Joint Research Center Report, PER 1133/86, Ispra, Italy (1986).

283. B. Putney, "WAMCOM, common-cause methodologies using large fault trees", Report EPRI NP-1851 (Electric Power Research Institute, Palo Alto, California, 1981).

284. T. Pyzdek, "Quality engineering handbook", in *Quality and Reliability Ser.: no. 29* (Marcel Dekker, 1991), 640pp.

285. A. Rangan and R. E. Grace, "A non-Markovian model for the optimun replacement of self-repairing system subject to shocks", *Journal of Applied Reliability* **25**, 375 (1978).

286. J. P. Rankin, "Identification of common-cause failures in instrumentation and control systems", *IEEE Trans. Nuclear Science* **29**, 979 (1982).

287. J. P. Rankin, "Common-cause hazard analysis for random glitches", in *Proc. of the Annual Reliability and Maintainability Symposium* (1982), pp. 1–4.

288. J. P. Rankin "Common-cause failure analysis of instrumentation and control systems", in *Proc. of the 8th Annual Reliability Engineering Conference for the Electrical Power Industry*, Portland Oregon (1981), pp. 195-204.

289. J. P. Rankin, "Common-cause failure analysis — Why interlocked redundant systems fail", Society of Automotive Engineers, SAE Technical Paper Series on Turbine Powered Executive Aircraft Meeting, Arizona, U.S.A., Paper No. 8000631 (1980), pp. 1-6.

290. D. M. Rasmuson, "Some practical considerations in treating dependencies in PRAs", *Reliability Engineering and System Safety* **34**, 327 (1991).

291. D. M. Rasmuson and D. H. Worledge, "Reflections on the NRC/EPRI common-cause failure analysis research", in *Proc. of PSA '89 International Topical Meeting on Probability, Reliability and Safety Assessment*, Pittsburgh, Pa. (1989), pp. 270-273.

292. D. M. Rasmuson *et al.*, "A common-cause failure analysis of McGuire unit 2 auxilliary feedwater system", Report 82-NE-17 (The American Society of Mechanical Engineers, 1982).

293. D. M. Rasmuson *et al.*, "Common-cause failure analysis techniques: A review and comparative evaluation", Report No. TREE-1349 (Idaho National Electric Laboratory, Idaho Falls, Idaho).

294. D. M. Rasmuson *et al.*, "COMCAN IIA — A computer program for automated common-cause failure analysis", Report No. TREE-1361, EG and G Inc. (Idaho National Electric Laboratory, Idaho Falls, Idaho).

295. D. M. Rasmuson *et al.*, "Use of COMCAN III in system design and reliability analysis", Report EGG-2187, EG and G Idaho Inc. (Idaho Falls, Idaho, 1982).

296. A. Sambhi *et al.*, "Common-cause failure analysis of multi-failure mode systems", in *Proc. of the 10th Annual Pittsburgh Conference on Modeling and Simulation*, part 2, Vol. 10 (1979), pp. 511–515.

297. R. Schilling and P. Dörre, "Consideration of common-cause aspects in the design of Siemens/KWU PWRs", *Reliability Engineering and System Safety*, Vol. 34 (1991), pp. 345–354.

298. G. C. Sharma *et al.*, "Stochastic analysis of a parallel system with common-cause failure, preventive maintenance", *Microelectronics and Reliability* 25, 1035 (1985).

299. C. Singh, "Reliability modeling of TMR computer systems with repair and common-mode failures", *Microelectronics and Reliability* 21, 259 (1981).

300. C. Singh and M. Reza Ebrahimian, "Non-Markovian models for common-mode failures in transmission systems", *IEEE Trans. Power Apparatus and Systems* 101, 1545 (1982).

301. C. Singh *et al.*, "Modeling common-mode failures in transmission systems", in *Proc. of the 11th Annual Pittsburgh Conference on Modeling and Simulation*, Vol. 11 (1980), pp. 863–867.

302. C. Singh, *et al.*, "Modeling common-mode failures in transmission lines when repair times are gamma distributed", in *13th Inter-RAM Conference for the Electric Power Industry*, Syracuse, NY (1986), pp. 339–346.

303. J. Singh, "A warm standby redundant system with common-cause failures", *Reliability Engineering and System Safety* 26, 135 (1989).

304. H. R. Singh *et al.*, "Common-mode failure consideration in a cold standby duplex system", *Microelectronics and Reliability* 29, 723 (1989).

305. A. M. Smith, "Common-cause failure — A mountain or molehill", in *Proc. of the 2nd National Reliability Conference*, Birmingham, U.K. (1979), pp. 5A/5/1–5A/5/3.

306. A. M. Smith and I. A. Watson, "Common-cause failures — A dilemma in perspective", in *Proc. of the Annual Reliability and Maintainability Symposium* (1980), pp. 332–339.

307. A. M. Smith *et al.*, "Defensive strategies for reducing susceptibility to common-cause failures, Vol. 1: Defensive strategies", EPRI NP-5777 (1988).

308. M. G. Stamatelatos, "Improved method for evaluating common-cause failure probabilities", *Trans. the American Nuclear Society on Probabilistic Risk Assessment* 43, 474 (1982).

309. K. Stecher, "Fault tree analysis, taking into account causes of common-mode failures", Siemens Forschungs-und Entwicklungs-berichte, R and D Reports, Vol. 13 (1984), pp. 184–191.

310. J. A. Steverson and C. L. Atwood, "Common-cause failure rate estimates for diesel generators in nuclear power plants", ANS/ENS Topical Meeting on Probabilistic Risk Assessment, Port chester, U.S.A. (1981).

311. C. Sundararajan, *Guide to Reliability Engineering: Data, Analysis, Applications, Implementation and Mangement* (Van Nostrand Reinhold, 1991), 496pp.

312. T. Taniguchi *et al.*, "Common-cause evaluations in applied risk analysis of nuclear power plants", Report No. ORNL/TM-8297 (Oak Ridge National Laboratory, Oak Ridge, Tennessee, 1983).

313. J. R. Taylor, "A study of failure causes based on U.S. power reactor abnormal occurrence reports", in *Proc. of the Symposium on Reliability of Nuclear Power Plants*, Innsbruck, Report No. IAEA-SM-195/16 (1975), pp. 119–130.

314. J. R. Taylor, "Common-mode and coupled failure", Report No. RISO-M-1826 (Danish Atomic Energy Commission, Research Establishment RISO, Electronics Department, 1975), pp. 1–60.

315. J. R. Taylor, "Comments on the subject of common-mode failures", in *Proc. of the 2nd National Reliability Conference*, Birmingham, U.K. (1979), pp. 5A/7/1.

316. J. R. Taylor, "Design Errors in Nuclear Power Plant", Ris-M-1742 (Danish Atomic Energy Commission, Roskilde, Denmark, 1974).

317. A. J. Unione and R. L. Ritzman, "Examination of common-mode failure analysis methodology", Report No. SAI/SR-141-PA (Science Applications, U.S.A., 1976).

318. U.S. Nuclear Regulatory Commission Reactor Safety Study, "An assessment of accident risks in U.S. commercial nuclear power plants", WASH 1400, NUREG-75/014, Appendices III and IV: Failure Data (1985).

319. J. Vaurio, "Availability of redundant safety syatems with common-mode and undetected failures", *Nuclear Engineering and Design* **58**, 415 (1980).

320. J. K. Vaurio, "Structures for common-cause failure analysis", International American Nuclear Society/European Nuclear Society Topical Meeting on Probabilistic Risk Assessment, Port Chester, U.S.A. (1981).

321. K. J. Vavrek and G. R. Andre, "Sensitivity study of common-mode failure rates for sizewell B", in *Proc. of the International Topical Meeting on Probabilistic Safety Methods and Applications*, EPRI NP-3912-SR, Vol. 2, sessions 9–16, California (1985), pp. 102/1–102/18.

322. S. M. Verma and A. A. Chari, "Availability and frequency of failures of a system in the presence of chance common-cause shock failures", *Microelectronics and Reliability* **31**, 265 (1991).

323. W. E. Vesely, "Estimating common-cause failure probabilities in reliability and risk analysis: Marshall Olkin specializations", in *Proc. of the International Conference on Nuclear Systems, Reliability Engineering and Risk Assessment*, Gatlinburg, Tennessee (1977), pp. 314–341.

324. B. Victor and F. Joyce, *Understanding System Failures* (Manchester University Press, Manchester, 1984), pp. 187–188.

325. R. Virolainen, "On common-cause failures, statistical dependence and calculation of uncertainty — Disagreement in interpretation of data", *Nuclear Engineering and Design* **77**, 103 (1984).

326. R. Virolainen, "On common-cause failure methods dealing with dependent failures: A comparative application", in *Proc. of the International Topical Meeting on Probabilistic Safety Methods and Applications*, EPRI NP-3912-SR, Vol. 1, sessions 1–8, California (1985), pp. 4/1–4/11.

327. G. Volta, "The common-mode failure analysis", Report No. CEC ISPRA, SR 76, No. 8, Italy (1976).

328. D. P. Wagner, "A procedure for qualitative common-cause failure analysis of complex systems", M.Sc Thesis, University of Tennessee, Knoxville, Tennessee (1977).

329. D. P. Wagner *et al.*, "Common-cause failure analysis methodology for complex systems", in *Proc. of the International Conference on Nuclear Systems Reliability Engineering and Risk Assessment*, Gatlinburg, Tennessee (1977), pp. 289–341.

330. I. A. Watson, "Analysis of dependent events and multiple unavailabilities with particular reference to common-cause", in *Nuclear Engineering and Design* **93**, 227 (1986).

331. I. A. Watson, "Common mode/cause failures — An overall review", in *Mechanical Reliability* (Science and Technology Press, U.K., 1980), pp. 136–162.

332. I. A. Watson, "The rare event dilemma and common-mode failures", in *Proc. of Annual Reliability and Maintainability Symposium* (1982), pp. 5–10.

333. I. A. Watson, "Introductory paper: Common-mode failures", *Proc. of the 2nd National Reliability Conference*, Birmingham, U.K. (1979), pp. 5A/1/1–5A/1/7.

334. I. A. Watson, "A study of common mode failures", Safety and Reliability Directorate Report, SRD R 146 (UKAEA, Culcheth, England, 1979).

335. I. A. Watson, "Review of common-cause failures", Report No. NCSR R 27 (United Kingdom Atomic Energy Authority, National Center of Systems Reliability, Warrington, U.K., 1981), pp. 1–31.

336. I. A. Watson and G. T. Edwards, "Common-mode failures in redundancy systems", *Nuclear Technology* **46**, 183 (1979).

337. I. A. Watson *et al.*, "CSNI task force on rare events research sub-group on common mode failures — Interim report" (UKAEA, SRD, U.K., 1977).

338. D. Whitehead *et al.*, "Recommended Improvements on data collection and reporting of single and multiple failure events", Report NUREG/CR-5471 (1980).

339. D. W. Williams, "Common-mode failures in U.S. commercial power reactors", M.Sc Thesis, University of Tennessee, Knoxville, Tennessee (1972).

340. J. R. Wilson and R. J. Crump, "Computer-aided common-cause analysis of an LMFBR system", *Trans. the American Nuclear Society on Safety Systems Reliability Methods and Applications* **22**, 474 (1975).

341. D. Worledge and I. B. Wall, "What has been learned about common-cause failures in the last 5 years" (International Topical Conference on Probabilistic Safety Assessment and Risk Management), Zurich, Switzerland (1987).

342. D. H. Worledge *et al.*, "Common-cause failure and systems interactions issues — An overview", in *Proc. of the International Topical Meeting on Probabilistic Safety Methods and Applications*, EPRI NP-3912-SR, Vol. 2 sessions 9–16, California (1985), pp. 97/1–97/10.

343. R. B. Worrell and G. R. Burdick, "Qualitative analysis in reliability and safety studies", *IEEE Trans. Reliability* **25**, 164 (1976).

344. R. B. Worrell and D. W. Stack, "Common-cause analysis using SETS", Report No. SAND 77-1832, Sandia Laboratory, Albuquerque, New Mexico, U.S.A. (1977).

345. R. B. Worrell and D. W. Stack, "A Boolean approach to common-cause analysis", in *Proc. of the Annual Reliability and Maintainability Symposium* (1980), pp. 363–366.

346. J. Wreathall, "Limits to reliability technology results", in *Proc. of the 2nd National Reliability Conference*, Birmingham, U.K. (1979), pp. 5A/4/1–5A/4/4.

347. R. I. Wright, "Some data on common-cause failures in redundancy industrial computer systems", *The Nuclear Engineer* **26**, 72 (1985).

348. J. Yuan, "Pivotal decomposition to find availability and failure frequency of systems with common-cause failures", *IEEE Trans. Reliability* **36**, 48 (1987).

349. J. Yuan, "A conditional probability approach to reliability with common-cause failures", *IEEE Trans. Reliability* **34**, 38 (1985).

350. J. Yuan *et al.*, "On 'inclusion and exclusion' and 'sum of mixed products' formulas to calculate system reliability with common-cause failures", *Reliability Engineering and System Safety* **27**, 219 (1990).

351. J. Yuan *et al.*, "Evaluation of system reliability with common-cause failures by pseudo-environments models", *IEEE Trans. Reliability* **38**, 328 (1989).

352. W. Y. Yun and D. S. Bai, "Optimal Numbers of redundant units for parallel systems with common-mode failures", *Reliability Engineering* **16**, 201 (1986).

353. S. Zacks, *Introduction to Reliability Analysis: Probability Models and Statistical Methods, Series in Statistics* (Springer-Verlag, New York, 1991), 212pp.

https://doi.org/10.1142/9789819812547_0017

Chapter 17

FAILURE TIME DISTRIBUTION UNDER A δ-SHOCK MODEL AND ITS APPLICATION TO ECONOMIC DESIGN OF SYSTEMS[#]

ZEHUI LI,[*,‡] LING-YAU CHAN,[†,§] and ZHIXIN YUAN[*]

Department of Mathematics, Lanzhou University
People's Republic of China
‡*E-mail: lizehui@lzu.edu.cn*

†*Department of Industrial and Manufacturing Systems Engineering*
The University of Hong Kong
E-mail: plychan@hku.hk

Suppose that shocks arrive and act on a system according to a Poisson distribution with mean rate of arrival equal to λ shock(s) per unit time. A δ-shock failure model is proposed in this paper, which assumes that when a system is acted on by a shock, it will recover fully in time $\delta(> 0)$, and after that it will function as if no shock had occurred before. If the time lag between two successive shocks is less than δ, the second shock will cause failure of the system. Theoretical expressions related to the distribution of the failure time of the system are derived. These results can be used to optimize the design of a system from a costing point of view.

Keywords: Shock Model; Reliability; Failure Time; Economic Design.

1. Introduction

When a system is subject to shocks, energy will accumulate in the system and the system will be damaged when the accumulated energy reaches a certain threshold level. This applies to individual components as well. Gaver[6] studied the survival of a system when it is subject to shocks, under the assumption that shocks arrive according to a Poisson random variable with a fixed incoming rate, and that each shock has the same probability of causing the system to fail (that is, the effect of a shock on the system is history-independent). A cumulative damage model (Ref. 1, Chap. 4, Sec. 3.1) is one which assumes that the system fails when the accumulative effect of k shocks of varying intensity exceed a certain threshold level. Esary *et al.*[5] studied various properties of the life distribution of a system under

§ *Corresponding author*: Dr. Ling-Yau Chan, Department of Industrial and Manufacturing Systems Engineering, The University of Hong Kong, Pokfulam Road, Hong Kong, *Fax Number*: (852) 2858 6535.

[#] This chapter appeared previously on the International Journal of Reliability, Quality and Safety Engineering. To cite this chapter, please cite the original article as the following: Z. Li, L. -Y. Chan and Z. Yuan, *Int. J. Reliab. Qual. Saf. Eng*, **6**, 237–247 (1999), doi:10.1142/S0218539399000231.

this failure model; Ross[11] obtained sufficient conditions, in terms of the accumulative damage, for the system failure time to have an increasing failure rate; Shanthikumar and Sumita[13,14] studied the distribution function of the failure time under the assumption that the magnitude of the nth shock and the time interval between the $(n-1)$th shock and the nth shock are correlated, and obtained results regarding monotonicity and being "new better than used" of the system failure time; Ghurye and Marshall[7] studied a system of two components, assuming that the failure of one component has an after-effect on the other component. When there are two or more components in the system, studies of the failure time of the system will involve bivariate and multivariate exponential distributions; for references, the readers may refer to Refs. 2–4, 9, 10 and 12.

The function of a protective device in an electrical or electronic system is to cut off a part of the circuit when damage is about to occur. A limitation of applying the above cumulative damage model is that, in practice it is usually not easy to measure the intensity of each shock, and it is even more difficult and expensive to measure the cumulative intensity of shocks.

In fact, after a shock, if the system is still not damaged, the accumulated energy caused by the shock will drain away as time passes, and after a certain length of time the residual energy will have an negligible effect on the system and the system will function as if no shocks had occurred. This phenomenon is not being taken into account in the cumulative damage model. If shocks arrive shortly one after another, however, the accumulated energy will easily reach a threshold level which will cause failure of the system.

In the present paper, a new shock model, called δ-*shock model*, will be proposed. In this δ-shock model, it is assumed that shocks of fixed intensity arrive as in a Poisson process, and that the after-effect of a shock on the system lasts for only a fixed period of time. If the time lag between two successive shocks is shorter than this period of time, the system will fail at the second shock; otherwise, the system will function as normal after the second shock. This scenario is not the same as that of the cumulative damage model where shocks of varying intensity act on the system and the system fails when the accumulative damage caused by k shock exceeds a threshold level. In fact, the concept of the δ-shock model has been used in the industry in the design of protective devices (e.g., time-delay switches) in electrical and electronic systems as well as in household appliances; this model is also used in the design of insulation materials and heat sink in electronic circuit systems. A similar shock model on traffic systems was used by Li[8] to study the distribution of occurrence of traffic accidents when the time span between arrivals of two consecutive vehicles travelling on a highway is less than a certain value.

In Sec. 2 of this paper, the δ-shock model will be defined, and theoretical expressions related to the life distribution of a system under this model will be derived. The ability of a system or a component to withstand shocks depends on factors such as product design, choice of insulation material, design of insulation and heat sink, and so on. In general, a component or system which has a higher ability to withstand shocks costs more than one with lower ability. In Sec. 3, the results obtained in Sec. 2 will be applied in the economic design of components based on a costing model.

2. The δ-shock Model

Throughout this paper, it is assumed that independent shocks of fixed intensity that arrive and act on a system follow a Poisson distribution. It is also assumed that the intensity of each shock does not exceed the threshold level of the system, so that the system will not be damaged by a single shock alone. When a shock acts on a system, however, the residual effect of the shock on the system will last for a time period of δ unit(s), so that the system will recover fully in δ unit(s) of time, and from then on it will function as if no shocks had acted on it previously. Based on this, throughout this paper it is assumed that if the time lag between two successive shocks is greater than or equal to δ unit(s) of time, the system will survive immediately after the second shock; if the time lag between two successive shocks is less than δ, the system has not yet recovered from the first shock when the second shock arrives, and the second shock, which will be called a *fatal shock*, will cause failure of the system. The following notation will be used throughout this paper:

$$\lambda \;=\; \text{the mean rate of arrival of incoming shocks, measured in shock(s) per unit time}$$

X_1 = a random variable, which is the time of arrival of the first shock

X_i = a random variable, which is the time lag between the $(i-1)$th shock and the ith shocks $(i = 2, 3, \ldots)$

δ = a constant, such that if $k \geq 2$ and $X_k < \delta$, the system will fail immediately after the kth shock, and if $X_k \geq \delta$, the system will survive immediately after the kth shock

Y = a discrete random variable, such that the system fails immediately after it is acted on by exactly Y shocks

T = the failure time of the system, which is a continuous random variable

$F(t)$ = the distribution function of the failure time T

$\bar{F}(t)$ = $1 - F(t)$ = the survival function of T (Ref. 1, Chap. 3)

$N(t)$ = the number of shocks that arrive and act on the system within the time interval $[0, t]$

$P_k(t)$ = the probability that exactly k shocks have acted on the system within the time interval $[0, t]$ and the system survives at time t $(k = 0, 1, \ldots)$

$\lfloor u \rfloor$ = the integer part of u, when u is a real number

a = $e^{-\lambda\delta}$

According to the above assumptions, the first shock alone will not cause failure of the system, no matter whether $X_1 < \delta$ or $X_1 \geq \delta$. Since T is the failure time of the system, the system survives at the instant t if $t < T$. It is clear that $T = X_1 + X_2 + \cdots + X_k$ for some integer $k \geq 2$, where $X_2 \geq \delta, \ldots, X_{k-1} \geq \delta$ and $X_k < \delta$, and X_1 can either be $\geq \delta$ or $< \delta$.

2.1. Probability distributions

Proposition 2.1.1. The probability function $\Pr(Y = h)$, the expectation $E[Y]$ and the variance $\text{var}[Y]$ of Y are given by

$$\Pr(Y = h) = a^{h-2}(1 - a) \quad (h = 2, 3, \ldots),$$

$$E[Y] = (2 - a)/(1 - a),$$

$$\text{var}[Y] = a/(1 - a)^2.$$

Proof. Since shocks are independent and arrive according to a Poisson process with mean rate of arrival λ, the following holds for $h = 2, 3, \ldots$:

$$\Pr(Y = h) = \Pr(T_2 \geq \delta) \times \cdots \times \Pr(T_{h-1} \geq \delta) \times \Pr(T_h < \delta) = a^{h-2}(1 - a).$$

Hence, the expectation of Y is

$$E[Y] = \sum_{h=2}^{\infty} ha^{h-2}(1 - a) = ((1 - a)/a)\sum_{h=2}^{\infty} ha^{h-1}$$

$$= ((1 - a)/a)(1/(1 - a)^2 - 1) = (2 - a)/(1 - a).$$

It follows from

$$E[Y^2] = \sum_{h=2}^{\infty} h^2 a^{h-2}(1 - a) = \sum_{h=2}^{\infty} h(h - 1)a^{h-2}(1 - a) + \sum_{h=2}^{\infty} ha^{h-2}(1 - a)$$

$$= 2/(1 - a)^2 + (2 - a)/(1 - a)$$

that $\text{var}[Y] = E[Y^2] - (E[Y])^2 = a/(1 - a)^2$. This proves Proposition 2.1.1.

Proposition 2.1.2. The survival function of T is given by

$$\bar{F}(t) = e^{-\lambda t}\left(1 + \sum_{k=1}^{[t/\delta]+1} \lambda^k(t - (k - 1)\delta)^k/k!\right).$$

Proof. Since shocks arrive according to a Poisson process and the system always survives at time t when $k = 0, 1$, it is clear that $P_0(t) = e^{-\lambda t}$ and $P_1(t) = \lambda t e^{-\lambda t}$ for any $t > 0$. If the system is acted on by $k \geq 2$ shocks within the time interval $[0, t]$ and the system survives at time t, the spacing between any two successive shocks within the time interval $[0, t]$ must be greater than or equal to δ, that is,

$X_2 \geq \delta, \ldots, X_k \geq \delta$. Therefore the number of shocks that can arrive within the time interval $[0, t]$ is at most $[t/\delta] + 1$, and thus $P_k(t) = 0$ when $k > [t/\delta] + 1$. When $2 \leq k \leq [t/\delta] + 1$, $P_k(t)$ is given by

$$P_k(t) = \Pr(\{X_1 + \cdots + X_k \leq t < X_1 + \cdots + X_{k+1}\} \cap \{X_2 \geq \delta, \ldots, X_k \geq \delta\})$$

$$= \Pr(X_1 + \cdots + X_k \leq t < X_1 + \cdots + X_{k+1} | X_2 \geq \delta, \ldots, X_k \geq \delta)$$

$$\times \Pr(X_2 \geq \delta, \ldots, X_k \geq \delta) = A \times B,$$

say. The transformation $X_i' = X_i - \delta(i = 2, \ldots, k)$ gives $X_1 + X_2' + \cdots + X_k' \leq t - (k-1)\delta$, and thus

$$A = \Pr(X_1 + X_2' + \cdots + X_k' \leq t - (k-1)\delta < X_1 + X_2' + \cdots + X_{k+1}')$$

$$= \Pr(N(t - (k-1)\delta) = k) = \lambda^k (t - (k-1)\delta)^k e^{-\lambda(t-(k-1)\delta)} / k!.$$

It is obvious that $B = (e^{-\lambda\delta})^{k-1}$. Hence $P_k(t) = \lambda^k (t-(k-1)\delta)^k e^{-\lambda t} / k!$. Therefore

$$\bar{F}(t) = \sum_{k=0}^{\infty} P_k(t) = e^{-\lambda t} + \sum_{k=1}^{[t/\delta]+1} \lambda^k (t - (k-1)\delta)^k e^{-\lambda t} / k!.$$

This proves Proposition 2.1.2.

Proposition 2.1.3. The mean and variance of T are given by

$$\mathrm{E}[T] = \lambda^{-1}(2 - a)/(1 - a), \quad \mathrm{var}[T] = \lambda^{-2}(1 + (1 + 2\lambda\delta a)/(1 - a)^2).$$

Proof. It follows from Proposition 2.1.2 that

$$\mathrm{E}[T] = \int_0^{\infty} \bar{F}(t) \, dt = \int_0^{\infty} e^{-\lambda t} \left(1 + \sum_{k=1}^{[t/\delta]+1} \lambda^k (t - (k-1)\delta)^k / k! \right) dt$$

$$= \int_0^{\infty} e^{-\lambda t} \, dt + \left(\int_0^{\delta} + \int_{\delta}^{2\delta} + \cdots + \int_{n\delta}^{(n+1)\delta} \right) \left(\sum_{k=1}^{[t/\delta]+1} \lambda^k (t - (k-1)\delta)^k / k! \right) dt$$

$$= \int_0^{\infty} e^{-\lambda t} \, dt + \int_0^{\delta} e^{-\lambda t} \lambda t \, dt + \int_{\delta}^{2\delta} e^{-\lambda t} \left(\lambda t + \lambda^2 (t - \delta)^2 / 2! \right) dt$$

$$+ \cdots + \int_{n\delta}^{(n+1)\delta} e^{-\lambda t} \left(\sum_{k=1}^{n+1} \lambda^k (t - (k-1)\lambda)^k / k \right) dt + \ldots$$

$$= \int_0^{\infty} e^{-\lambda t} \, dt + \int_0^{\infty} e^{-\lambda t} \lambda t \, dt + \int_{\delta}^{\infty} e^{-\lambda t} \lambda^2 (t - \delta)^2 / 2! \, dt$$

$$+ \cdots + \int_{n\delta}^{\infty} e^{-\lambda t} \lambda^{n+1} (t - n\delta)^{n+1} / (n+1)! \, dt + \ldots.$$

It follows from the substitution $u = \lambda(t - n\delta)$ and the well-known result $\int_0^\infty e^{-u}u^{n+1}$ $du = (n+1)!$ that the last series is equal to

$$\lambda^{-1} + \lambda^{-1} + \lambda^{-1}e^{-\lambda\delta} + \cdots + \lambda^{-1}e^{-n\lambda\delta} + \cdots = \lambda^{-1}(1 + 1/(1-a)).$$

By the same token, it is readily proved that

$$E[T^2] = \int_0^\infty 2t\bar{F}(t)\,dt = \int_0^\infty 2te^{-\lambda t}\left(1 + \sum_{k=1}^{[t/\delta]+1} \lambda^k(t - (k-1)\delta)^k/k!\right) dt$$

$$= \int_0^\infty 2te^{-\lambda t}\,dt + \int_0^\delta 2te^{-\lambda t}\lambda t\,dt + \int_\delta^{2\delta} 2te^{-\lambda\delta}\left(\lambda t + \lambda^2(t-\delta)^2/2!\right) dt$$

$$+ \cdots + \int_{n\delta}^{(n+1)\delta} 2te^{-\lambda t}\left(\sum_{k=1}^{n+1} \lambda^k(t - (k-1)\lambda)^k/k!\right) dt + \cdots$$

$$= 2\lambda^{-2} + 2\delta\lambda^{-1}\sum_{n=0}^\infty ne^{-n\lambda\delta} + 2\lambda^{-2}\sum_{n=0}^\infty (n+2)e^{-n\lambda\delta}$$

$$= 2\lambda^{-2} + 2(\delta\lambda^{-1} + \lambda^{-2})a/(1-a)^2 + 4\lambda^{-2}/(1-a).$$

Then, the stated expression for $\mathrm{var}[T]$ can be obtained from $\mathrm{var}[T] = E[T^2] - (E[T])^2$ and simplification of the result. This proves Proposition 2.1.3.

Comparison of the results of Propositions 2.1.1 and 2.1.3 shows that $E[Y] = \lambda E[T]$. The last relationship has the obvious intuitive meaning that the average number of shocks the system experiences before it fails is equal to the rate of arrival of shocks multiplied by the average failure time of the system. These propositions also indicate that for a fixed λ, the smaller is the value of δ, the larger are the values of $E[Y]$ and $E[T]$. When $\delta \to 0$, the system takes zero time to recover from a shock, and both $E[Y]$ and $E[T]$ approach infinity which intuitively means that incoming shocks will not cause failure of the system. When $\delta \to \infty$, the system takes an infinite length of time to recover from the first shock; this means a second shock that arrives within a finite length of time will definitely cause failure of the system, in which case, Y will be constantly equal to 2, T has a gamma distribution with probability density function $f(t) = \lambda te^{-\lambda t}(t > 0)$, mean $E[T] = 2/\lambda$, and variance $\mathrm{var}[T] = 2/\lambda^2$.

2.1.1. Estimation and hypothesis testing

Using Propositions 2.1.1 and 2.1.3, the parameters λ and δ can be estimated from sample observations $Y_1, \ldots, Y_n$ of Y, the number of shocks the system has experienced immediately before it fails, or from sample observations $T_1, \ldots, T_n$ of the failure time T of the system.

By equating $\bar{Y} = (Y_1 + \cdots + Y_n)/n$ with $E[Y] = (2-a)/(1-a)$, it is easy to obtain a moment estimate of $a = e^{-\lambda\delta}$, and $-\log a$ is then an estimate of $\lambda\delta$. If both λ and δ are to be estimated, $\lambda\delta$ may be first estimated, and then an estimate of λ will be obtained by equating $\bar{T} = (T_1 + \cdots + T_n)/n$ with $\mathrm{E}[T] = \lambda^{-1}(2-a)/(1-a)$. Then an estimate of δ is the ratio of the estimates of $\lambda\delta$ and λ.

It follows from the central limit theorem that the statistics

$$U_Y = (\bar{Y} - \mathrm{E}[Y])/(\mathrm{var}[Y]/n)^{1/2}, \quad U_T = (\bar{T} - \mathrm{E}[T])/(\mathrm{var}[T]/n)^{1/2}$$

are approximately a standard normal distribution when the sample size n is large, and they can be used for testing hypotheses regarding λ and δ.

3. Design of System or Component from a Costing Approach

Reliability of systems and components is an important issue in the industry. From the user's point of view, it is desirable to have systems with large values of $\mathrm{E}[Y]$ or $\mathrm{E}[T]$, or small values of δ. From the manufacturer's point of consideration, it will be more costly to produce components (or systems) with small values of δ than those with large values, because a small value of δ means the requirement of better material, better insulation, better design of heat sink and perhaps larger size of the component. Let $C(\delta)$ be the cost of producing a component which can withstand two successive shocks separated by a time span of at least δ. The results obtained in Sec. 2 can be used to optimize the design of components in several ways. Given an upper limit of $C(\delta)$, the manufacturer may attempt to produce components with the smallest possible δ for a certain value of λ, or with the highest possible λ for a certain value of δ. Attempts may also be made to maximize $\mathrm{E}[Y]$ or $\mathrm{E}[T]$ for certain values of λ and δ, given an upper limit of $C(\delta)$. Alternatively, the following average cost per unit operating time $\mathrm{cost}(\delta)$ of the component may be used as an objective function for optimization:

$$\mathrm{cost}(\delta) = C(\delta)/\mathrm{E}[T] = \lambda C(\delta)(1-a)/(2-a). \tag{3.1}$$

An economical value of δ, say δ_0, may be defined as one that minimizes $\mathrm{cost}(\delta)$ given by Eq. (3.1). For a given form of $C(\delta)$, δ_0 can be found by minimizing $\mathrm{cost}(\delta)$ analytically or numerically, or simply by plotting the curve of $\mathrm{cost}(\delta)$ against δ and locate the minimum point graphically. This can be facilitated by using readily available software such as *Mathematica*.[15] Generally speaking, whether δ_0 takes a large or a small value depends on whether $C(\delta)$ decreases quickly or slowly compared with $\mathrm{E}[T]$ as δ increases. The following examples, in which $\lambda = 1$ is assumed for simplicity, illustrate three different scenarios.

Example 1. Suppose that $C(\delta) = A_0 + A_1/(1 + \delta^c)$. Here A_0 is a fixed overhead cost of producing and installing a component, and $A_1 = /(1 + \delta^c)$ is the additional cost of making the component being able to withstand two successive shocks that are separated by at least a time span of δ. Let $A_0 = 1$, $A_1 = 0.9$, and $c = 4$. Then $\mathrm{cost}(\delta) = C(\delta)/\mathrm{E}[T] = (1 + 0.9/(1 + \delta^4))(1-a)/(2-a)$. The graph of $\mathrm{cost}(\delta)$ against δ is shown in Fig. 1. It is clear that $\mathrm{cost}(\delta)$ approaches the smallest possible

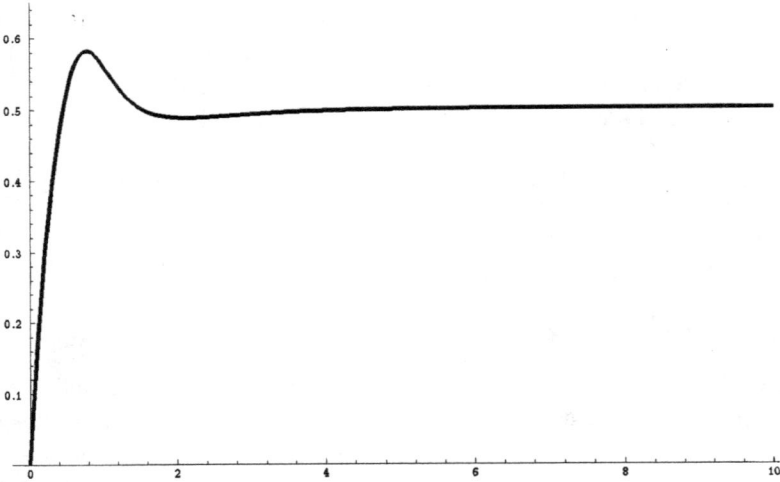

Fig. 1. The graph of cost$(\delta) = (1 + 0.9/(1 + \delta^4))(1 - a)/(2 - a)$ against δ.

limit 0, if and only if, $\delta \to 0$. Hence $\delta_0 = 0$, and it is most economical to produce components with $\delta = 0$. Incoming shocks will not cause failure of such components. The cost of producing such a component is $C(\delta = 0) = A_0 + A_1 = 1.9$.

If $C(\delta) = A_0 + A_1 f(\delta)$ for some function $f(\delta)$ as in Example 1, and the optimal value δ_0 is found to be nonzero, then the component will fail at time T with a positive probability. If the failed component will be replaced, the replacement cost will have to be added to the overhead cost A_0, and the average cost function cost(δ) with this value of A_0 will have to be minimized.

Example 2. In this example, suppose that $C(\delta) = 1 + 2/(e^\delta - 1)$. Hence cost$(\delta) = (1 + 2/(e^\delta - 1))(1 - a)/(2 - a)$. The graph of cost$(\delta)$ against δ is shown in Fig. 2. As the derivative of cost(δ) with respect to δ is $-3e^\delta/(2e^\delta - 1)^2 < 0$, cost$(\delta)$ decreases steadily to $1/2$ as $\delta \to \infty$. Hence $\delta_0 = \infty$, and it is most economical to produce components with $\delta = \infty$. Such components will take an infinite length of time to recover from a shock, that is, the second shock will surely cause failure of the component. A component with this design will have to be replaced immediately after it has been acted on by two shocks.

Example 3. In this example, consider $C(\delta) = 1 + 1/(e^{2\delta} - 1)$. The graph of cost$(\delta) = (1 + 1/(e^{2\delta} - 1))(1 - a)/(2 - a)$ against δ is shown in Fig. 3. The function cost(δ) has a derivative equal to $(e^\delta - 2)e^{2\delta}(e^\delta + 1)^{-2}(2e^\delta - 1)^{-2}$, and attains the global minimum value $4/9$ at $\delta = \log 2 = 0.6931 = \delta_0$. Hence, it is most economical to produce components with this value of δ. This δ_0 corresponds to $a = 1/2$ and $E[Y] = (2 - a)/(1 - a) = 3$, that is, on the average such a component will fail after it has been acted on by three shocks. This provides the maintenance personnel with information on what should the safety stock of the component be.

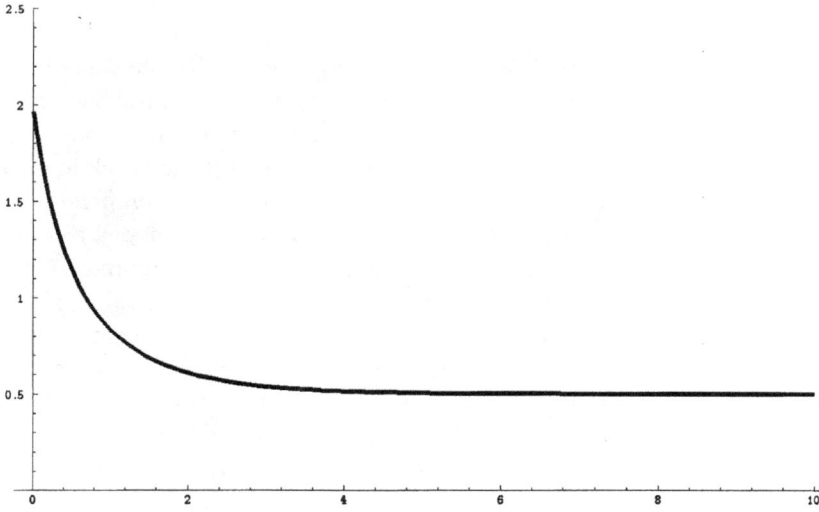

Fig. 2. The graph of $\text{cost}(\delta) = (1 + 2/(e^{\delta} - 1))(1 - a)/(2 - a)$ against δ.

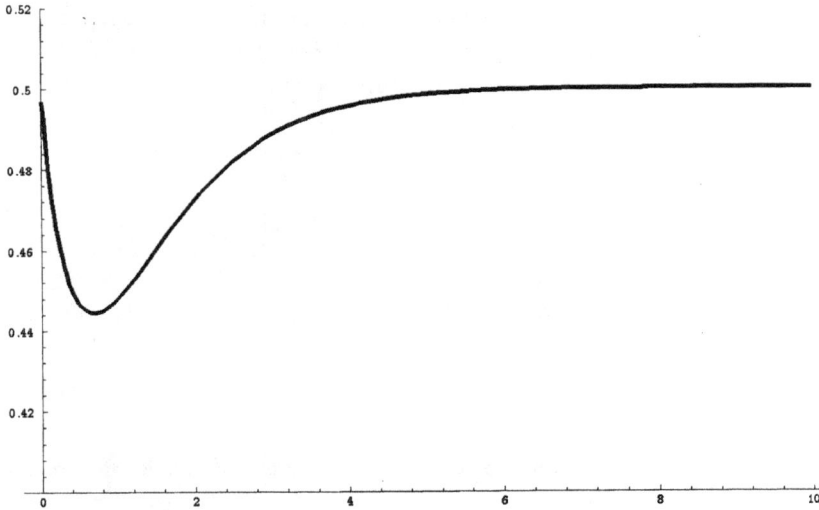

Fig. 3. The graph of $\text{cost}(\delta) = (1 + 1/(e^{2\delta} - 1))(1 - a)/(2 - a)$ against δ.

If a system consists of more than one component, and each component has a specific value of δ, optimizing the whole system will be more complicated than optimizing each individual component separately. However, the concept of cost optimization still applies, but simulation or other heuristic methods may have to be used.

4. Discussion

In this paper, a δ-shock model is proposed to study the life distribution of a system or a component when it is subject to shocks. It is assume that shocks arrive according to a Poisson random variable with a constant rate of arrival λ. A shock is called a fatal shock if the time span between this shock and the shock immediately before it is less than δ. It is assumed that the system or the component will fail immediately when a fatal shock occurs. Based on this δ-shock model, a component with a small value of δ has a longer failure time, but is usually more costly to produce, and vice versa. If the cost function is known, an optimal value of δ can be found to minimize the cost per unit operating time of the component or system.

The studies in this paper can be generalized in several ways. The constant δ can be extended to a random variable. This means that the probability of the system being damaged by a shock which arrives at a time u after the preceding shock is a function $Q(u)$. In this paper, it was assumed that $Q(u) = 1$ when $0 \le u < \delta$ and $Q(u) = 0$ when $u \ge \delta$. Another generalization is to consider the case when a component fails after it has been acted on by N fatal shocks, where N is a positive integer (which could be a random variable). The case when multiple components in a system are subjected to multiple shocks may also be studied. The very simple model of cost(δ) in Eq. (3.1) illustrates how the results established in Sec. 2 can be used in economic design of a component, and economic design of a system which consists of many inter-related components with different values of δ follows the same principle. Results of investigation in these directions will be reported elsewhere.

Acknowledgments

This research is supported by

(1) Grant 19871035 from National Natural Science Foundation of China;
(2) Grant B4 from Natural Science Foundation of Gansu Province of China;
(3) CRCG Grant 10201989/04902/14400/301/01 from The University of Hong Kong.

References

1. R. E. Barlow and F. Proschan, *Statistical Theory of Reliability and Life Testing* (Holt, Rinehart and Winston, New York, 1975).
2. H. W. Block, "A characterization of a bivariate exponential distribution," *Ann. Statist.* **5** (1977), pp. 808–812.
3. H. W. Block and A. P. Basa, "A continuous bivariate exponential extension," *J. Am. Statist. Assoc.* **69** (1974), pp. 1031–1037.
4. F. Downton, "Bivariate exponential distributions in reliability theory," *J. R. Statist. Soc., Ser. B* **32** (1970), pp. 408–417.
5. J. D. Esary, A. W. Marshall, and F. Proschan, "Shock models and wear processes," *Ann. Probability* **1**(4) (1973), pp. 627–649.
6. D. P. Gaver, Jr., "Random hazard in reliability problems," *Technometrics* **5** (1963), pp. 211–226.

7. S. G. Ghurye and A. W. Marshall, "Shock processes with after effects and multivariate lack of memory," *J. Appl. Probability* **21** (1984), pp. 786–801.

8. Z. Li, "Some distributions related to Poisson processes and their application in solving the problem of traffic jam," *J. Lanzhou University* **20** (1984), pp. 127–136.

9. A. W. Marshall and I. Olkin, "A multivariate exponential distribution," *J. Am. Statist. Assoc.* **62** (1967a), pp. 30–40.

10. A. W. Marshall and I. Olkin, "A generalized bivariate exponential distribution," *J. Appl. Probability* **4** (1967b), pp. 291–302.

11. S. M. Ross, "Generalized Poisson shock models,' *Ann. Statist.* **9**(5) (1981), pp. 896–898.

12. K. Rye, "An extension of Marshall and Olkin's bivariate exponential distribution," *J. Am. Statist. Assoc.* **88** (1993), pp. 1458–1465.

13. J. G. Shanthikumar and U. Sumita, "General shock models associated with correlated renewal sequences," *J. Appl. Probability* **20** (1983), pp. 600–614.

14. J. G. Shanthikumar and U. Sumita, "Distribution properties of the system failure time in a general shock model," *Adv. Appl. Probability* **16** (1984), pp. 363–377.

15. S. Wolfram, *Mathematica — A System for Doing Mathematics by Computer* (Addison-Wesley, Redwood City, California, 1991).

About the Authors

Zehui Li graduated from the Department of Mathematics in Lanzhou University in China in 1966, and is now a professor in the department and Chairman of Reliability and Statistics Institute in Lanzhou University. Her other official posts include Director of Institute of Industrial and Applied Mathematics in China, and Director of Chinese Association for Applied Statistics. Her main research areas include mathematical theory of reliability, queuing theory, applied probability, and applied statistics. She has published three books and more than 30 papers in these areas.

Ling-Yau Chan has a B.Sc. in Mathematics and a Ph.D. in Statistics, both obtained at The University of Hong Kong. He is now an associate professor in the Department of Industrial and Manufacturing Systems Engineering at The University of Hong Kong. His main research areas include optimal design of experiments, reliability analysis and statistical quality control. He has published more than 50 papers in these areas. He is a Fellow of The Royal Statistical Society and a reviewer of Mathematical Reviews.

Zhixin Yuan has obtained an M.A. in Mathematics from Lanzhou University in China in 1996. He is now a research officer in Agriculture Bank of China in Qihai province in China. His main research interests are in reliability analysis and computing.

Chapter 18

WARRANTY AND PREVENTIVE MAINTENANCE*

I. DJAMALUDIN and D. N. P. MURTHY

Department of Mechanical Engineering, The University of Queensland, Australia

C. S. KIM

Department of Industrial Engineering, Sangji University, Wonju, Korea

For products sold with warranty, preventive maintenance actions by manufacturers and/or buyers have an impact on the total costs for both parties. This paper develops a framework to study preventive maintenance actions when items are sold under warranty and reviews the models that have appeared in the literature. It then develops a new model and carries out its analysis.

Keywords: Preventive Maintenance; Product Warranty; Condition-Based Maintenance; Optimization.

1. Introduction

All products are unreliable in the sense that they eventually fail. An item failure can occur early in its life due to manufacturing defects or late in its life due to degradation of the item. The degradation is dependent on age and usage. Most products are sold with a warranty that offers protection to buyers against early failures over the warranty period. The warranty period offered has been progressively getting longer. For example, the warranty period for cars was three months in the early thirties and this changed to one year in the sixties and currently it varies from three to five years. With extended warranties, an item is covered for a significant part of its useful life. This implies that failures due to degradation can occur within the warranty period. The degradation of item can be controlled by preventive maintenance and this reduces the likelihood of failures. This implies that preventive maintenance becomes important when warranty periods are long.

Offering warranty implies additional costs to the manufacturer. This is the cost of repairing item failures (through corrective maintenance) over the warranty period. Preventive maintenance during the warranty period can reduce this cost. Since the buyer pays nothing for repairs during the warranty period, there is no incentive for him/her to invest any effort into preventive maintenance. It is

*This chapter appeared previously on the International Journal of Reliability, Quality and Safety Engineering. To cite this chapter, please cite the original article as the following: I. Djamaludin, D. N. P. Murthy and C. S. Kim, *Int. J. Reliab. Qual. Saf. Eng*, **8**, 89–107 (2001), doi:10.1142/S0218539301000396.

worthwhile for the manufacturer to carry out preventive maintenance only if the reduction in the warranty servicing cost is greater than the extra cost incurred with preventive maintenance.

However, from the buyer's perspective, investment in preventive maintenance during the warranty period and after the warranty has expired can have a significant impact on the maintenance cost after the warranty has expired which is borne by the buyer. As a result, buyer's preventive maintenance actions (during the warranty period and afterwards) needs to be determined in the life cycle context.

In this paper, we develop a framework to study preventive maintenance for items sold with warranty from both buyer and manufacturer perspective. Such a framework allows one to build alternate models to determine optimal preventive maintenance strategies. We carry out a literature review of models dealing with warranty and maintenance against this framework and then propose a new model formulation and carry out its analysis.

The outline of the paper is as follows. In Sec. 2, we give a brief overview of product warranty and maintenance so as to set the background for the main contribution of the paper. Following this, we develop a framework to study preventive maintenance for items sold with warranty in Sec. 3. In Sec. 4, we carry out a review of the literature dealing with warranty and maintenance against this framework. Section 5 deals with a new model formulation and its analysis. Finally, we conclude with some discussions on topics for research in the future.

2. Warranties and Maintenance

2.1. *Product warranties*

A warranty is a contract between a buyer and a manufacturer that becomes effective on the sale of an item. The purpose of a warranty is basically to establish liability in the event of a premature failure of an item, where "failure" is meant as the inability of the item to perform its intended function. The contract specifies the promised performance and if this is not met, the means for the buyer to be compensated. The contract also specifies the buyer's responsibilities with regards to due care and operation of the purchased item.

There are many different types of warranties and they depend on the type of product. Products can be broadly divided into three categories — consumer durables, commercial and industrial products, and defence acquisition. For consumer durables, the most common warranties are the free replacement warranty (where failed items are either repaired or replaced by new ones at no cost to the buyer), the pro-rata warranty (which involve replacement at pro-rated cost) and, combinations of these two policies. The above policies are also offered to commercial and industrial products sold individually. However, when they are sold in lots, then cumulative and fleet warranties cover the lot as a whole as opposed to separate warranties for each item. The advantage of this is that it reduces the cost of administering the warranty. Warranties for items procured by the government include all of the above plus some special warranties, particularly in acquisition of defence products. The best known of these special warranties is the Reliability Improvement Warranty, which includes provisions for product development and improvement subsequent to the sale. For a taxonomy for warranty policies, see Blischke and Murthy.[1]

There are many aspects to warranty. These have been studied extensively by researchers from different disciplines. Blischke and Murthy[30] deal with these different aspects for consumer durables and industrial and commercial products. For more on RIW policies see Blischke and Murthy.[1] In this section we focus our attention on the cost analysis of warranties as they are relevant for later sections of the paper.

There are many issues involved in the cost analysis of a warranty. Two of these are — the perspective (buyer or manufacturer as the costs are different for each) and the basis on which the costs are to be assessed. There are a number of approaches to the costing of warranty. The following are some of the methods for calculating costs:

(1) Cost to the manufacturer, per item sold. This per unit cost may be calculated as the total cost of warranty, as determined by general principles of accounting, divided by number of items sold.

(2) Cost per item to the buyer, averaging over all items purchased plus those obtained free or at reduced price under warranty.

(3) Life cycle cost of ownership of an item with or without warranty, including purchase price, operating and maintenance cost, etc., and finally including cost of disposal.

(4) Life cycle cost of an item and its replacements, whether purchased at full price or replaced under warranty, over a fixed time horizon.

(5) Cost per unit of time.

The selection of an appropriate cost basis depends on the product, the context and perspective. The type of customer, individual, corporation or government is important, as are many other factors. Cost models must be developed separately for manufacturer and buyer. Murthy and Blischke[1] deal with the cost analysis for several warranty policies.

2.2. *Maintenance*

Maintenance can be defined as actions to (i) control the deterioration process leading to failure of a system and (ii) restore the system to its operational state through corrective actions after a failure. The former is called "preventive" maintenance and the latter "corrective" maintenance.

Corrective maintenance (CM) actions are unscheduled actions intended to restore a system from a failed state to a working state. This involves either repair or replacement of failed components. In contrast, preventive maintenance actions are scheduled actions carried out to either reduce the likelihood of a failure or prolong the life of the component. Preventive maintenance (PM) actions are divided into the following categories:

- *Clock-based maintenance*: PM actions are carried out at set times. An example of this is the "Block replacement" policy.
- *Age-based maintenance*: PM actions are based on the age of the component. An example of this is the "Age replacement" policy.
- *Usage-based maintenance*: PM actions are based on usage of the product. This is appropriate for items such as tires, components of an aircraft, and so forth.
- *Condition-based maintenance*: PM actions are based on the condition of the component being maintained. This involves monitoring of one or more variables characterizing the wear process (e.g., crack growth in a mechanical component). It is often difficult to measure the variable of interest directly and in this case, some other variable may be used to obtain estimates of the variable of interest. For example, the bearing wear in an engine can be measured by dismantling the bearing case of the engine. However, measuring the vibration, noise or temperature of the bearing case provides information about wear since there is a strong correlation between these variables and bearing wear.
- *Opportunity-based maintenance*: This is applicable for multi-component systems, where maintenance actions (PM or CM) for a component provides an opportunity for carrying out PM actions on one or more of the remaining components of the system.
- *Design-out maintenance*: This involves carrying out modifications through re-designing the component. As a result, the new component has better reliability characteristics.

In general, preventive maintenance is carried out at discrete time instants. In cases where they are done fairly frequently, (the mean time between maintenance actions is ≪ the life of the item) then they can be approximated as occurring continuously over time. This results in the modeling and analysis becoming simpler. Many different types of model formulations have been proposed to study the effect of preventive maintenance on the degradation and failures of items to derive optimal preventive maintenance strategies. See, Murthy[15] and Murthy and Hwang.[16]

Several review papers on maintenance have appeared over the last 30 years. These include McCall,[13] Pierskalla and Voelker,[20] Sherif and Smith,[27] Monahan,[14] Jardine and Buzzacot,[11] Thomas,[28] Pham and Wang,[19] Valdez-Flores and Feldman,[29] Pintelon and Gelders,[21] Dekker[6] and Scarf.[26] Cho and Parlar[4] and Dekker *et al.*[7] deal with the maintenance of multi-component systems. These contain references to the large number of papers and books dealing with maintenance.

3. Framework for the Study of Warranty and Maintenance

As mentioned earlier, the cost analysis for the manufacturer is different from that for the buyer. For the manufacturer, it is the cost of servicing the warranty over the warranty period. This depends on the type of warranty offered. In the case of non-renewing warranty, the warranty period is fixed. In the case of renewing warranty, the warranty period is a random variable. The warranty servicing cost to the manufacturer is the cost of rectifying all failures over the warranty period. Blischke and Murthy[1] deal with this topic in great detail.

From the buyer's perspective, the time interval of interest is from the instant an item is purchased to the instant when it is disposed or replaced. This interval includes the warranty period and the post-warranty period. The cost of rectification over the warranty period depends on the type of warranty. It can vary from no cost (in the case of free replacement warranty) to cost sharing (in the case of pro rata warranty). The cost of rectification during the post-warranty period is borne completely by the buyer. As such, the variable of interest to the buyer is the cost of maintaining an item over its useful life.

Preventive maintenance actions are carried out either to reduce the likelihood of a failure or to prolong the life of an item. Preventive maintenance can be perfect (restoring the item to "good-as-new") or imperfect (restoring the item to a condition that is between as "good-as-new" and as "bad-as-old"). Corrective maintenance can be either minimal (repairing to back-as-old), imperfect or perfect as indicated earlier.

Preventive maintenance over the warranty period has an impact on the warranty servicing cost. It is worthwhile for the manufacturer to carry out this maintenance only if the reduction in the warranty cost exceeds the cost of preventive maintenance. From a buyer's perspective, a myopic buyer might decide not to invest in any preventive maintenance over the warranty period as failures over this period are rectified by the manufacturer at no cost to the buyer. Investing in maintenance is viewed as an additional unnecessary cost. However, from a life cycle perspective, the total life cycle cost to the buyer is influenced by maintenance actions during the warranty period and the post warranty period. This implies that the buyer needs to evaluate the cost under different scenarios for preventive maintenance actions.

This raises several interesting questions. These include the following:

(1) Should preventive maintenance be used during the warranty period?
(2) If so, what should be the optimal maintenance effort? Should the buyer or the manufacturer pay for this or should it be shared?
(3) What level of maintenance should the buyer use during the post warranty period?

Preventive maintenance actions are normally scheduled and carried out at discrete time instant. When the preventive maintenance is carried out frequently and the time between the two successive maintenance actions is small, then one can treat the maintenance effort as being continuous over time. This leads to two different ways (discrete and continuous) of modeling maintenance effort.

Another complicating factor is the information aspect. This relates to various issues such as the state of item, type of distribution function appropriate for modeling failures, parameters of the distribution function etc. The two extreme situations are complete information and no information. Often, the information available to manufacturer and buyer is somewhere in between these two extremes and can differ. This raises several interesting issues such as the adverse selection and moral hazard problems. Quality variations (so that all items are not statistically similar) add yet another dimension to the complexity.

As such, effective study of preventive maintenance for products sold under warranty requires a framework that incorporates the factors discussed above. The number of factors considered and the nature of their characterisation results in many different model formulations linking preventive maintenance and warranty. In the next section we review the models that have appeared in the literature.

4. Review of Literature Linking Warranty and Maintenance

In this section we carry out a chronological review of models dealing with warranty and preventive maintenance.

Ritchken and Fuh[23] discuss an age replacement policy for a non-repairable item. The warranty offered is the pro-rata policy, hence any failure within the warranty period results in a replacement by a new one with the associated cost shared by the producer and the buyer. At the end of the warranty period, the item in use is preventively replaced after a period T (measured from the end of the warranty period) or on failure should it occur earlier. The optimal T^* is selected by minimizing the buyer's asymptotic expected cost per unit time using the renewal reward theorem.

Chun and Lee[3] consider a model of a system with an increasing failure rate and subjected to periodic preventive maintenance actions during warranty period and after the warranty expired. They assume that the preventive maintenance is imperfect, that is the failure rate after maintenance is lower than that before maintenance but not as good-as-new. The hazard rate reduction is assumed to be equivalent to the reduction of the age of the system at preventive maintenance. The reduction is assumed to be the same for each maintenance action regardless of the age of the system and of a warranty period. The costs to the buyer consist of price of the system, cost during warranty period (portion of the preventive maintenance cost) and cost after the warranty has expired (all the maintenance costs). Any failures between preventive maintenance actions during the warranty period are repaired minimally by the manufacturer at no cost to the buyer. The optimal period between preventive maintenance actions is obtained by minimising the buyer's asymptotic expected cost per unit time over an infinite period. An example is given for a system with Weibull failure distribution.

Chun[2] deals with a model similar to the one in Chun and Lee[3] but the focus is on the warranty cost to the manufacturer as opposed to the buyer. The optimal number of preventive maintenance actions, N^*, is obtained by minimizing the warranty cost over a finite horizon.

Jack and Dagpunar[10] deal with the model studied by Chun.[2] They show that when the product has an increasing failure rate, a strict periodic policy for preventive maintenance action is not the optimal strategy. As a result, time intervals between successive preventive maintenance actions should not be identical. They derive the optimal preventive maintenance strategies to minimize the manufacturer's expected warranty cost over the warranty period. They show that for the policy to be strictly periodic, the preventive maintenance action must result in the product being restored to as good as new.

Dagpunar and Jack[5] deal with a model similar to that in Jack and Dagpunar.[10] The cost of each preventive maintenance action is a function of the operating age and the effective age reduction resulting from the action. This cost is an increasing function of the age reduction. In this case, the optimal maintenance can result in the product not being restored to as good-as-new. The optimal number of preventive maintenance actions, N^*, operating age s^*, and age reduction x^* are obtained by minimizing the manufacturer's expected warranty cost.

Sahin and Polatoglu[24] discuss a preventive replacement policy for repairable item following the expiration of warranty. Failures over the warranty period are minimally repaired at no cost to the buyer. The item is kept for a period T after the expiration of the warranty and replaced by a new item. Failures over this period are rectified minimally with the buyer paying the costs. They consider stationary and non-stationary strategies that minimize the long run average cost to the buyer.

Monga and Zuo[12] deal with a model formulation where the components of a system are replaced under preventive maintenance action when their failure rate reaches some specified value. The model formulation includes warranty period and preventive maintenance action in addition to system design and burn-in period. The cost of rectifying failures under warranty is borne by the manufacturer and post warranty costs are borne by the buyer. The various decision variables (including preventive maintenance) are optimally selected by minimizing the system life cycle cost which is the sum of the manufacturing cost (including burn-in cost), installation and setup costs, warranty cost and post warranty cost.

Finally, extended warranties can be viewed as maintenance service contracts. Padmanabhan[18] and Murthy and Padmanabhan[17] deal with extended warranties and Murthy and Ashgarizadeh[8,9] deal with maintenance service contracts.

As can be seen from the models reviewed in this section, preventive maintenance is performed at discrete time instants. The cost of preventive maintenance actions during warranty period is borne by the manufacturer in order to reduce the warranty servicing cost.

5. New Model Formulation

The following notation is used in this section.

L : life of product
W : warranty period

m : preventive maintenance level (decision variable) $[0 \leq m \leq M]$
       [$m = 0$ implies no preventive maintenance]
M : Upper limit on maintenance level
$F_0(t)$: failure distribution function with no preventive maintenance
$f_0(t)$: failure density function with no preventive maintenance
$r_0(T)$: failure rate with no preventive maintenance
$r(t; m)$: failure rate function with preventive maintenance
$r_m(t)$: failure rate with preventive maintenance over $[0, L]$
$C_R(m)$: cost of each repair
$C_m(t)$: maintenance cost per unit time with fixed maintenance level m
β : shape parameter for Weibull distribution
θ_m : scale parameter for Weibull distribution
C_{BX} : buyer's expected lifecycle cost under Option X
C_{MX} : expected warranty cost per unit to manufacturer under Option X
X : Options: $= A$ [no preventive maintenance],
             $= B$ [preventive maintenance over L],
             $= C$ [no preventive maintenance during the warranty period]

The model considers the following three options:

- Option A : No preventive maintenance action over the life of the item.
- Option B : Continuous preventive maintenance over $[0, L)$
- Option C : No preventive maintenance over the warranty period and continuous preventive maintenance over $[W, L)$

It examines the implications of this to the manufacturer's warranty servicing cost and the buyer's life cycle cost and then discusses the optimal strategies for both the manufacturer and buyer. The details of the model formulation are as follows.

5.1. *Product warranty*

The product is repairable and sold with a non-renewing free replacement warranty policy with a warranty period W. All failures in the warranty period $[0, W)$ are rectified (through corrective maintenance actions) by the manufacturer at no cost to the buyer. The product has a useful life L and the cost of rectifying failures (through corrective maintenance actions) in the interval $[W, L)$, subsequent to the expiry of the warranty is borne by the buyer.

5.2. *Item failures*

Since the product is repairable, any failure can be rectified through repair. We confine our attention to rectification through minimal repair. Under minimal repair the failure rate after repair is considered to be the same as that just before failure. We assume that the time to rectify a failure is small in relation to the mean time between failures and that it can be ignored.

We first consider the case without preventive maintenance [Option A]. Let $F_0(t)[f_0(t)]$ denote the product failure distribution [density] function. From Barlow and Hunter[25] the failures over time occur according to a non-homogeneous Poisson process with intensity function given by the failure rate function

$$r_0(t)\frac{f_0(t)}{1 - F_0(t)} .$$ (1)

5.3. *Preventive maintenance actions*

We assume that preventive maintenance actions are carried out at short regular interval so that it can be modeled as occurring continuously over time. As indicated earlier we consider two different options [Options B and C] as indicated earlier.

The maintenance level m is constrained by $0 \leq m \leq M$ where M denotes the upper limit. Larger value of m corresponds to greater maintenance effort (more frequent inspection and/or inspections of more components). One can model m as a continuous or discrete variable and we treat it as continuous valued. We assume constant maintenance level whenever it is used over the life of the item. Relaxing of this assumption is discussed in the concluding section.

Let $r(t; m)$ denote the failure rate for a given history of maintenance with level m. The characterization under different options is as follows:

Option A: In this case, $m = 0$ and as a; result the failure rate is given by

$$r(t; 0) = r_0(t)$$ (2)

Option B: The failure rate is given by

$$r(t; m) = r_m(t)$$ (3)

$[r_0(t) - r_m(t)]$ is an increasing function of m implying that the reduction in the failure rate increases as the maintenance level increases. For $m = M$, $r_m(t)$ is an increasing function of t implying that even with maximum maintenance effort, the failure rate is increasing with age.

One can model the relationship between $r_m(t)$ and $r_0(t)$ in many different ways. We indicate two of these. In the first, the linking is done through the scale parameter of the distribution function. Let θ_m denote the scale parameter with maintenance level m and θ_0 with no maintenance. Then, the relationship is given by

$$\theta_m = g(\theta_0, m)$$ (4)

with $\partial g(\theta_0, m)/\partial m > 0$ implying that the scale parameter increases with the maintenance level effort. This implies that the mean time to first failure increases with m. The second approach to modeling is as follows:

$$r(t; m) = \phi(m)r(t; 0)$$ (5)

with $\phi(m)$ decreasing with m increasing.

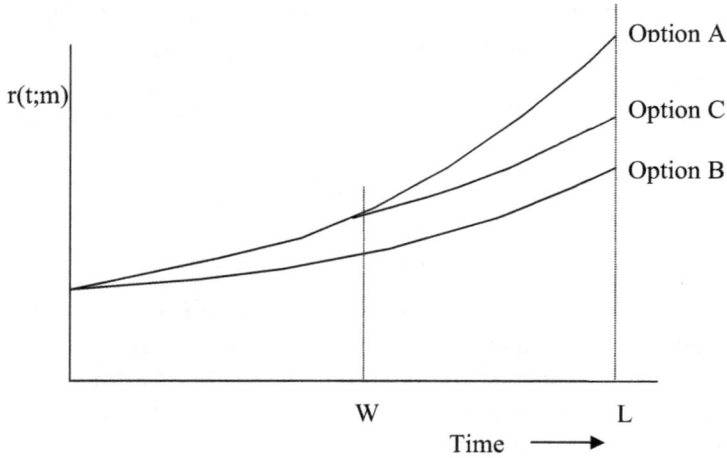

Fig. 1.   The failure rates for options A, B and C.

Option C: The failure rate is given by

$$r(t; m) = \begin{cases} r_0(t) & \text{for } 0 \leq t < W \\ [r_0(W) - r_m(W)] + r_m(t) & \text{for } W \leq t < L \end{cases}. \tag{6}$$

Figure 1 shows a plot of $r(t; m)$ for the three different options for a fixed m. For a given t, the failure rate under Option B is less than under Option C which in turn is less than under Option A.

5.4. Maintenance cost

Let C_m denote the cost per unit time with maintenance level m and this increases with m. Let C_R denote the average cost of each repair and the cost of corrective maintenance depends on the number of failures over the interval $[W, L)$ and C_R.

6. Model Analysis

In this section we derive expressions for the expected life cycle cost to the buyer over the life of an item and the expected warranty cost to the manufacturer for the three options.

Since failures are rectified through minimal repair and the repair times are negligible, the expected number of failures over any interval is given by the integral of the failure rate function over the interval. As a result, the expected number of failures over warranty period is given by $\int_0^W r(t; m)dt$ and over the post warranty period by $\int_W^L r(t; m)dt$. $r(t; m)$ for Options A–C are given by Eqs. (2), (3) and (6) respectively.

The preventive maintenance costs are zero, $C_m L$ and $C_m(L - W)$ under Options A, B and C respectively.

6.1. *Buyer's expected life cycle cost*

The expected life cycle cost to the buyer is the sum of the preventive maintenance cost and the expected cost of rectifying failures in the post warranty period. As such, we have the following results. Under Option A, it is given by

$$C_{BA} = C_R \int_W^L r_0(t) \, dt \tag{7}$$

as there is no preventive maintenance. Under Option B, the expected life cycle cost is given by

$$C_{BB} = C_R \int_W^L r_m(t) \, dt + C_m L . \tag{8}$$

Finally, under Option C the expected life cycle cost is given by,

$$C_{BC} = C_R \int_W^L (r_0(W) - r_m(W) + r_m(t)) \, dt + C_m(L - W) . \tag{9}$$

The optimal preventive maintenance level m for Options B and C can be obtained by minimizing the expected life cycle cost. If m is modeled as a continuous variable, this can be obtained by the usual first order conditions for optimality if the optimal m is an interior point of the interval $[0, M]$. If not, then the optimal m is either zero (no preventive maintenance) or M (maximum maintenance level). If m is modeled as a discrete variable, then the optimal m is determined by evaluating the costs for the different values of m and then carrying out a relative comparison.

From the buyer's point of view, the decision to carry out preventive maintenance (either over the life of the item [Option B] or only in the post warranty period [Option C]) or not [Option A] depends on the relative values of the expected life cycle costs for the three scenarios.

Note that for a given maintenance effort $m(0 \leq m \leq M)$, the optimal choice between the three options is determined by a relative comparison of the expected life cycle costs. The one that yields the lowest cost is the optimal strategy. Also, the optimal maintenance efforts for Options B and C are obtained by minimizing the expected life cycle costs. The optimal maintenance effort achieves an optimal trade-off between the preventive and corrective maintenance costs incurred by the buyer.

One can derive various relationships for comparing the options. One of them (comparing Options A and B) is as follows. Define $\Delta(m) = C_R \int_W^L (r_0(t) - r_m(t)) dt$. This is the saving in corrective maintenance cost by using preventive maintenance at level m. Option B is better than Option A if $\Delta(m) > C_m L$ for some m, for $0 < m < M$.

6.2. *Expected warranty cost per unit to manufacturer*

The cost to the manufacturer is the warranty servicing cost (the cost of rectifying failures under warranty). Under Options A and C, the expected warranty cost per unit is given by

$$C_{MA} = C_{MC} = C_R \int_0^W r_0(t)\, dt\,. \tag{10}$$

Under Option B, it is given by

$$C_{MB} = C_R \int_0^W r_m(t)\, dt\,. \tag{11}$$

Since $r_m(t) < r_0(t)$, the expected warranty cost per unit to the manufacturer under Option B is smaller than that under Options A and C. Hence, the manufacturer would always prefer to have the buyer carry out preventive maintenance during the warranty period.

6.3. *Discussion*

As mentioned earlier, the optimal decision of the buyer with regards to preventive maintenance depends on the parameter values of the model. A non-myopic buyer would choose between all three options whereas a myopic buyer would only choose between Options A and C.

For a non-myopic buyer if Option B is the optimal strategy, then the buyer carries out preventive maintenance effort over the life of the item. This not only yields the lowest life cycle cost to the buyer but also results in a saving to the manufacturer. In contrast, a myopic buyer would choose Option C that does not result in a reduction in the cost to the manufacturer. In this case, the manufacturer might like to explore some incentive scheme (for example, monetary compensation) to induce the buyer to carry out preventive maintenance over the warranty period. As long as the cost of this is less than the savings in the manufacturer's cost, it results in a win–win situation for both. However, the manufacturer needs to be aware of the resulting moral hazard problem — the buyer might collect the compensation and not carry out the preventive maintenance actions over the warranty period when there is no way for the manufacturer to observe the actions of the buyer.

With modern technology, in some cases it is possible for the manufacturer to observe the maintenance effort during warranty period by building in proper tamper proof data logging sensors and recorders. In this case, the manufacturer can modify the warranty policy so that the warranty period is extended from W to W_1 should the buyer carry out preventive maintenance during warranty period. This implies that non-myopic buyers are rewarded and myopic buyers tempted to carry out preventive maintenance over the warranty period through this inducement. This fits well with the incentive theory of warranties proposed by Priest[22] in the sense that a warranty must act as an incentive to buyers to carry out preventive maintenance and for the manufacturer to produce a more reliable product. We will discuss this further in the next section.

7. Special Case: Weibull Distribution

Let $F_0(t)$ be a Weibull distribution with shape parameter β and scale parameter θ_0. We model the effect of preventive maintenance on the failure distribution through a change in the scale parameter. As a result, if a constant maintenance effort m is used over the life of the item, then the failure distribution is given by a Weibull distribution with shape parameter β and scale parameter θ_m. The relationship between θ_m and θ_0 is assumed to be of the form

$$\theta_m = \theta_0 \left(\frac{10}{10 - m} \right)^{\gamma} \tag{12}$$

with $0 \le m < 10$ and $\gamma > 0$. This implies that θ_m increases as m increases.

$F(t; m)$ is given by

$$F(t, m) = 1 - e^{-(t/\theta_m)\beta} . \tag{13}$$

7.1. *Buyer's expected life cycle cost*

From Eqs. (7)–(9), the expected life cycle cost is given by

$$C_{BA} = (C_R/\theta_0^\beta)(L^\beta - W^\beta) \tag{14}$$

under Option A, by

$$C_{BB} = (C_R/\theta_m^\beta)(L^\beta - W^\beta) + C_m L \tag{15}$$

under Option B, and by

$$C_{BC} = C_R[\beta W^{\beta-1}(1/\theta_0^\beta - 1/\theta_m^\beta)(L - W) + (1/\theta_m^\beta)(L^\beta - W^\beta)] + C_m(L - W) \tag{16}$$

under Option C.

7.2. *Expected warranty cost per unit to manufacturer*

From Eqs. (10) and (11), the expected warranty cost under Options A and C are given by

$$C_{MA} = C_{MC} = C_R(W/\theta_0)^\beta \tag{17}$$

and under Option B by

$$C_{MB} = C_R(W/\theta_m)^\beta . \tag{18}$$

8. Numerical Example

8.1. *Buyer's expected life cycle cost*

Let $W = 2$, $L = 5$, $\beta = 3$, $\gamma = 1$ and $\theta_0 = 2$. This implies that the mean time to first failure is 1.8 years. We assume that m takes on only discrete values and these are indicated in Table 1. Also shown is the cost per unit time for each of the different levels of preventive maintenance.

Table 1.   Maintenance level m, θ_m and C_m.

Maintenance Level m	θ_m	C_m ($)
0	2.00	–
1	2.22	50
2	2.50	100
3	2.86	150
4	3.33	240
5	4.00	350

Table 2.   Expected life cycle costs for Options A, B and C [$\beta = 3$ and C_R varying].

	Option A	Option B					Option C				
C_R	$m = 0$	$m = 1$	$m = 2$	$m = 3$	$m = 4$	$m = 5$	$m = 1$	$m = 2$	$m = 3$	$m = 4$	$m = 5$
20	292.5	463.2	649.8	850.3	1263.2	1786.6	387.6	493.7	609.5	853.7	1165.3
40	585.0	676.5	799.5	950.7	1326.4	1823.1	625.2	687.4	768.9	987.5	1280.6
60	877.5	889.7	949.3	1051.0	1389.5	1859.7	862.9	881.0	928.4	1121.2	1395.9
80	1170.0	1102.9	1099.0	1151.3	1452.7	1896.3	1100.5	1074.7	1087.8	1255.0	1511.3
100	1462.5	1316.2	1248.8	1251.6	1515.9	1932.8	1338.1	1268.4	1247.3	1388.7	1626.6
120	1755.0	1529.4	1398.6	1352.0	1579.1	1969.4	1575.7	1462.1	1406.7	1522.4	1741.9
240	3510.0	2808.8	2297.1	1953.9	1958.2	2188.8	3001.5	2624.2	2363.5	2324.9	2433.8
250	3656.3	2915.4	2372.0	2004.1	1989.8	2207.0	3120.3	2721.0	2443.2	2391.8	2491.4
340	4972.5	3875.0	3045.9	2455.6	2274.1	2371.6	4189.6	3592.6	3160.8	2993.6	3010.3
400	5850.0	4514.7	3495.2	2756.6	2463.6	2481.3	4902.5	4173.6	3639.2	3394.8	3356.3
420	6142.5	4727.9	3645.0	2856.9	2526.8	2517.8	5140.1	4367.3	3798.6	3528.5	3471.6
500	7312.5	5580.8	4244.0	3258.2	2779.5	2664.1	6090.6	5142.0	4436.4	4063.5	3932.8

Table 2 shows the expected life cycle cost to the buyer under Options A, B and C with C_R varying from \$20 to \$500.

The result shows that for low corrective maintenance cost (< \$60), the optimal decision is Option A (no preventive maintenance during the life of the item). With the corrective maintenance cost increasing (but < \$120), the optimal decision is Option C, that is to carry out preventive maintenance after the warranty period has expired. The decision on the optimum preventive maintenance level depends on the ratio of the preventive maintenance and corrective maintenance costs.

For still higher repair costs (> \$120), the optimal decision is Option B (carry out preventive maintenance over the whole interval). The optimal preventive maintenance level depends on the ratio of the preventive maintenance and corrective maintenance costs.

The corresponding results for the case when β changes from 3 to 2 are shown in Table 3.

Table 3. Expected life cycle costs for Options A, B and C [$\beta = 2$ and C_R varying].

C_R	Option A $m = 0$	Option B $m = 1$	$m = 2$	$m = 3$	$m = 4$	$m = 5$	Option C $m = 1$	$m = 2$	$m = 3$	$m = 4$	$m = 5$
20	105.0	335.1	567.2	801.5	1237.8	1776.3	246.5	388.8	532.1	796.2	1121.3
40	210.0	420.1	634.4	852.9	1275.6	1802.5	342.9	477.6	614.1	872.4	1192.5
60	315.0	505.2	701.6	904.4	1313.4	1828.4	439.4	566.4	696.2	948.6	1263.8
80	420.0	590.2	768.8	955.8	1351.2	1855.0	535.8	655.2	778.2	1024.8	1335.0
100	525.0	675.3	836.0	1007.3	1389.0	1881.3	632.3	744.0	860.3	1101.0	1406.3
120	630.0	760.3	903.2	1058.7	1426.8	1907.5	728.7	832.8	942.3	1177.2	1477.5
240	1260.0	1270.6	1306.4	1367.4	1653.6	2065.0	1307.4	1365.6	1434.6	1634.4	1905.0
250	1312.5	1313.1	1340.0	1393.1	1672.5	2078.1	1355.6	1410.0	1475.6	1672.5	1940.6
260	1365.0	1355.7	1373.6	1418.9	1691.4	2091.3	1403.9	1454.4	1516.7	1710.6	1976.3
280	1470.0	1440.7	1440.8	1470.3	1729.2	2117.5	1500.3	1543.2	1598.7	1786.8	2047.5
300	1575.0	1525.8	1508.0	1521.8	1767.0	2143.8	1596.8	1632.0	1680.8	1863.0	2118.8
320	1680.0	1610.8	1575.2	1573.2	1804.8	2170.0	1693.2	1720.8	1762.8	1939.2	2190.0
340	1785.0	1695.9	1642.4	1624.7	1842.6	2196.3	1789.7	1809.6	1844.9	2015.4	2261.3
400	2100.0	1951.0	1844.0	1779.0	1956.0	2275.0	2079.0	2076.0	2091.0	2244.0	2475.0
420	2205.0	2036.1	1911.2	1830.5	1993.8	2301.3	2175.5	2164.8	2173.1	2320.2	2546.3
500	2625.0	2376.3	2180.0	2036.3	2145.0	2406.3	2561.3	2520.0	2501.3	2625.0	2831.3

Table 4. Expected warranty servicing cost per unit for Options A (or C), and B [$\beta = 3$ and C_R varying from \$20 to \$500].

C_R	Option A (or C) $m = 0$	Option B $m = 1$	$m = 2$	$m = 3$	$m = 4$	$m = 5$
20	20.0	14.6	10.2	6.9	4.3	2.5
40	40.0	29.2	20.5	13.7	8.6	5.0
60	60.0	43.7	30.7	20.6	13.0	7.5
80	80.0	58.3	41.0	27.4	17.3	10.0
100	100.0	72.9	51.2	34.3	21.6	12.5
120	120.0	87.5	61.4	41.2	25.9	15.0
240	240.0	175.0	122.9	82.3	51.8	30.0
250	250.0	182.3	128.0	85.8	54.0	31.3
340	340.0	247.9	174.1	116.6	73.4	42.5
400	400.0	291.6	204.8	137.2	86.4	50.0
420	420.0	306.2	215.0	144.1	90.7	52.5
500	500.0	364.5	256.0	171.5	108.0	62.5

The results are similar. The expected costs are smaller as to be expected. Note that the optimal strategy is Option A for $C_R < 240$ and Option B for $240 < C_R < 500$.

8.2. *Manufacturer's expected warranty servicing cost per unit*

For the same parameter values as in Sec. 8.1 and $\beta = 3$, the expected warranty cost per unit to the manufacturer under the three Options A (or C) and B are given in Table 4 for C_R varying from \$20 to \$500.

The result show that as the corrective maintenance cost increases, the expected warranty cost per unit to the manufacturer also increases. On the other hand, greater preventive maintenance effort on the part of the buyer implies smaller warranty servicing cost.

8.3. *Comment*

The results of Secs. 8.1 and 8.2 show the effect of preventive maintenance on the buyer's life cycle cost and manufacturer's warranty servicing cost. As the buyer's effort on preventive maintenance increases, the manufacturer warranty servicing cost decreases.

For the case $C_R = \$500$ and $\beta = 3$, from Table 2 it is seen that optimal preventive maintenance results in lower costs to both the manufacturer and buyer. This is a win–win situation.

If the buyer is myopic and does not invest in preventive maintenance during the warranty period then the manufacturer warranty servicing cost is higher. In this case, the manufacturer might get the buyer to invest in maintenance effort during the warranty period by offering monetary incentive as long as it is less than \$438.5 and the buyer chooses the highest level of maintenance.

9. Conclusion

In this paper, we have proposed a framework to study warranty and preventive maintenance and formulated a simple model involving continuous preventive maintenance. The model can be extended in several ways and we indicate a few.

(1) The effect of maintenance often leads to the life of the item being extended. This implies that L increases with m. (i.e., L increasing with m).
(2) We have confined our analysis to the free replacement policy and failed item being always repaired minimally. The analysis of other types of warranty policies (for example, pro-rata, combination) is yet to be carried out.
(3) We have not studied the different incentive schemes and the related moral hazard issues. This is a topic for considerable new research.
(4) We have assumed that the maintenance level is constant. Often, this is not realistic and the maintenance effort changes with the age of the item — less when it is new and more as it ages.
(5) We have confined our attention to continuous preventive maintenance effort. Often, one employs both continuous and discrete (overhaul) preventive maintenance actions. This makes the problem more difficult and also interesting.

Some of these problems are currently under investigation by the authors.

References

1. W. R. Blischke and D. N. P. Murthy, *Warranty Cost Analysis* (Marcel Dekker, New York, 1994).
2. Y. H. Chun, "Optimal number of periodic preventive maintenance operations under warranty," *Reliability Engineering and System Safety* **37** (1992), pp. 223–225.
3. Y. H. Chun and C. S. Lee, "Optimal replacement policy for a warranted system with imperfect preventive maintenance operations," *Microelectronics Reliability* **32** (1992), pp. 839–843.
4. D. Cho and M. Parlar, "A survey of maintenance models for multi-unit systems," *European J. Operational Research* **51** (1991), pp. 1–23.
5. J. S. Dagpunar and N. Jack, "Preventive maintenance strategy for equipment under warranty," *Microelectronics Reliability* **34** (1994), pp. 1089–1093.
6. R. Dekker, "Applications of maintenance optimization models: A review and analysis," *Reliability Engineering and System Safety* **51** (1996), pp. 229–240.
7. R. Dekker, R. E. Wildeman, and F. A. van der Duyn Schouten, "A review of multi-component maintenance models with economic dependence," *Mathematical Methods of Operations Research* **45** (1997), pp. 411–435.
8. D. N. P. Murthy and E. Ashgarizadeh, "A stochastic model for service contract," *Int. J. of Reliability Quality and Safety Engineering* **5** (1998) pp. 29–45.
9. D. N. P. Murthy and E. Ashgarizadeh, "Optimal decision making in a maintenance service operation," *European J. Operational Res.* **116** (1999) pp. 259–273.
10. N. Jack and J. S. Dagpunar, "An optimal imperfect maintenance policy over a warranty period," *Microelectronics Reliability* **34** (1994), pp. 529–534.
11. A. K. S. Jardine and J. A. Buzacott, "Equipment reliability and maintenance," *European J. Operational Research* **19** (1985), pp. 285–296.
12. A. Monga and M. J. Zuo, "Optimal system design considering maintenance and warranty," *Computers Operations Research* **9** (1998), pp. 691–705.
13. J. J. McCall, "Maintenance policies for stochastically failing equipment: A survey," *Management Science* **11** (1965), pp. 493–524.
14. G. E. Monahan, "A survey of partially observable Markov decision processes: Theory, models and algorithms," *Management Science* **28** (1982), pp. 1–16.
15. D. N. P. Murthy, "Optimal maintenance and sale date of a machine," *International J. Systems Science* **15** (1984), pp. 277–292.
16. D. N. P. Murthy and M. C. Hwang, "Optimal discrete and continuous maintenance policy for a complex unreliable machine," *International Journal of System Science* **27** (1996), pp. 483–495.
17. D. N. P. Murthy and V. Padmanabhan, *A Dynamic Model of Product Warranty with Consumer Moral Hazard*, Research paper No. 1263 (Graduate School of Business, Stanford University, Stanford, CA, 1993).
18. V. Padmanabhan, "Usage heterogeneity and extended warranty," *Journal of Economics and Management Strategy* **4** (1995), pp. 33–53.
19. H. Pham and H. Wang, "Imperfect maintenance," *European J. Operations Research* **94** (1996), pp. 425–438.
20. W. P. Pierskalla and J. A. Voelker, "A survey of maintenance models: The control and surveillance of deteriorating systems," *Naval Research Logistics Quarterly* **23** (1976), pp. 353–388.
21. L. M. Pintelon and L. F. Gelders, "Maintenance Management Decision Making," *European Journal of Operational Research* **58** (1992), pp. 301–317.
22. G. L. Priest, "A theory of the consumer product warranty," *Yale Law Journal* **90** (1981), pp. 1297–1352.

23. P. H. Ritchken and D. Fuh, "Optimal replacement policies for irreparable warrantied items," *IEEE Transactions on Reliability* **R-35** (1986), pp. 621–623.
24. I. Sahin and H. Polatoglu, "Maintenance strategies following the expiration of warranty," *IEEE Transactions on Reliability* **45** (1996), pp. 220–228.
25. R. E. Barlow, and L. Hunter, "Optimum preventive maintenance policies," *Operations Research* **8** (1960), pp. 90–100.
26. P. A. Scarf, "On the application of mathematical models in maintenance," *European Journal of Operational Research* **99** (1997), pp. 493–506.
27. Y. S. Sherif and M. L. Smith, "Optimal maintenance models for systems subject to failure — A review," *Naval Research Logistics Quarterly* **23** (1976), pp. 47–74.
28. L. C. Thomas, "A survey of maintenance and replacement models for maintainability and reliability of multi-item systems," *Reliability Engineering* **16** (1986), pp. 297–309.
29. C. Valdez-Flores and R. M. Feldman, "A survey of preventive maintenance models for stochastically deteriorating single-unit systems," *Naval Research Logistics Quarterly* **36** (1989), pp. 419–446.
30. W. R. Blischke and D. N. P. Murthy, eds., *Product Warranty Handbook* (Marcel Dekker Inc., New York, 1996).

About the Authors

I. Djamaludin received her BE degree from the Institute Technology of Bandung, Indonesia, Master of Engineering and Ph.D. degrees from the University of Queensland, Brisbane, Australia. She has published a number of papers in international journals, presented several papers at international conferences and contributed chapters to several books. Her research areas include Total Quality Management, Product Warranty, Reliability, Maintenance, Production and Engineering Management, Process Improvement, Data Analysis and Operation Research. She has supervised a number of undergraduate and postgraduate projects in the application of Quality Management and Improvement in industry.

D. N. P. Murthy obtained B.E. and M.E. degrees from Jabalpur University and the Indian Institute of Science in India and M.S. and Ph.D. degrees from Harvard University. He is currently Professor of Engineering and Operations Management in the Department of Mechanical Engineering at the University of Queensland and a Senior Scientific Advisor to the Norwegian University of Science and Technology. He has held visiting appointments at several universities in the USA, Europe and Asia. His current research interests include various aspects of technology management (new product development, strategic management of technology), operations management (lot sizing, quality, reliability, maintenance), and post-sale support (warranties, service contracts). He has authored or co-authored over 125 journal papers and 110 conference papers. He is co-author of three books (*Mathematical Modeling* [Pergamon Prsss, London] and *Warranty Cost Analysis* [Marcel Dekker, New York], *Reliability: Modelling, Prediction and Optimization* [Wiley, New York]) and co-editor of *Product Warranty Handbook* (Marcel Dekker, New York). He is currently working on two books (Weibull Models and, Reliability Case Studies) to be published by Wiley, New York. He is a member of several professional societies and is on the editorial boards of seven international journals. He has run short courses for industry on various topics in technology management, operations management and post-sale support in Australia, Asia, Europe and the USA.

Chesoong Kim received his masters and doctoral degrees in Engineering from the Department of Industrial Engineering at the Seoul National University in 1989 and 1993 respectively. He was a visiting scholar in the Department of Mechanical Engineering at the University of Queensland, Brisbane, Australia from September of 1998 to August of 1999. He is currently Associate Professor of the Department of Industrial Engineering at Sangji University. His current research interests include system reliability and maintenance, warranties, wireless communication network and system performance analysis.

Author Index

www.ingramcontent.com/pod-product-compliance
Lightning Source LLC
Chambersburg PA
CBHW081509190326
41458CB00015B/5328